Springer Theses

Recognizing Outstanding Ph.D. Research

Aims and Scope

The series "Springer Theses" brings together a selection of the very best Ph.D. theses from around the world and across the physical sciences. Nominated and endorsed by two recognized specialists, each published volume has been selected for its scientific excellence and the high impact of its contents for the pertinent field of research. For greater accessibility to non-specialists, the published versions include an extended introduction, as well as a foreword by the student's supervisor explaining the special relevance of the work for the field. As a whole, the series will provide a valuable resource both for newcomers to the research fields described, and for other scientists seeking detailed background information on special questions. Finally, it provides an accredited documentation of the valuable contributions made by today's younger generation of scientists.

Theses are accepted into the series by invited nomination only and must fulfill all of the following criteria

- They must be written in good English.
- The topic should fall within the confines of Chemistry, Physics, Earth Sciences, Engineering and related interdisciplinary fields such as Materials, Nanoscience, Chemical Engineering, Complex Systems and Biophysics.
- The work reported in the thesis must represent a significant scientific advance.
- If the thesis includes previously published material, permission to reproduce this must be gained from the respective copyright holder.
- They must have been examined and passed during the 12 months prior to nomination.
- Each thesis should include a foreword by the supervisor outlining the significance of its content.
- The theses should have a clearly defined structure including an introduction accessible to scientists not expert in that particular field.

More information about this series at http://www.springer.com/series/8790

Edoardo Vescovi

Perturbative and Non-perturbative Approaches to String Sigma-Models in AdS/CFT

Doctoral Thesis accepted by
the Humboldt University Berlin, Berlin, Germany

Author
Dr. Edoardo Vescovi
Institute of Physics
University of São Paulo
São Paulo, São Paulo
Brazil

Supervisor
Prof. Valentina Forini
Institute of Physics, IRIS Adlershof
Humboldt University Berlin
Berlin
Germany

ISSN 2190-5053 ISSN 2190-5061 (electronic)
Springer Theses
ISBN 978-3-319-87551-4 ISBN 978-3-319-63420-3 (eBook)
DOI 10.1007/978-3-319-63420-3

Softcover reprint of the hardcover 1st edition 2017

Printed on acid-free paper

This Springer imprint is published by Springer Nature
The registered company is Springer International Publishing AG
The registered company address is: Gewerbestrasse 11, 6330 Cham, Switzerland

Supervisor's Foreword

This thesis, which encompasses results from the six peer-reviewed publications that Dr. Vescovi co-authored in his 3 years at Humboldt University Berlin, deals with the most relevant string model in the framework of the gauge/gravity duality—one of the most far-reaching recent ideas in theoretical physics. This work is in my opinion remarkable for three aspects. First, it contains some of the technically hardest calculations in this framework (next-to-leading order results for string excitations which are highly coupled, as well as next-to-next-to-leading order results, the last one setting the current limit of perturbative string worldsheet analysis). Second, it contains the results of a new, highly interdisciplinary project—involving various themes in mathematical and high-energy physics—which initiates an entirely new way to analyse worldsheet string theory models: the use of Monte Carlo simulations for their lattice-discretized version. Finally, the thesis is very well written and highly pedagogical, thus enabling the reader to follow easily (and I believe with pleasure) the complex subjects treated and the analytic and numerical results reported.

Berlin, Germany
April 2017

Prof. Valentina Forini

Abstract

This thesis discusses perturbative and non-perturbative aspects of type II superstring theories in $AdS_5 \times S^5$ and $AdS_4 \times \mathbb{CP}^3$ backgrounds relevant for the AdS/CFT correspondence. We present different approaches to compute observables in the context of the duality and we test the quantum properties of these two superstring actions. Our methods of investigation span from the traditional perturbative techniques for the worldsheet sigma-model at large string tension, consisting in expanding around minimal area surfaces, to the development of a novel non-perturbative analysis, based upon numerical methods borrowed from lattice field theory.

We review the construction of the supercoset sigma-model for strings propagating in the $AdS_5 \times S^5$ background. When applied to the $AdS_4 \times \mathbb{CP}^3$ case, this procedure returns an action that cannot consistently describe the general quantum dynamics of the superstring. This can be attained instead by an alternative formulation based on the double-dimensional reduction of the supercoset action for a supermembrane moving in $AdS_4 \times S^7$.

We then discuss a general and manifestly covariant formalism for the quantization of string solutions in $AdS_5 \times S^5$ in semiclassical approximation, by expanding the relevant sigma-model around surfaces of least area associated to BPS and non-BPS observables amenable to a dual description within the gauge/gravity duality. The novelty of our construction is to express the bosonic and fermionic semiclassical fluctuation operators in terms of intrinsic and extrinsic invariants of the background geometry for given *arbitrary* classical configuration.

We proceed with two examples in the more general class of quantum small fluctuations, governed by nontrivial matrix-valued differential operators and so far explored only in simplifying limits. Our results stem from the exact solution of the spectral problem for a generalization of the Lamé differential equation, which falls under a special class of fourth-order operators with coefficients being doubly periodic in a complex variable. Our exact semiclassical analysis applies to two-spin folded closed strings: the (J_1, J_2)-string in the $SU(2)$ sector in the limit described by a quantum Landau–Lifshitz model and the bosonic sector of the (S, J)-string rotating in AdS_5 and S^5. In both situations, we write the one-loop contribution to the string energy in an analytically closed integral expression that involves nontrivial nested combinations of Jacobi elliptic functions.

Similar techniques allow to address the strong-coupling behaviour of 1/4-BPS latitude Wilson loops in planar $SU(N)$ $\mathcal{N}=4$ supersymmetric Yang–Mills (SYM) theory. These operators are holographically mapped to fundamental strings in $AdS_5 \times S^5$. To compute the first correction to their classical values, we apply a corollary of the Gel'fand-Yaglom method for the functional determinants to the matrix-valued operators of the relevant semiclassical fluctuations. To avoid ambiguities due to the absolute normalization of the string partition function, we consider the ratio between the generic latitude and the maximal 1/2-BPS circular loop. Our regularization procedure reproduces the next-to-leading order predicted by supersymmetric localization in the dual gauge theory, up to a certain remainder function that we comment upon and that was later confirmed in a different setup by other authors. We also study the AdS light-cone gauge-fixed string action in $AdS_4 \times \mathbb{CP}^3$ expanded around the null cusp background, which is dual to a light-like Wilson cusp in the planar $\mathcal{N}=6$ Chern–Simons matter (ABJM) theory. The fluctuation Lagrangian has constant coefficients, thus it allows to extend the computation of the free energy associated to such string solution up to two loops, from which we derive the null cusp anomalous dimension $f(\lambda)$ of the dual ABJM theory at strong coupling to the same loop order. The comparison between this perturbative result for $f(\lambda)$ and its integrability prediction results in the computation of the nontrivial ABJM interpolating function $h(\lambda)$, which plays the role of effective coupling in all integrability-based calculations in the AdS_4/CFT_3 duality. The perturbative result is in agreement with the strong-coupling expansion of an all-loop conjectured expression of $h(\lambda)$.

The last part of the thesis is devoted to a novel and genuinely field-theoretical way to investigate the $AdS_5 \times S^5$ superstring at finite coupling, relying on lattice field theory methods. Deeply inspired by a previous study of Roiban and McKeown, we discretize the $AdS_5 \times S^5$ superstring theory in the AdS light-cone gauge and perform lattice simulations employing a Rational Hybrid Monte Carlo algorithm. We measure the string action, from which we extract the null cusp anomalous dimension of planar $\mathcal{N}=4$ SYM as derived from AdS/CFT, as well as the mass of the two AdS excitations transverse to the relevant null cusp classical solution. For both observables, we find good agreement in the perturbative regime of the sigma-model at large 't Hooft coupling. For small coupling, the expectation value of the action exhibits a deviation compatible with the presence of quadratic divergences. After their non-perturbative subtraction, the continuum limit can be taken and it suggests a qualitative agreement with the non-perturbative expectation from AdS/CFT. For small coupling, we also detect a phase in the fermionic determinant that leads to a sign problem not treatable via standard reweighting. We explain its origin and also suggest an alternative fermionic linearization.

Acknowledgements

I am immensely grateful to Prof. Dr. Valentina Forini, for her patient guidance, constant supervision and fruitful collaboration over the last 3 years. Most of the achievements collected in the present thesis would not have been possible without her and my other collaborators. I wish to express my gratitude to Lorenzo Bianchi, Marco Stefano Bianchi, Valentina Giangreco Marotta Puletti, Luca Griguolo, Björn Leder, Domenico Seminara and Michael Pawellek for having promptly responded to my numerous questions with competence and shared their expertise with me. In particular, I owe special thanks to Luca Griguolo and Domenico Seminara for having accepted to support my postdoctoral applications last year and to Björn Leder, Michael Pawellek and especially Valentina Forini for many helpful comments on the draft of this thesis. I am indebted to Arkady A. Tseytlin for always having made himself available with his deep expertise in a recently started collaboration. I acknowledge fruitful collaborations with the master students Alexis Brès and Philipp Töpfer.

I am grateful to Matteo Beccaria, Gerald Dunne and Arkady A. Tseytlin for earlier collaborations with two of my coauthors (V. Forini and M. Pawellek) on the topic presented in Chap. 4. I would like to thank Dmytro Volin for having shared a Mathematica script for the numerical solution of the BES equation, and also Benjamin Basso and Pedro Vieira for one related to the spectrum of the GKP string. All this has been useful for the theme of Chap. 7, which was made possible thanks to a long and fruitful collaboration with Mattia Bruno in the initial stages. I am particularly thankful to Radu Roiban and Rainer Sommer for having agreed to be members of my doctoral committee and for several discussions that eventually contributed to the achievements in Chap. 7. I thank Radu Roiban also for suggesting some corrections that improved the quality of the published version of this thesis.

I would like to thank Prof. Dr. Jan Plefka and Prof. Dr. Matthias Staudacher, together with the members of their groups, for several scientific interactions and the pleasant stay at Humboldt University Berlin. My research has benefited from invaluable discussions with a number of people: Marisa Bonini, Alessandra

Cagnazzo, Xinyi Chen-Lin, Stefano Cremonesi, Amit Dekel, Francesco Di Renzo, Harald Dorn, Nadav Drukker, Giovanni Eruzzi, Alberto Faraggi, Davide Fioravanti, Sergey Frolov, Ilmar Gahramanov, Simone Giombi, Jaume Gomis, Vasco Gonçalves, Nikolay Gromov, Ben Hoare, Yunfeng Jiang, George Jorjadze, Thomas Klose, Shota Komatsu, Martin Kruczenski, Matias Leoni, Florian Löbbert, Fedor Levkovich-Maslyuk, Tristan McLoughlin, Daniel Medina Rincón, Marco Meineri, Carlo Meneghelli, Vladimir Mitev, Dennis Müller, Hagen Münkler, Edvard Musaev, Fabrizio Nieri, David Schaich, Leopoldo Pando Zayas, Yi Pang, Sara Pasquetti, Alessandro Pini, Antonio Pittelli, Jonas Pollock, Michelangelo Preti, Israel Ramírez, Luca Romano, Matteo Rosso, Marco Sanchioni, Amit Sever, Alessandro Sfondrini, Christoph Sieg, Guillermo Silva, Stijn van Tongeren, Diego Trancanelli, Pedro Vieira, Gang Yang, Konstantin Zarembo and Stefan Zieme. A special acknowledgement goes to Ilmar Gahramanov for the useful tips that smoothened the bureaucracy for me over the last months and to Björn Leder and Hagen Münkler for translating the abstract of this thesis.

I feel obliged to give my special thanks to the Director of the Department of Physics Prof. Dr. Norbert Koch, the leader group of the "Quantum Field and String Theory Group" Prof. Dr. Jan Plefka and my doctoral advisor Prof. Dr. Valentina Forini for supporting the nomination and endorsement of this scientific work for publication in the Springer Theses collection. I also take the chance to express thanks to Prof. Diego Trancanelli and Prof. Victor de Oliveira Rivelles for the postdoctoral position offered to me in their group.

My doctoral studies have been funded by DFG via the Emmy Noether Program "Gauge Fields from Strings", the Research Training Group GK 1504 "Mass, Spectrum, Symmetry", the GATIS Initial Training Network, IRIS Adlershof of Humboldt University Berlin and the Marie Curie Initial Training Network UNIFY. Over the last 3 years, a significant part of my research has been done at the Galileo Galilei Institute for Theoretical Physics (Florence, Italy), Nordita (Stockholm, Sweden) and Perimeter Institute for Theoretical Physics (Waterloo, Canada), which I would like to warmly thank for the kind hospitality. I also want to thank the Max Planck Institute for Gravitational Physics (Potsdam, Germany) for the inspiring atmosphere provided in many occasions while this thesis was in preparation.

Contents

Chapter 1
Introduction

Modern theoretical physics has been written in the language of two major scientific paradigms: theory of general relativity and quantum field theory.

The Einstein's theory of gravitation provides an elegant geometric interpretation of gravitational attraction as a dynamical effect of the curvature of space and time, seen as interwoven in a single four-dimensional "fabric" called spacetime, determined by the distribution of energy and momentum carried by the matter and radiation filling the universe. Over the last century, a number of physical phenomena has been derived from this principle and found consistent with experimental data at the current level of accuracy [1]. The first direct detections of gravitational waves, travelling as "ripples" of spacetime, has been confirmed recently by the LIGO and Virgo collaborations [2, 3]. Despite these successes, general relativity still defies all efforts to reconcile them with a microscopical description at a quantum level.

On a parallel route, non-gravitational forces have been incorporated into a theoretical framework where special relativity fits together with quantum mechanics and the concept of field quanta supersedes the classical idea of single particles. Quantum field theory (QFT) has evolved to start new trends in condensed matter physics, leading to the study of critical phenomena in connection with phase transitions using the renormalization group flow [4], with the benefit of providing a new viewpoint on renormalization in particle physics [5–7] ([8] for a review). The same symbiosis has developed in connection to special QFTs with spacetime conformal invariance following earlier studies in two-dimensional critical systems [9].

The original focus of QFT arose within the first attempts to quantize gauge theories. The formulation of quantum theory of electrodynamics (QED) served as model for the development of quantum chromodynamics (QCD), which unravelled the puzzle behind the growing list of hadrons discovered in the late 1960s in terms of the strong interaction among constituents particles called quark and gluons. Subsequent efforts to describe weak interactions as the exchange of heavy bosons culminated in the foundation of the best theoretical tool to investigate nature at short distances as we

E. Vescovi, *Perturbative and Non-perturbative Approaches to String Sigma-Models in AdS/CFT*, Springer Theses, DOI 10.1007/978-3-319-63420-3_1

know it today, the *Standard Model of elementary particle physics*. Free of quantum anomalies and arguably theoretically self-consistent, it describes the dynamics of matter particles as the exchange of the force carriers of a non-abelian (Yang–Mills) theory with local (gauge) symmetry group $SU(3) \times SU(2) \times U(1)$ partially broken by the Higgs mechanism [10–12]. One of its greatest successes is the interpretation of the mysterious Feynman-Björken scaling as an effect of asymptotic freedom in non-abelian gauge theories [13], when quarks behave as non-interacting constituents in deep inelastic scattering. Since then, theoretical predictions have shown agreement with the experimental data with spectacular precision [14]. The process of experimental validation continues and recently led to the discovery of the last elusive particle, the Higgs boson, at the Large Hadron Collider [15, 16]. That being said, the Standard Model cannot be the last word on physical reality. The next future will likely shed light on many known inadequacies and unanswered questions, for instance the hierarchy problem of the fundamental forces, the phenomenon of neutrinos oscillations and cosmic observations hinting at the existence of dark matter and dark energy.

Most of the predictive power of the Standard Model is due to perturbative approximations around the free theory by means of Feynman diagrams. However, the hope of resumming loop expansions vanishes as soon as one realizes that they are typically asymptotic expansions with zero radius of convergence. Moreover, perturbation theory breaks down when applied to inherently strongly coupled quantum phenomena, e.g. solitons and bound states. Of course, there are direct attempts to quantitatively understand the mechanism of quark confinement and arrive at reasonable approximations for the hadronic spectra, but they are the product of numerical simulations of effective theories, which may obscure a microscopic description in terms of the elementary constituents.

An alternative step consists in engineering a "toy model" that abandons the immediate ambition to describe the real world. The first step is to reduce the complexity of the problem and exploit enlarged number of symmetries to make nontrivial analytical statements. Second, the simplified model can be enriched with more features in order to transfer some of its properties back to the original system to some degree. This strategy has proven to be extremely useful in countless occasions throughout the history of science. For instance, it happened at the dawn of quantum mechanics when the development of the Hartree–Fock method (e.g. in [17]) to calculate wavefunctions for multi-electron atoms and small molecules was guided by earlier semi-empirical methods based on the exact Schrödinger solution for the hydrogen atom.

In order to gain a better theoretical understanding of QCD physics and to develop new computational tools in QFT in general, theoretical physicists have looked for the "most symmetric" interacting gauge theory in four dimensions. This role is arguably played by $\mathcal{N} = 4$ *supersymmetric Yang–Mills (SYM) theory* [18]: it describes a Minkowskian universe containing scalars and fermions interacting via non-abelian gluons. It possesses the maximal amount of $\mathcal{N} = 4$ supercharges to be renormalizable in four dimensions, which fully constrains the precise form of the interactions, and it does not display any parameter other than the coupling constant and the gauge group.[1] In addition to supersymmetry, the model exhibits exact conformal symmetry at the quantum level and it is conjectured to have an "electric-magnetic" Montonen–Olive $SL(2, \mathbb{Z})$ duality [20–22], one of the earliest instances of S-duality. Of course, we cannot expect to draw heavily on this analogy, as it is clear from the fact that $\mathcal{N} = 4$ SYM has massless mass spectrum and no running coupling constant [23–26]—meaning neither a characteristic scale nor asymptotic freedom—leaving aside the fact that supersymmetry is not a feature of the Standard Model. However, there exist quantitative features of $\mathcal{N} = 4$ SYM found to survive in QCD, for instance in the conformal dimension of local gauge-invariant operators[2] and in the derivation of tree-level QCD scattering amplitudes from $\mathcal{N} = 4$ SYM [30]. In the remainder of the chapter, we will show that there are also other reasons that make $\mathcal{N} = 4$ SYM a theoretical laboratory worth to be studied in its own right.

1.1 The AdS_5/CFT_4 and AdS_4/CFT_3 Correspondences

One of the major breakthroughs of the recent years is the *Anti-de Sitter/Conformal Field Theory (AdS/CFT) correspondence* [31–33].[3] The conjecture asserts the exact equivalence between a pair of models. On one side, there is a QFT with conformal spacetime symmetry in d dimensions. On the other one, we have a superstring theory where strings move in the target space $AdS_{d+1} \times \mathcal{M}_{9-d}$ including an anti de-Sitter space, a $(d+1)$-dimensional manifold with constant negative curvature, and a compact manifold $\mathcal{M}_{9-d}$ in $9-d$ dimensions. The d-dimensional boundary of the background is a conformally flat space on which the CFT is formulated. Note that the dimensions of the two factors (AdS and $\mathcal{M}$) add up to yield a string theory with fermions in ten dimensions, which is the critical dimension to ensure the cancellation of conformal anomaly on the worldsheet. In this context, the term "equivalence" is a synonym of one-to-one correspondence between aspects of the two models (e.g. global symmetries, operator observables, states, correlation functions). The claim that the dynamics of the string degrees of freedom can be encoded in a lower dimensional (non-gravity) theory at its boundary suggests to see it as

---

[1] In principle, one can also consider the instanton angle θ which combines with the YM coupling constant into a complex coupling $\tau = \frac{\theta}{2\pi} + \frac{4\pi i}{g^2_{\text{YM}}}$. Through the AdS/CFT correspondence (Sect. 1.1), the angle θ equals the expectation value of the axion field in the spectrum of the dual Type IIB superstring, e.g. [19].

[2] Twist-two (Wilson) operators play an important role in deep inelastic scattering in QCD as much as in $\mathcal{N} = 4$. Their anomalous dimension for large spin is governed by the so-called *scaling function* of the theory in question, see Sect. 7.1. The *maximal transcendentality principle* conjectured in [27] states that the $\mathcal{N} = 4$ SYM scaling function has uniform degree of transcendentality $2l - 2$ at loop order l and can be extracted from the QCD expression by removing the terms that are not of maximal transcendentality. A brief account of the subject and references are in [28, 29].

[3] Among the many reviews on the topic, we suggest [19, 34–38] and the excellent textbook [39].

a realization of *holographic duality*. Since we will encounter CFTs that are gauge theories in this thesis (summarized in (1.1) and (1.5) below), we often refer to the correspondence also as *gauge/gravity duality*.

The first example [31] at the spotlight since 1997—later named *AdS_5/CFT_4 correspondence*—relates

$\mathcal{N}=4$ super Yang–Mills in flat space $\mathbb{R}^{1,3}$
with Yang-Mills coupling constant $g_{\rm YM}$ and gauge group $SU(N)$

and (1.1)

type IIB superstring theory with string tension T and coupling constant g_s
on $AdS_5 \times S^5$ with curvature radii $R_{AdS_5} = R_{S^5} \equiv R$
and N units of Ramond–Ramond five-form flux through S^5.

Here, T is an overall factor in the string action and g_s is the genus-counting variable in the perturbative expansion over topologies of string theory. The AdS/CFT dictionary relates the gauge/string parameters through the dimensionless 't Hooft coupling λ

$$\lambda = g^2_{\rm YM} N\,, \qquad \lambda = 4\pi^2 T^2 = 4\pi N g_s = \frac{R^4}{\alpha'^2}\,. \tag{1.2}$$

The constant α' is the square of the string characteristic length and historically the *slope parameter* in the linear relationship between energy/angular momentum of rotating relativistic bosonic strings in flat space. The motivation behind the correspondence (1.1) arose from the investigation of a stack of N parallel Dirichlet branes (D3-branes), three-dimensional objects sweeping out a (1+3)-dimensional volume, separated by a distance d and embedded in type IIB string theory in $\mathbb{R}^{1,9}$. D-branes can be viewed in two equivalent ways, fundamentally linked to the *open/closed string duality*, where $\mathcal{N}=4$ SYM theory and type IIB supergravity in $AdS_5 \times S^5$ emerge as two (arguably equivalent) low-energy descriptions of the same physics in *Maldacena limit* for α', $d \to 0$ while holding α'/d fixed. Relaxing the supergravity limit $\alpha' \to 0$, the claim [31] is that the two models in (1.1) continue to be dual for any values of the parameters.

An immediate "check" of the duality is the fact that the two models in (1.1) have the same global symmetry group $PSU(2,2|4)$, namely the super-Poincaré and conformal invariance of $\mathcal{N}=4$ SYM and the superisometry group of the string theory in $AdS_5 \times S^5$. On operative level, one establishes the equivalence of the superstring partition function, subject to sources ϕ for string vertex operators with boundary value ϕ_0, and the partition function in the CFT side with sources ϕ_0 for local operators

$$Z_{\rm string}\left[\phi|_{\partial(AdS_5)} = \phi_0\right] = Z_{\rm CFT}[\phi_0]\,. \tag{1.3}$$

The strongest version of the conjecture puts no restriction on the parameter space, but it is hard to check its validity if we do not work in certain simplifying limits to enable a perturbative approach. A unique parameter (λ) turns out to be a useful choice when considering the *'t Hooft limit* [40]

$$g_{\rm YM} \to 0\,, \qquad N \to \infty\,, \qquad \lambda = \text{constant}\,. \tag{1.4}$$

The Yang–Mills theory becomes a free non-abelian theory ($g_{\rm YM} \to 0$) for infinitely many "colors" ($N \to \infty$) where the class of *planar graphs* is dominant in the diagrammatical expansions. In the partner model, the joining and splitting of strings is suppressed ($g_s \to 0$) and only lowest genus surfaces survive. For small λ, the string is subject to large quantum mechanical fluctuations ($T \to 0$) on a highly curved $AdS_5 \times S^5$ ($R \ll \sqrt{\alpha'}$); conversely for large λ, the string behaves semiclassically ($T \to \infty$) in a flat-space limit ($R \gg \sqrt{\alpha'}$). For the latter interpretation, we recall that T is an overall factor of the string action and thus can be assimilated to a sort of inverse Planck constant. Conventional perturbative calculations on the gauge theory side are possible to a certain extent if we impose that λ is small (*weak coupling*), while semiclassical methods can probe the string corrections to the classical supergravity theory ($\alpha' = 0$) when we adjust λ to be large (*strong coupling*). This observation enables to make precise statements about a strongly coupled regime of a gauge theory, typically lacking systematic quantitative tools previous to the AdS/CFT correspondence, as long as it admits a higher dimensional string theory.

The seminal paper by Maldacena [31] sparked a quest for other realizations of AdS/CFT duality. Following earlier works [41, 42], Aharony, Bergman, Jafferis and Maldacena (ABJM) [43] ([44] for a review) established the equivalence between a theory of M2-branes in 11 dimensions and a certain three-dimensional gauge theory. The two parameters k and N (defined below) allow for a somewhat richer structure than the AdS_5/CFT_4 system. In this thesis we will limit ourselves to consider the duality between

$\mathcal{N} = 6$ super Chern–Simons theory with matter in flat space $\mathbb{R}^{1,2}$
with integer Chern–Simons levels k and $-k$
and gauge group $SU(N)_k \times SU(N)_{-k}$

and (1.5)

type IIA superstring theorywith string tension T and coupling constant g_s
on $AdS_4 \times \mathbb{CP}^3$ with curvature radii $2R_{AdS_4} = R_{\mathbb{CP}^3} \equiv R$
and N units of Ramond–Ramond four-form flux through AdS_4
and k units of Ramond–Ramond two-form flux through $\mathbb{CP}^1 \subset \mathbb{CP}^3$,

provided the identifications through the 't Hooft coupling λ[4]

[4] We will make clear the distinction between the 't Hooft parameter $\lambda_{\rm YM}$ of $\mathcal{N} = 4$ SYM and the one $\lambda_{\rm ABJM}$ of ABJM when necessary, namely in Chap. 6.

$$\lambda = \frac{N}{k}, \qquad \lambda = \frac{R^6_{AdS_4}}{32\pi^2 k^2 \ell_P^6}. \tag{1.6}$$

The duality (1.5) holds only in the analogue [43] of the 't Hooft limit (1.4)

$$N,\, k \to \infty, \qquad \lambda = \text{constant}. \tag{1.7}$$

On the gauge theory side, the ABJM theory is a supersymmetric extension of pure Chern–Simons theory, which is a broad subject with applications to three-dimensional gravity theory [45] and knot theory [46]. The addition of $\mathcal{N} = 6$ supercharges[5] renders ABJM a non-topological theory, still retaining conformal invariance. The global symmetry group of the ABJM theory and the dual string theory is the orthosymplectic supergroup $OSp(6|4)$.

The original "dictionary" proposal [43] for the string tension in terms of the 't Hooft coupling λ reads

$$T = \frac{R^2}{2\pi\alpha'} = 2\sqrt{2\lambda}, \qquad g_s \propto \frac{N^{1/4}}{k^{5/4}}. \tag{1.8}$$

As suggested in [48] and later quantified in [49], the relation between T and λ receives quantum corrections. The geometry of the background (and also the flux, in the ABJ theory [50], generalization of the ABJM theory with gauge group $U(N) \times U(M)$) induces higher order corrections to the radius of curvature in the Type IIA description, which reads in the planar limit (1.7) of interest in this thesis

$$T = \frac{R^2}{2\pi\alpha'} = 2\sqrt{2\left(\lambda - \frac{1}{24}\right)}. \tag{1.9}$$

The *anomalous radius shift* by $-\frac{1}{24}$ in (1.9) is important at strong coupling, because it affects the corrections to the energy and anomalous dimensions of giant magnons and spinning strings starting from worldsheet two-loop order $O(\lambda^{-1/2})$. It will also turn out to be crucial in Chap. 6 to translate the string tension T into the gauge coupling λ.

Another instance of holography is the AdS_3/CFT_2 correspondence between superstring theories on backgrounds involving the AdS_3 space and two-dimensional superconformal field theories. The supersymmetric backgrounds of interest, especially because of their integrable properties, are the $AdS_3 \times S^3 \times S^3 \times S^1$ and $AdS_3 \times S^3 \times T^4$ supergravity backgrounds which preserve 16 real supercharges. However, in light of the work done in the next chapters, we will be mostly concerned with the other two dualities spelt out above, referring the reader to [51] (also [52]) and references therein for an account of the subject.

[5]Supersymmetry is enhanced to $\mathcal{N} = 8$ at Chern–Simons level is $k = 1, 2$ [43, 47]. We can disregard this exception since we will be working in planar limit.

1.2 Integrable Systems in AdS/CFT

Since its discovery, the AdS/CFT correspondence prompted a new interest in $\mathcal{N}=4$ SYM and offered a (strong coupling) perspective to study this gauge theory. Ideally, the aim of *solving* a QFT means to express arbitrary n-point correlation functions of any combination of fields in terms of elementary functions or integral/differential equations involving the parameters of the model. When this happens, it signals the presence of an infinite number of conserved charges and the theory in question is called *classically integrable*, and *quantum integrable* if the property persists at the quantum level.[6] It is clear that this requirement is extraordinarily difficult to satisfy, save for a few exceptions typically relegated to two-dimensional models. A less trivial occurrence, the first in four dimensions, emerges in high-energy QCD scattering [57–59].

Evidence of integrable structures in planar $\mathcal{N}=4$ SYM later emerged in relation to single-trace operators (the only relevant ones at $N \to \infty$) and certain spin-chain models.[7] Since the theory is conformal, the dynamical information is contained in the two- and three-point functions of local gauge-invariant operators.[8] Conformal symmetry fixes their two-point correlators in terms of their eigenstates under the action of the dilatation operator $D \in \mathfrak{psu}(2,2|4)$, namely the spectra of scaling dimensions of all operators.

The breakthrough of [61] was realizing that single-trace operators in the flavour sector $SO(6)$ (i.e. traces of a product of any of the scalars of $\mathcal{N}=4$ SYM) are mapped to states of a periodic spin-chain and the (one-loop) dilatation operator to the Hamiltonian of the spin-chain system. The spectrum of scaling dimensions at weak-coupling one-loop order was set equivalent to the diagonalization problem of the auxiliary $SO(6)$ spin-chain Hamiltonian which, since it was known to be integrable, could be solved exactly using Bethe ansatz techniques [63–66]. Dropping the restriction to scalar operators, integrability was established for all operators at the one-loop order [67] in terms of an integrable $PSU(2,2|4)$ super spin-chain, later diagonalized in [68]. A further development concerned the generalization of the Hamiltonian method to two and three loops [69].

This program was pushed further to reveal classical integrability on the string theory side of AdS/CFT by explicitly rewriting the equations of motion of the non-linear sigma-model on $AdS_5 \times S^5$ background [70] into a zero-curvature condition for a Lax pair operator [71]. Following the same approach the analogous set of non-local conserved charges was constructed in [72] in the pure spinor formulation of the $AdS_5 \times S^5$ action [73–75] and the same Lax pair was found in [76].[9]

---

[6] We suggest [53, 54] for an extensive discussion of integrable systems and also [55, 56] for a focus on AdS/CFT.

[7] A transparent and concise introduction to the subject is in [60–62].

[8] Higher point functions decompose into these elementary constituents [9].

[9] The integrability of the string in the $AdS_5 \times S^5$ background has been mostly studied in the supercoset description [70] (e.g. in [77]) than in the pure spinor version. Some integrable properties in the former formalism will be discussed to some extent in Sects. 2.1.1, 2.1.2 and 2.1.3, while we

With integrability becoming a solid fact at both weak and strong coupling, the focus shifted to speculate about this property holding true at all loops. In [83] a direct relationship between Bethe equations and classical string integrability was reinforced using the language of algebraic curves, interpreted as a sort of continuum version of Bethe equations. On the *assumption* of exact *quantum* integrability of the AdS_5/CFT_4 system, a set of Bethe equations valid at all loop order was formulated [84] for all *long* local operators [85]. These results were complemented by the study of the so-called *dressing factor* [86–91][10] and collectively referred to as all-loop *asymptotic Bethe ansatz (ABA)*, as their validity is limited to asymptotically long chains in the auxiliary picture. In principle this enabled to solve the spectral problem for the anomalous dimension of all long single-trace operators in planar $\mathcal{N} = 4$ SYM.

The understanding of the conjectured integrability has steadily advanced towards the inclusion of finite-size effects (*wrapping effects*) [93–96]. This ambitious program included the development of an infinite set of coupled integral equations called *Thermodynamic Bethe ansatz (TBA)* [94, 97–102] (also in [103, 104]) which are solvable in some cases for scattering amplitudes [105] and cusped Wilson lines [106–108]. This served as a basis for the so-called *Y-system* [99] (an infinite set of nonlinear functional equations) and its successor *FiNLIE* (acronym for "finite system of non-linear integral equations"). The state of the art in elegance and computational efficiency in solving the spectral problem seems to be achieved in the form of a set of Riemann–Hilbert equations that defines the *quantum spectral curve (QSC)* approach (or *Pμ-system*) [109, 110], where the so-called Q functions are a sort of quantum generalization of pseudo-momenta in the algebraic curve construction. The potential of this machinery extends beyond the original scope of computing spectrum of anomalous dimensions, e.g. in the high-precision and non-perturbative numerical computation of the generalized cusp anomalous dimension of a cusped Wilson line [111, 112].

Almost all relevant statements that have been made about integrability for the planar AdS_5/CFT_4 system have been reworked almost in parallel for the lower dimensional correspondence AdS_4/CFT_3 in the planar limit, see [113] for a comprehensive overview. The investigation started perturbatively at planar two-loop order for scalar operators by constructing the corresponding integrable spin-chain Hamiltonian [114, 115]. The extension to all operators (at two-loop order) was derived in [116, 117]. One of the most distinguishing differences between the integrable structure of ABJM and the one of $\mathcal{N} = 4$ is the fact that the transition from weak to strong coupling is more intricate due to the presence of the nontrivial *(ABJM) interpolating function* $h(\lambda)$, introduced and analysed in Chap. 6. This function plays the crucial role of a "dressed" coupling constant that absorbs the dependence on the 't Hooft coupling λ in all integrability-based computations, e.g. in the set of ABA equations for the

(Footnote 9 continued)
refer the reader to a non-exhaustive selection of relevant references in the latter formalism in [78, 79] and in the reviews [80, 81]. Arguments that support the quantum integrability of the pure spinor action were given in [82].

[10]This is a function undetermined by the symmetries of the theory, but constrained by physical requirements such as crossing symmetry and unitarity, see [92] for a short review.

complete spectrum of all long single-trace operators proposed in [118]. At strong coupling, the classical spectral curved was constructed in [119] and integrability was demonstrated for the supercoset action at classical level [120].[11] Echoing the developments in $\mathcal{N} = 4$ SYM, the Y-system was proposed in [99] along with the analogue one for AdS_5/CFT_4 system. The infinite set of nonlinear integral TBA equations encoding the anomalous dimensions spectrum was derived in [121, 122]. The QSC formalism was set up in [123] and used to put forward a conjecture for the exact form of $h(\lambda)$ in the ABJM model [124][12] and in its generalization, the ABJ model, in [125].

The concept of integrability has been rephrased in several contexts and its facets detected in a wide range of observables. Another realization is *Yangian symmetry* [126], a sort of enhancement of the Lie algebra symmetry $\mathfrak{psu}(2, 2|4)$ of the theory, which benefited from the previous discovery of the duality [127] mapping scattering amplitudes of n gluons to polygonal Wilson loops with n light-like segments, see also [128–131] for some later developments. This duality was proposed at strong coupling and later noticed in perturbative computations at weak coupling [132] (also [133]) where it inspired the discovery of a hidden *dual superconformal symmetry* [134, 135]. Soon after, the latter and the conventional conformal symmetry were shown to combine into the Yangian symmetry [136]. This symmetry has been seen in colour-ordered scattering amplitudes at tree level [136] and in loop quantum corrections [137–139], the dilaton operator [140] and supersymmetric extensions of Wilson loops [141, 142].

A further area rich of developments is the study of the dual polygonal light-like Wilson loops at any coupling through a pentagon-block decomposition in the form of an OPE-like expansion [143–147] which can be determined again on the basis of integrability arguments. An integrability-based framework to compute structure constants of higher point correlation functions was recently established in [148].

1.3 Quantization of Strings in AdS/CFT

In the previous section, we have seen that integrability offers a wide range of techniques to make quantitative predictions about the spectral problem and other observables in many AdS/CFT systems, by means of analytical and high-accuracy numerical methods. All these statements are based on the conjectured all-loop integrability of the model. Without this assumption, a restricted class of supersymmetry-protected observables can still be computed at finite coupling and beyond the planar limit via supersymmetric localization techniques [149], which are however only defined on the field theory side. Leaving aside the ambition of *proving the assumptions* of integrability from first principles, the natural question arising is whether one can *check*

[11] More references on the subject are below (2.35) in Sect. 2.2.

[12] We will test the strong-coupling expansion of the interpolating function in Chap. 6.

their predictions against perturbative results and non-perturbative ones obtained with different methods. Field theory computations maintain a crucial role in detecting the precise pattern of such functions of coupling and charges, as well as in checking the proposed all-loop formalisms. This viewpoint shifts the attention from exact methods to the development of computational tools, in principle flexible enough to work in different frameworks when neither integrability nor localization is available.

We shall pursue this goal in the AdS/CFT systems (1.1) and (1.5) considered in the respective 't Hooft limit (1.4) and (1.7) of their parameter spaces, exclusively working on the string theory side. Put in simple words, *strings* are objects spatially extended in one dimension, at variance with point-like particles of ordinary QFTs, and are embedded in a higher dimensional ambient manifold (*target space*), for us the ten-dimensional $AdS_5 \times S^5$ or $AdS_4 \times \mathbb{CP}^3$. Note that quantum mechanical consistency guarantees the absence of conformal anomaly when the dimensionality of the spacetime is 10. Strings sweep out a (1+1)-dimensional surface Σ (*worldsheet*) in their time evolution. Since all computations are in a limit where N is put to infinity, scattering of two or more strings does not occur and worldsheets are genus-0 surfaces (i.e. without "holes" and "handles"). All observables depend on the single tunable parameter T (or equivalently λ) of the model under investigation. The fields of a string theory consist of the bosonic embedding coordinates of Σ into the target space and their fermionic supersymmetric partners. From the worldsheet viewpoint, they are bosonic (collectively denoted by X) and fermionic fields (Ψ) propagating in the two-dimensional curved manifold Σ.

From now on we shall focus on the prototypical duality (1.1), as the following statements hold for (1.5) after the necessary changes having been made. Since the string theory in $AdS_5 \times S^5$ includes a Ramond–Ramond five-form flux, the Neveu-Schwarz-Ramond (NSR) formalism [150, 151] is not applicable in a straightforward way. As we will see in Chap. 2, the $AdS_5 \times S^5$ is 10d supersymmetric background of type IIB supergravity [152]. The Green–Schwarz (GS) approach [153, 154] seems to be adequate when the RR fields are not vanishing and would endow the string action with invariance under supersymmetry (manifestly realized as a target-space symmetry) and κ-symmetry (a local fermionic symmetry that ensures the correct number of physical fermionic degrees of freedom), but it is not very practical for finding the explicit form of the action in terms of the coordinate fields. For $AdS_5 \times S^5$ the superstring action is formulated [70] as a sigma-model on a supercoset target space. This is an highly interacting two-dimensional field theory for which a first-principle quantization is a hard theoretical problem.[13]

[13]For a general curved target space, the string equations of motion are nonlinear and the right and left oscillator modes of the string interact with themselves and with each other [155], see also [156] for further issues. As for the quantization of a generic field theory in curved spacetime, a good initial reference is the textbook [157].

The quantization is more straightforward if one picks a suitable string vacuum[14] (whose properties and/or quantum numbers depend on the particular string observable to study), fixes the gauge symmetries (two-dimensional diffeomorphisms and κ-symmetry) and expands the degrees of freedom of the superstring in terms of fluctuation fields around such vacuum.

As in ordinary quantum field theory, the fundamental object is the *string partition function*

$$Z_{\text{string}} = \int \mathcal{D}g\, \mathcal{D}X\, \mathcal{D}\Psi\, e^{-S_{\text{IIB}}[g,X,\Psi]} \,. \tag{1.10}$$

We work with S_{IIB} being the sigma-model action of [70], where one has to integrate over the two-dimensional metric g_{ij} and fix the diffeomorphism-Weyl invariance of the action with the Faddeev–Popov procedure.[15] The *fluctuation string action* is written in terms of the fluctuations $\delta X = X - X_{\text{cl}}$ and $\delta\Psi = \Psi$ around the nontrivial vacuum $(X_{\text{cl}}, \Psi = 0)$, where X_{cl} is the chosen classical solution of the string equations of motion and fermions are set to zero on a classical configuration. The expansion of the action (1.10) delivers an infinite tower of complicated-looking interaction vertices organized in increasing inverse powers of T. Note that we have not made any assumption on the (small or finite) value of the coupling constant T^{-1} up to this point.

One can proceed with *perturbation theory* for large string tension $T \sim \sqrt{\lambda} \gg 1$, which indeed corresponds to the nearly free regime of the sigma-model at small T^{-1}. To access the non-perturbative regime of the full quantum superstring, one can resort to techniques of lattice field theory and evaluate numerically the string observable of interest. One main objective of this thesis will be to present evidence that this route is indeed viable and that the data collected so far (Chap. 7) is consistent with the expectations based on the integrability of the AdS_5/CFT_4 system.

Before addressing this important methodological distinction, we recall a few facts about the properties of classical backgrounds X_{cl}.

Solutions that are translationally invariant in the time and space coordinates (τ, σ) of the worldsheet, namely with constant derivatives of the background X_{cl}, are called *homogeneous*. In this case the effective action $\Gamma \equiv -\log Z_{\text{string}}$ is an extensive quantity—proportional to the area of the classical worldsheet—and the semiclassical analysis is highly simplified since the action turns out to have constant coefficients. Then the kinetic/mass-operator determinants entering the one-loop partition function are expressed in terms of characteristic frequencies which are relatively simple to calculate. We will see that computation of quantum corrections can be pushed to higher loop order by standard diagrammatic methods. In this context, generalized

[14]The string sigma-model of [70] is *nonlinear* because the curvature of the target space brings *field-dependent* coefficients of the kinetic terms. Expanding the path-integral around a classical solution generates standard quadratic kinetic terms (and interaction terms) for the fluctuation fields that make the sigma-model tractable.

[15]One should also remember that a rigorous definition takes into account some factors associated with conformal Killing vectors and/or Teichmüller moduli. We defer the discussion to [158, 159] and the textbooks [156, 160].

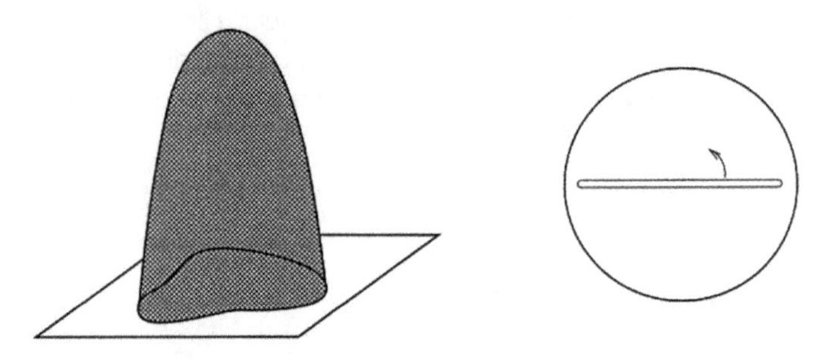

Fig. 1.1 Sketchy depiction of two distinctive classes of string configurations that will be of interest in the next chapters: an open string protruding in the bulk of the *AdS* space and ending on the path of a supersymmetric Wilson loop operator at the *AdS* boundary (*left panel*, from [178]) and a closed string that is folded upon itself and rotating in a subregion of the $AdS_5 \times S^5$ space (*right panel*, from [170])

unitarity techniques are a promising way to reproduce loop-level worldsheet amplitudes in terms of lower loop ones [161–163]. Instances of such homogeneous cases are the rational rigid string solutions in [164–168]. Other cases can still fall under this category if it is possible to redefine coordinates and fields to make the coefficients in the fluctuation action constant, as in Chaps. 6 and 7, as well as in [169] (Fig. 1.1).

Next-to-simplest cases are *inhomogeneous solutions*, namely nontrivial solutions of the string sigma-model that are not translationally invariant in either the τ- or σ-direction. Beyond the leading order, direct computations are generally difficult and one-loop corrections are already a daunting task that requires the diagonalization of many 2d matrix-valued differential operators using functional methods based on the notion of spectral zeta-function. The rigid spinning string elliptic solutions rotating with spin S in AdS_5 and momentum J in S^5 [170, 171] (and Chap. 4) are well-known examples of inhomogeneous backgrounds. Another non-negligible difficulty to face is the appearance of nontrivial special elliptic functions in the fluctuation spectrum (and thus in the propagator) in [172–175] which depend on the worldsheet coordinates. However, there are non-homogeneous cases that become homogeneous in certain limits, as for the example above in the limit $S/\sqrt{\lambda} \gg 1$ with $J/(\sqrt{\lambda} \log S)$ fixed [176, 177].

The classical backgrounds studied the next Chaps. 4–7 assume a special relevance in the AdS/CFT correspondence since they will be dual to two types of gauge theory observables.

One class of observables comprises gauge-invariant, non-local observables called Wilson loops. In ordinary (conformal or not) gauge theories they are obtained [179] from the holonomy of the gauge connection around a closed spacetime path and carry information on the potential between static quarks when defined on a rectangular contour. In $\mathcal{N} = 4$ SYM "quarks" are modelled by infinitely massive W-bosons arising from a Higgs mechanism, and Wilson loops admit a supersymmetric extension

(*Maldacena-Wilson loops*) locally invariant under half of the supercharges.[16] The AdS/CFT formulation of the duality between Wilson loops and open strings [183, 184] states that the expectation value of a supersymmetric Wilson loop $\mathcal{W}[\mathcal{C}]$ defined along a contour $\mathcal{C} \subset \mathbb{R}^4$ equals the string partition function $Z_{\text{string}}[\mathcal{C}]$ where the string embedding ends on $\mathcal{C}$

$$\langle \mathcal{W}[\mathcal{C}] \rangle = Z_{\text{string}}[\mathcal{C}] \equiv \int \mathcal{D}g\, \mathcal{D}X\, \mathcal{D}\Psi\, e^{-S_{\text{IIB}}[g,X,\Psi]}\,. \tag{1.11}$$

The reader can consult the review papers [185, 186] for a (not latest though) collection of related works.

The second example comprises local gauge-invariant operators made of traces of fully contracted products of fields of $\mathcal{N}=4$ SYM. The AdS/CFT correspondence conjectures a relation between their conformal dimension and the energy of rotating string states in $AdS_5 \times S^5$ with the same quantum numbers. Since the Cartan subalgebra of $\mathfrak{su}(2,2) \times \mathfrak{su}(4) \subset \mathfrak{psu}(2,2|4)$ has six commuting generators, an operator is labelled by a sextuplet of charges: the *scaling dimension* Δ under spacetime dilatations, the two *spins* S_i $(i=1,2)$ of the Lorentz group and the J_i $(i=1,2,3)$ associated to the three commuting R-symmetry generators. In AdS/CFT the symmetry group $SU(2,2) \times SU(4)$ is identified with the isometry group of $AdS_5 \times S^5$ and the six charges above correspond to the energy E conjugated to the AdS global time t, the AdS_5 spins S_i and the S^5 angular momenta J_i of the string.

Not all parameters are independent due to the Virasoro conditions, so we can express the worldsheet energy $E = E(S_1, S_2, J_1, J_2, J_3)$ of closed spinning strings as a function of the given remaining charges. We will be interested in *rigid* strings, i.e. for which the shape does not change in time. Computing the expectation value of the energy, including its quantum corrections in the coupling constant (the string tension T), is conveniently done by taking a "thermodynamical" approach to connect the semiclassical computation of the energy to the perturbative evaluation of the worldsheet effective action $\Gamma = -\log Z_{\text{string}}$ expanded around the relevant classical configuration. This relation was elucidated in [187, 188] and synthetically reexplained in the review [189].

1.4 Perturbation Theory for Sigma-Models

The perturbative approach to string quantization has proven to be an extremely useful tool for investigating the structure of the AdS/CFT correspondence [189, 190]. As a matter of fact, the first attempt in this direction was the determination of the strong-coupling correction [191] to the quark–antiquark potential of [183], although

[16] See Sect. 5.1. Subsequent steps were made to construct further generalization called *super Maldacena-Wilson loops* [180]. The field theoretical description is explored in [142, 181] while the complementary view at strong coupling in [182].

obstructed by an issue of UV divergences. The study of semiclassical partition functions was systematically set up in [192] and it has played an important role for spinning string states [164–168, 170–172, 193], worldsheet S-matrices [161–163, 194, 195], scattering amplitudes [130] and Wilson loops [173–175, 196–199].

In semiclassical quantization, observables are computed in worldsheet-loop series in T^{-1}, as we will do in Chaps. 3–6. In the example of (1.10), this means that we can truncate the fluctuation action at quadratic (aiming at a one-loop result in Chaps. 4–5), at quartic (in the two-loop example of Chap. 6) or higher order in δX and $\delta\Psi$, depending on the accuracy sought in the final result, and evaluate the path-integral in saddle-point approximation. The *effective action* takes into account semiclassical corrections around the background solution as

$$\Gamma \equiv -\log Z_{\text{string}} = \Gamma^{(0)} + \Gamma^{(1)} + \Gamma^{(2)} + \dots . \tag{1.12}$$

A covariant formalism for the one-loop semiclassical quantization of the action will be the topic of Chap. 3. The supercoset action is not a necessary starting point because the complete (Nambu–Goto and Polyakov) bosonic action was well-known before and the covariant derivative in the Green–Schwarz action at quadratic order in fermions has already appeared in the Killing spinor equation of type IIB supergravity [200].[17] The action at quadratic level is a free model for eight physical bosons ($y^{\underline{i}}$) and 16 κ-symmetry fixed fermionic degrees of freedom (encapsulated in a new spinor that we call Ψ again)

$$S_{\text{IIB}}[X, \Psi] = S_{\text{IIB}}[X = X_{\text{cl}}, \Psi = 0] + \int d\tau d\sigma \sqrt{h}\left(y_{\underline{i}}(\mathcal{O}_B)^{\underline{i}}{}_{\underline{j}} y^{\underline{j}} + 2\bar{\Psi}\mathcal{O}_F\Psi\right) + \dots \tag{1.13}$$

and the one-loop approximation of the path-integral (1.10) can be put into the form (1.12)[18]

$$\Gamma^{(0)} = S_{\text{IIB}}\left[X = X_{\text{cl}}, \Psi = 0\right], \qquad \Gamma^{(1)} = -\frac{1}{2}\log\frac{\text{Det}\,\mathcal{O}_F}{\text{Det}\,\mathcal{O}_B}. \tag{1.14}$$

When the background solution is homogeneous, the one-loop effective action can be formally expressed as a summation (integral) over the discrete (continuous) eigenvalues—usually called *characteristic frequencies*—the operators $\mathcal{O}_B$ and $\mathcal{O}_F$. The matching of the bosonic and fermionic degrees of freedom and a sum rule for their masses (dictated by the geometry of the worldsheet and the target-space supersymmetry of the Green–Schwarz action) guarantee that the result is eventually finite. In the next Sect. 1.4.1 we explain how one has to proceed to quantify the one-loop correction around inhomogeneous solutions. Then, in Sect. 1.4.2 we will

---

[17] The Green–Schwarz action is known to quadratic order in fermions for any general type II supergravity background [201] and recently up to fourth order [202].

[18] The net contribution comes only from the kinetic operators $\mathcal{O}_B$ and $\mathcal{O}_F$ because we suppose that the determinant of the diffeomorphism ghosts cancels the one of the "unphysical" (longitudinal) bosons, see comments below (3.46) and (3.98), and we also ignore the caveats in footnote 15.

make some comments on higher loop corrections $\Gamma^{(\ell)}$ with $\ell \geq 2$ in homogeneous backgrounds, which consist of vacuum Feynman diagrams computable via standard diagrammatical techniques.

1.4.1 Two-Dimensional Fluctuation Operators

A fully two-dimensional definition of (the finite and divergent part of) a determinant can be achieved with the notion of *heat kernel propagator* [203–208] of a $r \times r$ matrix operator $\mathcal{O}$ defined on the classical worldsheet with Riemannian metric h_{ij}. This object contains all spectral information on the operator (eigenvalues and eigenfunctions) [209–211] and it is defined as the unique solution $K_{\mathcal{O}}\left(\tau,\sigma;\tau',\sigma';t\right)$ of the *heat equation*, namely the evolutionary Schrödinger-type equation for the "Hamiltonian" $\mathcal{O}$ evolving in the "Wick-rotated time" $t>0$

$$(\partial_t+\mathcal{O}(\tau,\sigma))\,K_{\mathcal{O}}\left(\tau,\sigma;\tau',\sigma';t\right)=0\,, \tag{1.15}$$

supplemented by an initial normalization à la Dirac delta ($\mathbb{I}_r$ is the $r\times r$ identity matrix)

$$\lim_{t\to 0^+} K_{\mathcal{O}}\left(\tau,\sigma;\tau',\sigma';t\right)=\frac{\delta(\tau-\tau')\,\delta(\sigma-\sigma')}{\sqrt{h}}\,\mathbb{I}_r \tag{1.16}$$

and boundary conditions set by an operator $\mathcal{B}$ if the space boundary is not empty

$$\mathcal{B}_{\tau,\sigma}K_{\mathcal{O}}\left(\tau,\sigma;\tau',\sigma';t\right)=0\,. \tag{1.17}$$

A vast literature has covered the Laplace [212, 213] and the (square of the) Dirac operator [214, 215] for Euclidean manifolds without boundary—flat spaces $\mathbb{R}^d$, spheres S^d and hyperboloids H^d[19] and products thereof—and with boundary, e.g. [216].

The asymptotics of the traced heat kernel for small t is known for all Laplace and Dirac operators. In Sect. 3.5.2 it will be used to estimate the infinite part of the functional determinants in (1.14). The cancellation of the logarithmic divergences proportional to the worldsheet Ricci curvature in the one-loop effective action was clarified in [192] via a careful account of the Seeley-DeWitt coefficients of the operators, which exactly compensate the divergences arising in the measure factors of the string path-integral. The counting of anomalies in the GS string goes essentially as adding together the central charges of all the fields for the theory in flat space.

Since UV divergences are associated with conformal anomalies, the same mechanism can be seen as an explicit verification of the conformal invariance of the type

[19]The three spaces listed here are *maximally symmetric* because their metrics possess the maximal number $d(d+1)/2$ of Killing vectors in d dimensions.

IIB string theory on $AdS_5 \times S^5$ background at one loop, which was actually argued to hold true to all order in the T^{-1} expansion [70].

If we look instead for extracting the regularized part of the determinants in (1.14) via zeta-function regularization [209, 211, 217], knowing only the small-time behaviour does not suffice. In this case the spectral information that we need is carried by the heat kernel as a function of t, but finding a solution of the heat equation is practically impossible for most of the worldsheet geometries of interests. Some notable exceptions are represented by the spectral problems for the string worldsheets in AdS_5 dual to the straight line and circular Wilson loop [218, 219].

In the examples considered in Chaps. 4 and 5, the geometric properties of the classical surface deliver fluctuation operators that are translationally invariant in one variable,[20] which we call τ in what follows. The same coordinate does not appear in the two-dimensional induced metric because the associated vector field ∂_τ generates an isometry of the classical surface. To proceed, let us call ϕ the (scalar or spinor) field acted upon by the one-loop operator $\mathcal{O}$ in the free action $\int d\tau d\sigma\, \phi^\dagger\, \mathcal{O}\, \phi$. Without loss in generality, let us suppose that the isometry acts on the worldsheet points as a $U(1)$ rotation with $\tau \in [0, 2\pi)$,[21] so we can decompose the field into discrete Fourier modes

$$\phi(\tau, \sigma) = \sum_{\omega} \phi(\omega, \sigma)\, \frac{e^{i\omega\tau}}{\sqrt{2\pi}}\,. \tag{1.18}$$

The frequencies ω are integer or half-integer according to the periodicity or anti-periodicity of the field. In the case of non-compact symmetry, acting like as a translation along the τ-direction,[22] the frequency label turns into a continuous variable $\tau \in \mathbb{R}$ and the Fourier integral takes the place of the Fourier series

$$\phi(\tau, \sigma) = \int_{-\infty}^{+\infty} d\omega\, \phi(\omega, \sigma)\, \frac{e^{i\omega\tau}}{\sqrt{2\pi}}\,. \tag{1.19}$$

Plugging formula (1.18) or (1.19) into the Lagrangian, one effectively replaces $\partial_\tau \to i\omega$ in the operator $\mathcal{O}(\tau, \sigma) \to \mathcal{O}(\omega, \sigma)$. This trick eventually allows to trade the two-dimensional spectral problem for $\mathcal{O}(\tau, \sigma)$ with infinite many 1d spectral problems for the Fourier-transformed differential operator $\mathcal{O}(\omega, \sigma)$. Our strategy can be summarized in two steps.

- At fixed Fourier mode ω, the evaluation of the determinant $\mathrm{Det}_\omega(\mathcal{O}(\omega, \sigma))$ is a one-variable eigenvalue problem on a certain line segment $\sigma \in [a, b]$, which we solve using the *Gel'fand-Yaglom method* [220], a technique based on zeta-

[20] We are referring to (4.11) and (4.27) for the two examples of spinning strings studied in this thesis, (5.38) and (5.53) for the worldsheet dual to a 1/4-BPS latitude Wilson loop.

[21] This geometrically manifests as an axial symmetry of the minimal surface (5.25) in Fig. 5.3.

[22] For the spinning strings of Chap. 4, this is the invariance of the string surface under shifts of the AdS global time t, see text above (4.1) and formula (4.22).

function regularization for one-dimensional operators.[23] More precisely, we shall refer to an extension of the theorem elaborated by Forman in [226, 227] that encompasses also the case of odd-order (e.g. Dirac-like) differential operators $\mathcal{O}$. The theorem outputs the value of the *ratio* between two determinants in terms of solutions of an homogeneous ordinary differential equation, so it requires some preliminary work to pair bosonic and fermionic fluctuation determinants. It is worth mentioning that both the Gel'fand-Yaglom method and its corollaries do not naturally cope with infinite or semi-infinite intervals and with operators having singular coefficients for some $\sigma \in [a, b]$. One usually needs some regulators on σ to deal with such situations in AdS/CFT applications, only then to prescribe an appropriate regularization scheme to eliminate or subtract them from the final result.

- The full two-dimensional determinant is given by the sum over all discrete frequencies ω

$$\log \mathrm{Det}(\mathcal{O}(\tau,\sigma)) = \left(\int_0^{2\pi} \frac{d\tau}{2\pi}\right) \sum_\omega \log \mathrm{Det}_\omega(\mathcal{O}(\omega,\sigma)) = \sum_\omega \log \mathrm{Det}_\omega(\mathcal{O}(\omega,\sigma)) \quad (1.20)$$

or by an integral if ω is continuous

$$\log \mathrm{Det}(\mathcal{O}(\tau,\sigma)) = \left(\int_{-\infty}^{+\infty} \frac{d\tau}{2\pi}\right) \int_{-\infty}^{+\infty} d\omega \, \log \mathrm{Det}_\omega(\mathcal{O}(\omega,\sigma)) \,. \quad (1.21)$$

The infinite prefactor in the latter case is often neglected since it drops once bosonic and fermionic contributions are plugged into (1.14).

In the following we shall make a notational distinction between the algebraic determinant det and the functional determinant Det, involving the determinant on the matrix indices as well as on the two-dimensional space with coordinates (τ, σ). We also introduced the functional determinant Det_ω over σ at a given Fourier mode ω. The issue of how to define an unambiguous infinite product (1.20) is addressed in the case of rotationally invariant operators in [228][24] and goes case by case in the main text.

We conclude the section with the observation that, while the algorithm described above is highly versatile and easy to implement either analytically or numerically, it does not treat the worldsheet directions on equal footing. On general grounds, an optimal algorithm for fluctuation determinants should preserve the classical symmetries (e.g. diffeomorphisms of the classical worldsheet and target-space supersymmetry of the Green–Schwarz action) at any steps of the regularization procedure.

---

[23]This is of course just one possible route. It is a well-known result [221] that the logarithm of the determinant is equivalently computed by the *on-shell vacuum energy*, as obtained by summing over the frequency spectrum of the operator, e.g. [198], under certain assumptions. Alternatively, one can also employ the *phaseshift method* [222–224], see [225] for a recent application. We are grateful to Xinyi Chen-Lin and Daniel Medina Rincón for long discussions about this point.

[24]We thank Amit Dekel for informing us about this reference.

The Gel'fand-Yaglom theorem, paired up with cutoff regularization on σ and followed by the summation over the angular modes, breaks the invariance under 2d reparametrization $(\tau, \sigma) \to (\tau'(\tau, \sigma), \sigma'(\tau, \sigma))$ at least.

The heat kernel method does not suffer from this issue, but in the context of one-loop spectral problems its current level of development is limited to a very selected class of background solutions [192, 218, 219]. A fully two-dimensional method, based upon heat kernel techniques, is being developed in [229] (see also Chap. 8) when the worldsheet metric is a "small" deformation of the maximally symmetric hyperbolic space H^2.

1.4.2 String Effective Action Beyond the Next-to-Leading Order

The accuracy of the semiclassical approximation improves by including higher interaction vertices in the fluctuation Lagrangian in (1.10). Computations can be technically involved due to the large number of fields—eight bosonic and 16 fermionic real off-shell degrees of freedom after gauge-fixing—and the lack of manifest two-dimensional Lorentz invariance, broken by the curved background $X_{\rm cl}$, in the interactions terms.

Consistency at quantum level requires the theory to be finite. Perturbative calculations are crucial to directly address the nontrivial problem of divergences cancellation in a field theory that is expected to be finite, although it is not manifestly power-counting renormalizable beyond the one-loop level (see comments in [177, 230] and also [177, 188, 231, 232]). In fact, as soon as one expands around a particular vacuum and picks a proper gauge-fixing, the derivatives of the curved background generate a Green–Schwarz kinetic term $\partial X_{\rm cl}\bar{\Psi}\partial\Psi$ and fermionic interactions that contain derivatives. While all fields are dimensionless and no dimensionful parameter appears in the action, derivatives of the background act as a dimensional scale that turns the fermion mass dimension from 0 to the canonical $1/2$ in two dimensions, therefore leading to non-renormalizable interactions and to the potential appearance of power-like and logarithmic divergences.

Loop computations with related tests of finiteness of the model may be a thorny issue. A particularly convenient choice to fix the diffeomorphism invariance and κ-symmetry of the coset action is the *AdS light-cone gauge* throughly discussed in [233, 234] and summarized in our Sects. 2.1.3 and 7.2.1. The resulting action has a relatively simple structure—where fermions are *only* quadratic and quartic—that was efficiently used, for instance, in [231] to compute the string effective action up to two loops in the *null cusp background*. This is a homogeneous solution lying only in AdS that can be viewed as the light-like limit of the space-like cusp solution of [235].

The AdS light-cone gauge turns out to be perfectly applicable to the $AdS_4 \times \mathbb{CP}^3$ Lagrangian of [236, 237] in Sect. 2.2. A side objective of Chap. 6 is to begin the investigation of this quantum action, expanded about the $AdS_4 \times \mathbb{CP}^3$ counterpart of the null cusp solution above. Similar to the $AdS_5 \times S^5$ case, here the AdS light-cone approach to the evaluation of the two-loop effective action turns out to be extremely efficient. Simplifications occur due to bosonic propagators being only diagonal, which reduces the number of Feynman graphs to be considered.[25]

As a preparation for the two-loop study in Chap. 6 following what was done in [231], let us emphasise that various non-covariant integrals with components of the loop momenta in the numerators originate from the combinations of vertices and propagators. This naturally brings us to the problem of handling potentially divergent loop integrations. As for the $AdS_5 \times S^5$ coset action, the action is it not renormalizable by power-counting and naively it seems to lead to potential divergences. In principle, the expected finiteness of the theory implies that all their divergences cancel against a careful account of the contributions from the path-integral measure and κ-symmetry ghosts (see [177]). However, one typically does not attempt an exact evaluation of the terms crucial for the complete divergence cancellation. Alternatively, the use of dimensional regularization[26] allows to automatically set all power-like divergences to zero. In doing so, we are effectively discarding these divergences in loop integrals, but we can nontrivially check the absence of logarithmically divergent integrals in the final result.

A strong indication of consistency—in which the finite parts and the divergence cancellation are some of the ingredients—comes from the comparison with the integrability predictions of Bethe ansatz [118] in the ABJM theory. We also flash that the $AdS_4 \times \mathbb{CP}_3$ action of [236, 237] in AdS light-cone gauge offers an efficient setting for computing the dispersion relation of worldsheet excitations on the same cusp background [239], which were found to essentially agree with the predictions from the Bethe ansatz of [240].

1.5 Lattice Field Theory for the $AdS_5 \times S^5$ String Sigma-Model

In the last sections we saw that the $AdS_5 \times S^5$ sigma-model is a nontrivial theory which, as virtually any interacting QFT, we do not know how to define rigorously without relying on perturbative expansions. The aim of this section is to set the ground for Chap. 7, where we will see that lattice methods provide a concrete mean to non-perturbatively define the theory and evaluate observables on the two-dimensional (Wick-rotated) Euclidean worldsheet from the gravity side of AdS/CFT.

In general, the study of lattice models has a long and well-established tradition in QCD and condensed matter systems [241–243]. Such non-perturbative investigations are formulated in discrete rather than continuous spacetime. They provide a

[25] In the first two-loop calculation of [238] the conformal gauge was used, in which propagators are non-diagonal, implying the evaluation of a larger number of two-loop diagrams.

[26] In regularizing our integrals all manipulations of tensor structures in the Feynman integrands are however carried out strictly in two dimensions.

mathematically well-defined regularization of the theory of interest by introducing a momentum UV cutoff of order a^{-1}, where a is the spacing between lattice sites in the spacetime grid. The central idea is to construct a discrete form of the action and the operators, which formally reduce to corresponding continuum counterparts when the regulator is removed. By working on a discrete spacetime, path-integrals defining observables become finite multi-dimensional integrals which can be evaluated through stochastic simulation techniques such as the Monte Carlo method. The aim is to eventually recover the values of the observables in the continuum model on finer and finer lattices, through an appropriate prescription on the continuum limit $a \to 0$ and the other technical parameters of the simulations.

The first finite-coupling calculation in the $AdS_5 \times S^5$ superstring sigma-model using purely (lattice) field theory methods was pioneered in [244] with the measurement of the *universal scaling function*[27] $f(\lambda)$, one of the most important observables in the AdS_5/CFT_4 duality. It governs the (the renormalization of) a light-like cusped Wilson loop in $\mathcal{N}=4$ SYM, while from the holographic viewpoint it is captured by the path-integral of an open string ending on two intersecting null lines at the AdS boundary. The convenient ground to set up such lattice investigation was the same of the two-loop perturbative analysis of [231] mentioned in Sect. 1.4.1, i.e. the AdS light-cone gauge-fixed action of [233, 234] with no fermionic interactions of order higher than four and expanded around the null cusp background. The crucial difference is the fact that now all interactions must be kept in the fluctuation Lagrangian to be discretized. The numerical results of [244] for the scaling function were found in agreement with its exact integrability prediction from the Beisert-Eden-Staudacher (BES) equation [245] within reasonable numerical accuracy.

Our work in Chap. 7 takes a deep inspiration from the route opened up by [244]. We will scrutinize the way one should extract the scaling function from the expectation value of the action, as well as inaugurate the study of the dispersion relations of worldsheet fields starting from the measurement of the physical mass of one bosonic field. We will also elucidate and enlarge the discussion related to many aspects related to the numerical simulations.

The investigation here and in [244] is *not* a non-perturbative definition of the worldsheet string model à la Wilson lattice-QCD. In this case, one should work with a Lagrangian which is invariant under the local symmetries of the model (bosonic diffeomorphisms and fermionic κ-symmetry), whereas we will make use of the action in the AdS light-cone gauge which fixes them all. However, there is a number of reasons that makes this model interesting for lattice investigations, potentially beyond the community interested in numerical holography [246].

- If the aim is a test of holography and integrability, it is computationally cheaper to simulate a two-dimensional model, rather than a four-dimensional one ($\mathcal{N}=4$ SYM). Additionally, all fields are assigned to sites, since no gauge degrees of freedom are present and only (commuting and anti-commuting) scalar fields appear in the string action.

[27]In literature *scaling function* and *cusp anomalous dimension* are essentially synonyms, as we will explain in Sect. 7.1.

- Although we deal with superstrings, there is no subtlety involved with supersymmetry on the lattice because in the Green–Schwarz formulation of the action supersymmetry is manifest only in the target space, and also because κ-symmetry is gauge-fixed.
- At the same time, this is a computational playground interesting on its own, allowing in principle for explicit investigations/improvements of algorithms. In fact, we work with a highly nontrivial two-dimensional model for which relevant observables have not only, through AdS/CFT, an explicit analytic strong-coupling expansion but also, through AdS/CFT and the assumption of integrability, an explicit numerical prediction at all couplings.
- In principle, the scope of numerical simulations is not limited to partition functions. An interesting program was outlined in Sect. 2.2 of [244]. The goal is to provide a concrete mean to extract anomalous dimensions of local gauge-invariant operators in $\mathcal{N} = 4$ SYM from the numerical evaluation of worldsheet two-point functions of the dual integrated vertex operators [247, 248]. Significant work is necessary to investigate this direction in the next future.

On a different note, we conclude by mentioning—albeit not as much as it would deserve—that recent years have also witnessed rapid progress in constructing lattice discretizations of supersymmetric gauge theories [246, 249, 250], in particular of the $\mathcal{N} = 4$ SYM theory [251–255]. Large-scale calculations [256] have studied the static "interquark" potential [257, 258] and the anomalous scaling dimension of the Konishi operator [259] (see also [246]).

Alternative numerical, non-lattice approaches include the study of $\mathcal{N} = 4$ SYM on $\mathbb{R} \times S^3$ as plane-wave (BMN) matrix model [260–266].

1.6 Plan of the Thesis

Chapter 2 gives a short introduction to the construction of superstring theory in $AdS_5 \times S^5$ and discusses the advantages and the limits of the supercoset formalism in $AdS_4 \times \mathbb{CP}^3$.

In Chap. 3, which is largely based on [267], we present a pedagogical discussion about the general structure of the quadratic fluctuation Lagrangian around arbitrary classical configurations in $AdS_5 \times S^5$, expressing the relevant differential operators in terms of geometric invariants of the background geometry.

Chapter 4 is based on [268] and exploits the analytical solution to the spectral problem of a type of fourth-order differential operators. This finds application in the exact evaluation of the semiclassical contributions to the one-loop energy of a class of two-spin spinning strings in $AdS_5 \times S^5$.

In Chap. 5, we illustrate the perturbative computation in [269] aimed at the strong-coupling correction to the expectation value of a family of supersymmetric Wilson loops in $\mathcal{N} = 4$ SYM. We comment on the unexpected discrepancy between our result and the all-loop prediction from supersymmetric localization.

Chapter 6 presents the derivation in [270] of the cusp anomalous dimension of the ABJM theory up to two loops at strong coupling, which provides support to a recent conjecture for the exact form of the interpolating function $h(\lambda)$ in this theory.

Chapter 7 shows recent developments [271, 272] in the construction of consistent lattice discretizations of the two-dimensional $AdS_5 \times S^5$ string sigma-model in AdS light-cone gauge. We discuss the numerical results for the observables studied so far, test them against semiclassical and integrability predictions and comment upon the difficulties encountered in our approach.

In Chap. 8, we summarize our results and conclude with an outlook on interesting directions for further research, in particular based on [229, 273–275].

Several appendices supplement the main text with important methodological and technical details. For this reason, we encourage the reader to consult them in parallel with the reading of the main text.

The structure of the individual chapters is outlined in their introductions. Sometimes the presentation has required to slightly deviate from the notation of the original references, but we will clearly point it out when confusion may arise.

References

1. C.M. Will, The confrontation between general relativity and experiment. Living Rev. Rel. **17**, 4 (2014). arXiv:1403.7377
2. Virgo, LIGO Scientific Collaboration, B.P. Abbott et al., Observation of gravitational waves from a binary black hole merger. Phys. Rev. Lett. **116**, 061102 (2016). arXiv:1602.03837
3. Virgo, LIGO Scientific Collaboration, B.P. Abbott et al., GW151226: observation of gravitational waves from a 22-solar-mass binary black hole coalescence. Phys. Rev. Lett. **116**, 241103 (2016). arXiv:1606.04855
4. K.G. Wilson, J.B. Kogut, The Renormalization group, the epsilon expansion. Phys. Rept. **12**, 75 (1974)
5. K. Symanzik, Small distance behaviour in field theory and power counting. Commun. Math. Phys. **18**, 227 (1970)
6. C.G. Callan Jr., Broken scale invariance in scalar field theory. Phys. Rev. D **2**, 1541 (1970)
7. K. Symanzik, Small-distance-behaviour analysis and Wilson expansions. Commun. Math. Phys. **23**, 49 (1971)
8. M.E. Peskin, D.V. Schroeder, *An Introduction to Quantum Field Theory*, (Addison-Wesley Publishing Company, 1995)
9. A.A. Belavin, A.M. Polyakov, A.B. Zamolodchikov, Infinite conformal symmetry in two-dimensional quantum field theory. Nucl. Phys. B **241**, 333 (1984)
10. F. Englert, R. Brout, Broken symmetry and the mass of gauge vector mesons. Phys. Rev. Lett. **13**, 321 (1964)
11. P.W. Higgs, Broken symmetries and the masses of gauge bosons. Phys. Rev. Lett. **13**, 508 (1964)
12. G.S. Guralnik, C.R. Hagen, T.W.B. Kibble, Global conservation laws and massless particles, 5. Phys. Rev. Lett. **13**, 585 (1964)
13. D.J. Gross, F. Wilczek, Ultraviolet behavior of nonabelian gauge theories. Phys. Rev. Lett. **30**, 1343 (1973)
14. K.A. Olive et al., Particle Data Group Collaboration. Review of Particle Physics. Chin. Phys. **C38**, 090001 (2014)

15. C.M.S. Collaboration, S. Chatrchyan et al., Observation of a new boson at a mass of 125 GeV with the CMS experiment at the LHC. Phys. Lett. B **716**, 30 (2012). arXiv:1207.7235
16. ATLAS Collaboration, G. Aad et al., Observation of a new particle in the search for the Standard Model Higgs boson with the ATLAS detector at the LHC. Phys. Lett. B **716**, 1 (2012). arXiv:1207.7214
17. K. Ramachandran, G. Deepa, K. Namboori, *Computational Chemistry and Molecular Modeling: Principles And Applications* (Springer Science & Business Media, 2008)
18. L. Brink, J.H. Schwarz, J. Scherk, Supersymmetric yang-mills theories. Nucl. Phys. B **121**, 77 (1977)
19. E. D'Hoker, D.Z. Freedman, Supersymmetric gauge theories and the AdS/CFT correspondence. arXiv:hep-th/0201253
20. C. Montonen, D. Olive, Magnetic monopoles as gauge particles? Phys. Lett. B **72**, 117 (1977)
21. P. Goddard, J. Nuyts, D.I. Olive, Gauge theories and magnetic charge. Nucl. Phys. B **125**, 1 (1977)
22. E. Witten, D.I. Olive, Supersymmetry algebras that include topological charges. Phys. Lett. B **78**, 97 (1978)
23. M.B. Green, J.H. Schwarz, L. Brink, $\mathcal{N} = 4$ Yang-Mills and $\mathcal{N} = 8$ supergravity as limits of string theories. Nucl. Phys. B **198**, 474 (1982)
24. S. Mandelstam, Light cone superspace and the ultraviolet finiteness of the $\mathcal{N} = 4$ model. Nucl. Phys. B **213**, 149 (1983)
25. L. Brink, O. Lindgren, B.E.W. Nilsson, The ultraviolet finiteness of the $\mathcal{N} = 4$ yang-mills theory. Phys. Lett. B **123**, 323 (1983)
26. N. Seiberg, Supersymmetry and nonperturbative beta functions. Phys. Lett. B **206**, 75 (1988)
27. A.V. Kotikov, L.N. Lipatov, DGLAP and BFKL equations in the $\mathcal{N} = 4$ supersymmetric gauge theory, Nucl. Phys. B **661**, 19 (2003). arXiv:hep-ph/0208220, [Erratum: Nucl. Phys. B685,405(2004)]
28. L. Freyhult, Review of AdS/CFT Integrability, Chapter III.4: twist states and the cusp anomalous dimension. Lett. Math. Phys. **99**, 255 (2012). arXiv:1012.3993
29. G.P. Korchemsky, Review of AdS/CFT Integrability, Chapter IV.4: Integrability in QCD and $\mathcal{N} < 4$ SYM. Lett. Math. Phys. **99**, 425 (2012). arXiv:1012.4000
30. L.J. Dixon, J.M. Henn, J. Plefka, T. Schuster, All tree-level amplitudes in massless QCD. JHEP **1101**, 035 (2011). arXiv:1010.3991
31. J.M. Maldacena, The Large N limit of superconformal field theories and supergravity. Adv. Theor. Math. Phys. **2**, 231 (1998). arXiv:hep-th/9711200
32. S.S. Gubser, I.R. Klebanov, A.M. Polyakov, Gauge theory correlators from non-critical string theory. Phys. Lett. B **428**, 105 (1998). arXiv:hep-th/9802109
33. E. Witten, Anti-de Sitter space and holography. Adv. Theor. Math. Phys. **2**, 253 (1998). arXiv:hep-th/9802150
34. O. Aharony, S.S. Gubser, J.M. Maldacena, H. Ooguri, Y. Oz, Large N field theories, string theory and gravity. Phys. Rept. **323**, 183 (2000). arXiv:hep-th/9905111
35. A. Zaffaroni, Introduction to the AdS/CFT correspondence. Class. Quant. Grav. **17**, 3571 (2000)
36. J. Plefka, Spinning strings and integrable spin chains in the AdS/CFT correspondence. Living Rev. Rel. **8**, 9 (2005). arXiv:hep-th/0507136
37. H. Nastase, Introduction to AdS/CFT. arXiv:0712.0689
38. N. Beisert et al., Review of AdS/CFT integrability: an overview. Lett. Math. Phys. **99**, 3 (2012). arXiv:1012.3982
39. M. Ammon, J. Erdmenger, *Gauge/Gravity Duality* (Cambridge University Press, Cambridge, UK, 2015)
40. G. 't Hooft, A planar diagram theory for strong interactions. Nucl. Phys. B **72**, 461 (1974)
41. A. Gustavsson, Algebraic structures on parallel M2-branes. Nucl. Phys. B **811**, 66 (2009). arXiv:0709.1260
42. J. Bagger, N. Lambert, Gauge symmetry and supersymmetry of multiple M2-branes. Phys. Rev. D **77**, 065008 (2008). arXiv:0711.0955

43. O. Aharony, O. Bergman, D.L. Jafferis, J. Maldacena, $\mathcal{N} = 6$ superconformal Chern-Simons-matter theories, M2-branes and their gravity duals. JHEP **0810**, 091 (2008). arXiv:0806.1218
44. I.R. Klebanov, G. Torri, M2-branes and AdS/CFT. Int. J. Mod. Phys. A **25**, 332 (2010). arXiv:0909.1580 (In: Crossing the boundaries: Gauge dynamics at strong coupling. Proceedings, Workshop in Honor of the 60th Birthday of Misha Shifman, (Minneapolis, USA, May 14–17, 2009), p. 332–350)
45. S. Deser, R. Jackiw, S. Templeton, Three-dimensional massive gauge theories. Phys. Rev. Lett. **48**, 975 (1982)
46. E. Witten, Quantum field theory and the Jones polynomial. Commun. Math. Phys. **121**, 351 (1989)
47. A. Gustavsson, S.-J. Rey, Enhanced $\mathcal{N} = 8$ supersymmetry of ABJM theory on $\mathbb{R}^8$ and $\mathbb{R}^8/\mathbb{Z}_2$. arXiv:0906.3568
48. T. McLoughlin, R. Roiban, A.A. Tseytlin, Quantum spinning strings in $AdS_4 \times \mathbb{CP}^3$: testing the Bethe Ansatz proposal. JHEP **0811**, 069 (2008). arXiv:0809.4038
49. O. Bergman, S. Hirano, Anomalous radius shift in AdS_4/CFT_3. JHEP **0907**, 016 (2009). arXiv:0902.1743
50. O. Aharony, O. Bergman, D.L. Jafferis, Fractional M2-branes. JHEP **0811**, 043 (2008). arXiv:0807.4924
51. A. Sfondrini, Towards integrability for AdS_2/CFT_3. J. Phys. A **48**, 023001 (2015). arXiv:1406.2971
52. A. Pittelli, Dualities and Integrability in Low Dimensional AdS/CFT, Ph.D Thesis. http://epubs.surrey.ac.uk/812577/
53. O. Babelon, D. Bernard, M. Talon, *Introduction to Classical Integrable Systems* (Cambridge University Press, 2003)
54. G. Mussardo, *Statistical Field Theory: An Introduction to Exactly Solved Models in Statistical Physics* (Oxford University Press, 2010)
55. N. Beisert, *Integrability in QFT and AdS/CFT*, Lecture notes. http://edu.itp.phys.ethz.ch/hs14/14HSInt/IntAdSCFT14Notes.pdf
56. A. Torrielli, *Lectures on Classical Integrability*. arXiv:1606.02946
57. L.N. Lipatov, High energy asymptotics of multi-colour QCD and exactly solvable lattice models. JETP Lett. **59**, 596 (1994). arXiv:hep-th/9311037, [Pisma Zh. Eksp. Teor. Fiz. 59, 571(1994)]
58. L.D. Faddeev, G.P. Korchemsky, High-energy QCD as a completely integrable model. Phys. Lett. B **342**, 311 (1995). arXiv:hep-th/9404173
59. V.M. Braun, G.P. Korchemsky, D. Mueller, The Uses of conformal symmetry in QCD. Prog. Part. Nucl. Phys. **51**, 311 (2003). arXiv:hep-ph/0306057
60. M. Staudacher, Review of AdS/CFT Integrability, Chapter III.1: Bethe Ansátze and the R-matrix formalism. Lett. Math. Phys. **99**, 191 (2012). arXiv:1012.3990
61. J.A. Minahan, K. Zarembo, The Bethe ansatz for $\mathcal{N} = 4$ superYang-Mills. JHEP **0303**, 013 (2003). arXiv:hep-th/0212208
62. F. Levkovich-Maslyuk, Lectures on the Bethe Ansatz. arXiv:1606.02950
63. H. Bethe, Zur Theorie der Metalle. I. Eigenwerte und Eigenfunktionen der linearen Atomkette. Z. Phys. **71**, 205 (1931). (in German)
64. L.D. Faddeev, L.A. Takhtajan, Spectrum and scattering of excitations in the one-dimensional isotropic Heisenberg model. J. Sov. Math. **24**, 241 (1984). [Zap. Nauchn. Semin. 109, 134(1981)]
65. N.Y. Reshetikhin, A method of functional equations in the theory of exactly solvable quantum systems. Lett. Math. Phys. **7**, 205 (1983)
66. N.Y. Reshetikhin, Integrable models of quantum one-dimensional magnets with $O(N)$ and $Sp(2k)$ symmetry. Theor. Math. Phys. **63**, 555 (1985)
67. N. Beisert, M. Staudacher, The $\mathcal{N} = 4$ SYM integrable super spin chain. Nucl. Phys. B **670**, 439 (2003). arXiv:hep-th/0307042
68. N. Beisert, The complete one loop dilatation operator of $\mathcal{N} = 4$ superYang-Mills theory. Nucl. Phys. B **676**, 3 (2004). arXiv:hep-th/0307015

69. N. Beisert, C. Kristjansen, M. Staudacher, The Dilatation operator of conformal $\mathcal{N}=4$ superYang-Mills theory. Nucl. Phys. B **664**, 131 (2003). arXiv:hep-th/0303060
70. R. Metsaev, A.A. Tseytlin, Type IIB superstring action in $AdS_5 \times S^5$ background. Nucl. Phys. B **533**, 109 (1998). arXiv:hep-th/9805028
71. I. Bena, J. Polchinski, R. Roiban, Hidden symmetries of the $AdS_5 \times S^5$ superstring. Phys. Rev. D **69**, 046002 (2004). arXiv:hep-th/0305116
72. B.C. Vallilo, Flat currents in the classical $AdS_5 \times S^5$ pure spinor superstring. JHEP **0403**, 037 (2004). arXiv:hep-th/0307018
73. N. Berkovits, Super Poincare covariant quantization of the superstring. JHEP **0004**, 018 (2000). arXiv:hep-th/0001035
74. N. Berkovits, O. Chandia, Superstring vertex operators in an $AdS_5 \times S^5$ background. Nucl. Phys. B **596**, 185 (2001). arXiv:hep-th/0009168
75. B.C. Vallilo, One loop conformal invariance of the superstring in an $AdS_5 \times S^5$ background. JHEP **0212**, 042 (2002). arXiv:hep-th/0210064
76. N. Berkovits, BRST cohomology and nonlocal conserved charges. JHEP **0502**, 060 (2005). arXiv:hep-th/0409159
77. G. Arutyunov, S. Frolov, Foundations of the $AdS_5 \times S^5$ Superstring. Part I. J. Phys. A **42**, 254003 (2009). arXiv:0901.4937
78. A. Mikhailov, S. Schafer-Nameki, Algebra of transfer-matrices and Yang-Baxter equations on the string worldsheet in $AdS_5 \times S^5$. Nucl. Phys. B **802**, 1 (2008). arXiv:0712.4278
79. R. Benichou, First-principles derivation of the AdS/CFT Y-systems. JHEP **1110**, 112 (2011). arXiv:1108.4927
80. N. Berkovits, ICTP lectures on covariant quantization of the superstring. arXiv:hep-th/0209059 (In: Superstrings and related matters. Proceedings, Spring School (Trieste, Italy, March 18–26, 2002), p. 57–107. http://www.ictp.trieste.it/~pub_off/lectures/lns013/Berkovits/Berkovits.pdf)
81. Y. Oz, The pure spinor formulation of superstrings. Class. Quant. Grav. **25**, 214001 (2008). arXiv:0910.1195 (In: Strings, supergravity and gauge theories. Proceedings, European RTN Winter School (CERN, Geneva, Switzerland, January 21–25, 2008), p. 214001)
82. N. Berkovits, Quantum consistency of the superstring in $AdS_5 \times S^5$ background. JHEP **0503**, 041 (2005). arXiv:hep-th/0411170
83. V. Kazakov, A. Marshakov, J. Minahan, K. Zarembo, Classical/quantum integrability in AdS/CFT. JHEP **0405**, 024 (2004). arXiv:hep-th/0402207
84. N. Beisert, V. Dippel, M. Staudacher, A Novel long range spin chain and planar $\mathcal{N}=4$ super Yang-Mills. JHEP **0407**, 075 (2004). arXiv:hep-th/0405001
85. N. Beisert, M. Staudacher, Long-range $\mathfrak{psu}(2,2|4)$ Bethe Ansatze for gauge theory and strings. Nucl. Phys. B **727**, 1 (2005). arXiv:hep-th/0504190
86. G. Arutyunov, S. Frolov, M. Staudacher, Bethe ansatz for quantum strings. JHEP **0410**, 016 (2004). arXiv:hep-th/0406256
87. R. Hernandez, E. Lopez, Quantum corrections to the string Bethe ansatz. JHEP **0607**, 004 (2006). arXiv:hep-th/0603204
88. R.A. Janik, The $AdS_5 \times S^5$ superstring worldsheet S-matrix and crossing symmetry. Phys. Rev. D **73**, 086006 (2006). arXiv:hep-th/0603038
89. N. Beisert, R. Hernandez, E. Lopez, A Crossing-symmetric phase for $AdS_5 \times S^5$ strings. JHEP **0611**, 070 (2006). arXiv:hep-th/0609044
90. Z. Bern, M. Czakon, L.J. Dixon, D.A. Kosower, V.A. Smirnov, The four-loop planar amplitude and cusp anomalous dimension in maximally supersymmetric Yang-Mills theory. Phys. Rev. D **75**, 085010 (2007). arXiv:hep-th/0610248
91. N. Beisert, T. McLoughlin, R. Roiban, The four-loop dressing phase of $\mathcal{N}=4$ SYM. Phys. Rev. D **76**, 046002 (2007). arXiv:0705.0321
92. P. Vieira, D. Volin, Review of AdS/CFT Integrability, Chapter III.3: the dressing factor, Lett. Math. Phys. **99**, 231 (2012). arXiv:1012.3992
93. C. Sieg, A. Torrielli, Wrapping interactions and the genus expansion of the 2-point function of composite operators. Nucl. Phys. B **723**, 3 (2005). arXiv:hep-th/0505071

94. J. Ambjorn, R.A. Janik, C. Kristjansen, Wrapping interactions and a new source of corrections to the spin-chain/string duality. Nucl. Phys. B **736**, 288 (2006). arXiv:hep-th/0510171
95. R.A. Janik, T. Lukowski, Wrapping interactions at strong coupling: the Giant magnon. Phys. Rev. D **76**, 126008 (2007). arXiv:0708.2208
96. R.A. Janik, Review of AdS/CFT Integrability, Chapter III.5: luscher corrections. Lett. Math. Phys. **99**, 277 (2012). arXiv:1012.3994
97. G. Arutyunov, S. Frolov, On string S-matrix. Bound states and TBA, JHEP **0712**, 024 (2007). arXiv:0710.1568
98. G. Arutyunov, S. Frolov, String hypothesis for the $AdS_5 \times S^5$ mirror. JHEP **0903**, 152 (2009). arXiv:0901.1417
99. N. Gromov, V. Kazakov, P. Vieira, Exact spectrum of anomalous dimensions of planar $\mathcal{N} = 4$ supersymmetric Yang-Mills theory. Phys. Rev. Lett. **103**, 131601 (2009). arXiv:0901.3753
100. D. Bombardelli, D. Fioravanti, R. Tateo, Thermodynamic Bethe Ansatz for planar AdS/CFT: a proposal. J. Phys. A **42**, 375401 (2009). arXiv:0902.3930
101. N. Gromov, V. Kazakov, A. Kozak, P. Vieira, Exact spectrum of anomalous dimensions of planar $\mathcal{N} = 4$ supersymmetric Yang-Mills theory: TBA and excited states. Lett. Math. Phys. **91**, 265 (2010). arXiv:0902.4458
102. G. Arutyunov, S. Frolov, Thermodynamic Bethe Ansatz for the $AdS_5 \times S^5$ mirror model. JHEP **0905**, 068 (2009). arXiv:0903.0141
103. Z. Bajnok, Review of AdS/CFT Integrability, Chapter III.6: Thermodynamic Bethe Ansatz. Lett. Math. Phys. **99**, 299 (2012). arXiv:1012.3995
104. S.J. van Tongeren, Introduction to the thermodynamic Bethe ansatz. arXiv:1606.02951
105. L.F. Alday, J. Maldacena, A. Sever, P. Vieira, Y-system for scattering amplitudes. J. Phys. A **43**, 485401 (2010). arXiv:1002.2459
106. D. Correa, J. Maldacena, A. Sever, The quark anti-quark potential and the cusp anomalous dimension from a TBA equation. JHEP **1208**, 134 (2012). arXiv:1203.1913
107. N. Drukker, Integrable Wilson loops. JHEP **1310**, 135 (2013). arXiv:1203.1617
108. N. Gromov, A. Sever, Analytic solution of Bremsstrahlung TBA. JHEP **1211**, 075 (2012). arXiv:1207.5489
109. N. Gromov, V. Kazakov, S. Leurent, D. Volin, Quantum spectral curve for planar $\mathcal{N} = 4$ super-Yang-Mills theory. Phys. Rev. Lett. **112**, 011602 (2014). arXiv:1305.1939
110. N. Gromov, V. Kazakov, S. Leurent, D. Volin, Quantum spectral curve for arbitrary state/operator in AdS_5/CFT_4. JHEP **1509**, 187 (2015). arXiv:1405.4857
111. N. Gromov, F. Levkovich-Maslyuk, Quantum spectral curve for a cusped Wilson line in $\mathcal{N} = 4$ SYM. JHEP **1604**, 134 (2016). arXiv:1510.02098
112. N. Gromov, F. Levkovich-Maslyuk, Quark-anti-quark potential in $\mathcal{N} = 4$ SYM. arXiv:1601.05679
113. T. Klose, Review of AdS/CFT Integrability, Chapter IV.3: $\mathcal{N} = 6$ Chern-Simons and strings on $AdS_4 \times \mathbb{C}P^3$. Lett. Math. Phys. **99**, 401 (2012). arXiv:1012.3999
114. J. Minahan, K. Zarembo, The Bethe ansatz for superconformal Chern-Simons. JHEP **0809**, 040 (2008). arXiv:0806.3951
115. D. Bak, S.-J. Rey, Integrable spin chain in superconformal Chern-Simons theory. JHEP **0810**, 053 (2008). arXiv:0807.2063
116. J.A. Minahan, W. Schulgin, K. Zarembo, Two loop integrability for Chern-Simons theories with $\mathcal{N} = 6$ supersymmetry. JHEP **0903**, 057 (2009). arXiv:0901.1142
117. B.I. Zwiebel, Two-loop integrability of planar $\mathcal{N} = 6$ superconformal Chern-Simons theory. J. Phys. A **42**, 495402 (2009). arXiv:0901.0411
118. N. Gromov, P. Vieira, The all loop AdS_4/CFT_3 Bethe ansatz. JHEP **0901**, 016 (2009). arXiv:0807.0777
119. N. Gromov, P. Vieira, The AdS_4/CFT_3 algebraic curve. JHEP **0902**, 040 (2009). arXiv:0807.0437
120. G. Arutyunov, S. Frolov, Superstrings on $AdS_4 \times \mathbb{CP}^3$ as a Coset Sigma-model. JHEP **0809**, 129 (2008). arXiv:0806.4940

121. D. Bombardelli, D. Fioravanti, R. Tateo, TBA and Y-system for planar AdS_4/CFT_3. Nucl. Phys. B **834**, 543 (2010). arXiv:0912.4715
122. N. Gromov, F. Levkovich-Maslyuk, Y-system, TBA and Quasi-Classical strings in $AdS_4 \times \mathbb{CP}^3$. JHEP **1006**, 088 (2010). arXiv:0912.4911
123. A. Cavaglia, D. Fioravanti, N. Gromov, R. Tateo, The quantum spectral curve of the ABJM theory. Phys. Rev. Lett. **113**, 021601 (2014). arXiv:1403.1859
124. N. Gromov, G. Sizov, Exact slope and interpolating functions in $\mathcal{N} = 6$ supersymmetric Chern-Simons theory. Phys. Rev. Lett. **113**, 121601 (2014). arXiv:1403.1894
125. A. Cavaglia, N. Gromov, F. Levkovich-Maslyuk, On the exact interpolating function in ABJ theory. JHEP **1612**, 086 (2016). arXiv:1605.04888
126. V.G. Drinfeld, Hopf algebras and the quantum Yang-Baxter equation. Sov. Math. Dokl. **32**, 254 (1985). [Dokl. Akad. Nauk Ser. Fiz. 283 1060(1985)]
127. L.F. Alday, J.M. Maldacena, Gluon scattering amplitudes at strong coupling. JHEP **0706**, 064 (2007). arXiv:0705.0303
128. N. Beisert, R. Ricci, A.A. Tseytlin, M. Wolf, Dual superconformal symmetry from $AdS_5 \times S^5$ superstring integrability. Phys. Rev. D **78**, 126004 (2008). arXiv:0807.3228
129. I. Adam, A. Dekel, Y. Oz, On integrable backgrounds self-dual under fermionic T-duality. JHEP **0904**, 120 (2009). arXiv:0902.3805
130. L.F. Alday, Review of AdS/CFT Integrability, Chapter V. 3: scattering amplitudes at strong coupling. Lett. Math. Phys. **99**, 507 (2012). arXiv:1012.4003
131. M.C. Abbott, J. Murugan, S. Penati, A. Pittelli, D. Sorokin, P. Sundin, J. Tarrant, M. Wolf, L. Wulff, T-duality of Green-Schwarz superstrings on $AdS_d \times S^d \times M^{10-2d}$. JHEP **1512**, 104 (2015). arXiv:1509.07678
132. J.M. Drummond, G.P. Korchemsky, E. Sokatchev, Conformal properties of four-gluon planar amplitudes and Wilson loops. Nucl. Phys. B **795**, 385 (2008). arXiv:0707.0243
133. J.M. Drummond, Review of AdS/CFT Integrability, Chapter V. 2: dual superconformal symmetry. Lett. Math. Phys. **99**, 481 (2012). arXiv:1012.4002
134. J.M. Drummond, J. Henn, G.P. Korchemsky, E. Sokatchev, Dual superconformal symmetry of scattering amplitudes in $\mathcal{N} = 4$ super-Yang-Mills theory. Nucl. Phys. B **828**, 317 (2010). arXiv:0807.1095
135. A. Brandhuber, P. Heslop, G. Travaglini, A Note on dual superconformal symmetry of the $\mathcal{N} = 4$ super Yang-Mills S-matrix. Phys. Rev. D **78**, 125005 (2008). arXiv:0807.4097
136. J.M. Drummond, J.M. Henn, J. Plefka, Yangian symmetry of scattering amplitudes in $\mathcal{N} = 4$ super Yang-Mills theory. JHEP **0905**, 046 (2009). arXiv:0902.2987 (In: Strangeness in quark matter. Proceedings, International Conference, SQM 2008 (Beijing, P.R. China, October 5–10, 2008), p. 046)
137. T. Bargheer, N. Beisert, W. Galleas, F. Loebbert, T. McLoughlin, Exacting $\mathcal{N} = 4$ superconformal symmetry. JHEP **0911**, 056 (2009). arXiv:0905.3738
138. N. Beisert, J. Henn, T. McLoughlin, J. Plefka, One-loop superconformal and Yangian symmetries of scattering amplitudes in $\mathcal{N} = 4$ super Yang-Mills. JHEP **1004**, 085 (2010). arXiv:1002.1733
139. S. Caron-Huot, S. He, Jumpstarting the all-loop S-matrix of planar $\mathcal{N} = 4$ super Yang-Mills. JHEP **1207**, 174 (2012). arXiv:1112.1060
140. L. Dolan, C.R. Nappi, E. Witten, A relation between approaches to integrability in superconformal Yang-Mills theory. JHEP **0310**, 017 (2003). arXiv:hep-th/0308089
141. D. Mueller, H. Muenkler, J. Plefka, J. Pollok, K. Zarembo, Yangian symmetry of smooth Wilson loops in $\mathcal{N} = 4$ super Yang-Mills theory. JHEP **1311**, 081 (2013). arXiv:1309.1676
142. N. Beisert, D. Mueller, J. Plefka, C. Vergu, Integrability of smooth Wilson loops in $\mathcal{N} = 4$ superspace. JHEP **1512**, 141 (2015). arXiv:1509.05403
143. B. Basso, A. Sever, P. Vieira, Spacetime and Flux tube S-matrices at finite coupling for $\mathcal{N} = 4$ supersymmetric Yang-Mills theory. Phys. Rev. Lett. **111**, 091602 (2013). arXiv:1303.1396
144. B. Basso, A. Sever, P. Vieira, Space-time S-matrix and Flux tube S-matrix II. Extracting and matching data. JHEP **1401**, 008 (2014). arXiv:1306.2058

145. B. Basso, A. Sever, P. Vieira, Space-time S-matrix and Flux-tube S-matrix III. The two-particle contributions. JHEP **1408**, 085 (2014). arXiv:1402.3307
146. B. Basso, A. Sever, P. Vieira, Space-time S-matrix and Flux-tube S-matrix IV. Gluons and fusion. JHEP **1409**, 149 (2014). arXiv:1407.1736
147. B. Basso, J. Caetano, L. Cordova, A. Sever, P. Vieira, OPE for all helicity amplitudes. JHEP **1508**, 018 (2015). arXiv:1412.1132
148. B. Basso, S. Komatsu, P. Vieira, Structure constants and integrable bootstrap in planar $\mathcal{N} = 4$ SYM theory. arXiv:1505.06745
149. V. Pestun, Localization of gauge theory on a four-sphere and supersymmetric Wilson loops. Commun. Math. Phys. **313**, 71 (2012). arXiv:0712.2824
150. A. Neveu, J.H. Schwarz, Factorizable dual model of pions. Nucl. Phys. B **31**, 86 (1971)
151. P. Ramond, Dual theory for free fermions. Phys. Rev. D **3**, 2415 (1971)
152. J.H. Schwarz, Covariant field equations of chiral $\mathcal{N} = 2$ $D = 10$ supergravity. Nucl. Phys. B **226**, 269 (1983)
153. M.B. Green, J.H. Schwarz, Covariant description of superstrings. Phys. Lett. B **136**, 367 (1984)
154. M.B. Green, J.H. Schwarz, Properties of the covariant formulation of superstring theories. Nucl. Phys. B **243**, 285 (1984)
155. H.J. de Vega, N. G. Sanchez, Lectures on string theory in curved space-times. arXiv:hep-th/9512074 (In: String gravity and physics at the Planck energy scale. Proceedings, NATO Advanced Study Institute, Erice, Italy, September 8–19, 1995)
156. J. Polchinski, String theory, in *An Introduction to the Bosonic String*, vol. 1 (Cambridge University Press, 2007)
157. S.A. Fulling, Aspects of quantum field theory in curved space-time. Lond. Math. Soc. Stud. Texts **17**, 1 (1989)
158. O. Alvarez, Theory of strings with boundaries: fluctuations, topology, and quantum geometry. Nucl. Phys. B **216**, 125 (1983)
159. H. Luckock, Quantum geometry of strings with boundaries. Ann. Phys. **194**, 113 (1989)
160. M. Nakahara, *Geometry, Topology and Physics* (CRC Press, 2003)
161. L. Bianchi, V. Forini, B. Hoare, Two-dimensional S-matrices from unitarity cuts. JHEP **1307**, 088 (2013). arXiv:1304.1798
162. O.T. Engelund, R.W. McKeown, R. Roiban, Generalized unitarity and the worldsheet S matrix in $AdS_n \times S^n \times M^{10-2n}$. JHEP **1308**, 023 (2013). arXiv:1304.4281
163. L. Bianchi, B. Hoare, $AdS_3 \times S^3 \times M^4$ string S-matrices from unitarity cuts. JHEP **1408**, 097 (2014). arXiv:1405.7947
164. S. Frolov, A.A. Tseytlin, Multispin string solutions in $AdS_5 \times S^5$. Nucl. Phys. B **668**, 77 (2003). arXiv:hep-th/0304255
165. S. Frolov, A.A. Tseytlin, Quantizing three spin string solution in $AdS_5 \times S^5$. JHEP **0307**, 016 (2003). arXiv:hep-th/0306130
166. S. Frolov, I. Park, A.A. Tseytlin, On one-loop correction to energy of spinning strings in S^5. Phys. Rev. D **71**, 026006 (2005). arXiv:hep-th/0408187
167. I. Park, A. Tirziu, A.A. Tseytlin, Spinning strings in $AdS_5 \times S^5$: One-loop correction to energy in SL(2) sector. JHEP **0503**, 013 (2005). arXiv:hep-th/0501203
168. N. Beisert, A.A. Tseytlin, K. Zarembo, Matching quantum strings to quantum spins: one-loop versus finite-size corrections. Nucl. Phys. B **715**, 190 (2005). arXiv:hep-th/0502173
169. B. Hoare, Y. Iwashita, A.A. Tseytlin, Pohlmeyer-reduced form of string theory in $AdS_5 \times S^5$: semiclassical expansion. J. Phys. A **42**, 375204 (2009). arXiv:0906.3800
170. S.S. Gubser, I.R. Klebanov, A.M. Polyakov, A semi-classical limit of the gauge/string correspondence. Nucl. Phys. B **636**, 99 (2002). arXiv:hep-th/0204051
171. S. Frolov, A.A. Tseytlin, Semiclassical quantization of rotating superstring in $AdS_5 \times S^5$. JHEP **0206**, 007 (2002). arXiv:hep-th/0204226
172. M. Beccaria, G.V. Dunne, V. Forini, M. Pawellek, A.A. Tseytlin, Exact computation of one-loop correction to energy of spinning folded string in $AdS_5 \times S^5$. J. Phys. A **43**, 165402 (2010). arXiv:1001.4018

173. V. Forini, Quark-antiquark potential in AdS at one loop. JHEP **1011**, 079 (2010). arXiv:1009.3939
174. N. Drukker, V. Forini, Generalized quark-antiquark potential at weak and strong coupling. JHEP **1106**, 131 (2011). arXiv:1105.5144
175. V. Forini, V.G.M. Puletti, O. Ohlsson, Sax, The generalized cusp in $AdS_4 \times \mathbb{CP}^3$ and more one-loop results from semiclassical strings. J. Phys. A **46**, 115402 (2013). arXiv:1204.3302
176. S. Frolov, A. Tirziu, A.A. Tseytlin, Logarithmic corrections to higher twist scaling at strong coupling from AdS/CFT. Nucl. Phys. B **766**, 232 (2007). arXiv:hep-th/0611269
177. R. Roiban, A. Tirziu, A.A. Tseytlin, Two-loop world-sheet corrections in $AdS_5 \times S^5$ superstring. JHEP **0707**, 056 (2007). arXiv:0704.3638
178. N. Drukker, D.J. Gross, H. Ooguri, Wilson loops and minimal surfaces. Phys. Rev. D **60**, 125006 (1999). arXiv:hep-th/9904191
179. K.G. Wilson, Confinement of quarks. Phys. Rev. D **10**, 2445 (1974) [,45(1974)]
180. H. Ooguri, J. Rahmfeld, H. Robins, J. Tannenhauser, Holography in superspace. JHEP **0007**, 045 (2000). arXiv:hep-th/0007104
181. N. Beisert, D. Mueller, J. Plefka, C. Vergu, Smooth Wilson loops in $\mathcal{N} = 4$ non-chiral superspace. JHEP **1512**, 140 (2015). arXiv:1506.07047
182. H. Muenkler, J. Pollok, Minimal surfaces of the $AdS_5 \times S^5$ superstring and the symmetries of super Wilson loops at strong coupling. J. Phys. A **48**, 365402 (2015). arXiv:1503.07553
183. J.M. Maldacena, Wilson loops in large N field theories. Phys. Rev. Lett. **80**, 4859 (1998). arXiv:hep-th/9803002
184. S.-J. Rey, J.-T. Yee, Macroscopic strings as heavy quarks in large N gauge theory and anti-de Sitter supergravity. Eur. Phys. J. C **22**, 379 (2001). arXiv:hep-th/9803001
185. G.W. Semenoff, K. Zarembo, Wilson loops in SYM theory: from weak to strong coupling. Nucl. Phys. Proc. Suppl. **108**, 106 (2002). arXiv:hep-th/0202156 (In: Light cone physics: Particles and strings. Proceedings, International Workshop (Trento, Italy, September 3–11, 2001), p. 106–112, [,106(2002)])
186. Y. Makeenko, A brief introduction to Wilson loops, large N, Phys. Atom. Nucl. **73**, 878 (2010). arXiv:0906.4487 (In: 12th International Moscow School of Physics and 37th ITEP Winter School of Physics Moscow (Russia, February 9–16, 2009), p. 878–894)
187. R. Roiban, A.A. Tseytlin, Strong-coupling expansion of cusp anomaly from quantum superstring. JHEP **0711**, 016 (2007). arXiv:0709.0681
188. S. Giombi, R. Ricci, R. Roiban, A.A. Tseytlin, C. Vergu, Generalized scaling function from light-cone gauge $AdS_5 \times S^5$ superstring. JHEP **1006**, 060 (2010). arXiv:1002.0018
189. T. McLoughlin, Review of AdS/CFT Integrability, Chapter II.2: quantum strings in $AdS_5 \times S^5$. Lett. Math. Phys. **99**, 127 (2012). arXiv:1012.3987
190. A.A. Tseytlin, Review of AdS/CFT Integrability, Chapter II.1: classical $AdS_5 \times S^5$ string solutions. Lett. Math. Phys. **99**, 103 (2012). arXiv:1012.3986
191. S. Forste, D. Ghoshal, S. Theisen, Stringy corrections to the Wilson loop in $\mathcal{N} = 4$ superYang-Mills theory. JHEP **9908**, 013 (1999). arXiv:hep-th/9903042
192. N. Drukker, D.J. Gross, A.A. Tseytlin, Green-Schwarz string in $AdS_5 \times S^5$: semiclassical partition function. JHEP **0004**, 021 (2000). arXiv:hep-th/0001204
193. M. Beccaria, V. Forini, A. Tirziu, A.A. Tseytlin, Structure of large spin expansion of anomalous dimensions at strong coupling. Nucl. Phys. B **812**, 144 (2009). arXiv:0809.5234
194. T. Klose, T. McLoughlin, R. Roiban, K. Zarembo, Worldsheet scattering in $\mathrm{AdS}_5 \times S^5$. JHEP **0703**, 094 (2007). arXiv:hep-th/0611169
195. R. Roiban, P. Sundin, A. Tseytlin, L. Wulff, The one-loop worldsheet S-matrix for the $AdS_n \times S^n \times T^{10-2n}$ superstring. JHEP **1408**, 160 (2014). arXiv:1407.7883
196. M. Kruczenski, A. Tirziu, Matching the circular Wilson loop with dual open string solution at 1-loop in strong coupling. JHEP **0805**, 064 (2008). arXiv:0803.0315
197. E.I. Buchbinder, A.A. Tseytlin, One-loop correction to the energy of a wavy line string in AdS_5. J. Phys. A **46**, 505401 (2013). arXiv:1309.1581
198. J. Aguilera-Damia, D.H. Correa, G.A. Silva, Semiclassical partition function for strings dual to Wilson loops with small cusps in ABJM. JHEP **1503**, 002 (2015). arXiv:1412.4084

199. T. Kameyama, K. Yoshida, Generalized quark-antiquark potentials from a q-deformed $AdS_5 \times S^5$ background. PTEP **2016**, 063B01 (2016). arXiv:1602.06786
200. R. Kallosh, J. Kumar, Supersymmetry enhancement of D-p-branes and M-branes. Phys. Rev. D **56**, 4934 (1997). arXiv:hep-th/9704189
201. M. Cvetic, H. Lu, C.N. Pope, K.S. Stelle, T duality in the Green-Schwarz formalism, and the massless/massive IIA duality map. Nucl. Phys. B **573**, 149 (2000). arXiv:hep-th/9907202
202. L. Wulff, The type II superstring to order θ^4. JHEP **1307**, 123 (2013). arXiv:1304.6422
203. V. Fock, The proper time in classical and quantum mechanics, Izv. Akad. Nauk USSR (Phys.) 4–5, 551 (1937)
204. J.S. Schwinger, On gauge invariance and vacuum polarization. Phys. Rev. **82**, 664 (1951)
205. B. DeWitt, *Dynamical Theory of Groups and Fields* (Gordon and Breach, 1965)
206. B.S. DeWitt, Quantum theory of gravity. 1. The canonical theory. Phys. Rev. **160**, 1113 (1967)
207. B.S. DeWitt, Quantum theory of gravity. 2. The manifestly covariant theory. Phys. Rev. **162**, 1195 (1967)
208. B.S. DeWitt, Quantum theory of gravity. 3. Applications of the covariant theory. Phys. Rev. **162**, 1239 (1967)
209. D.V. Vassilevich, Heat kernel expansion: User's manual. Phys. Rept. **388**, 279 (2003). arXiv:hep-th/0306138
210. V. Mukhanov, S. Winitzki, *Introduction to Quantum Effects in Gravity* (Cambridge University Press, 2007)
211. D. Fursaev, D. Vassilevich, *Operators, Geometry and Quanta: Methods of Spectral Geometry in Quantum Field Theory* (Springer, Verlag, 2011)
212. R. Camporesi, Harmonic analysis and propagators on homogeneous spaces. Phys. Rept. **196**, 1 (1990)
213. R. Camporesi, A. Higuchi, Spectral functions and zeta functions in hyperbolic spaces. J. Math. Phys. **35**, 4217 (1994)
214. R. Camporesi, The Spinor heat kernel in maximally symmetric spaces. Commun. Math. Phys. **148**, 283 (1992)
215. R. Camporesi, A. Higuchi, On the eigenfunctions of the Dirac operator on spheres and real hyperbolic spaces. J. Geom. Phys. **20**, 1 (1996). arXiv:gr-qc/9505009
216. O. Calin, D.-C. Chang, K. Furutani, C. Iwasaki, *Heat Kernels for Elliptic and Sub-Elliptic Operators: Methods and Techniques* (Springer, 2011)
217. E. Elizalde, *Zeta Regularization Techniques with Applications* (World Scientific, 1994)
218. E.I. Buchbinder, A.A. Tseytlin, $1/N$ correction in the D3-brane description of a circular Wilson loop at strong coupling. Phys. Rev. D **89**, 126008 (2014). arXiv:1404.4952
219. R. Bergamin, A.A. Tseytlin, Heat kernels on cone of AdS_2 and k-wound circular Wilson loop in $AdS_5 \times S^5$ superstring. J. Phys. A **49**, 14LT01 (2016). arXiv:1510.06894
220. I.M. Gelfand, A.M. Yaglom, Integration in functional spaces and it applications in quantum physics. J. Math. Phys. **1**, 48 (1960)
221. R. Camporesi, A. Higuchi, Stress-energy tensors in anti-de Sitter spacetime. Phys. Rev. D **45**, 3591 (1992)
222. R.F. Dashen, B. Hasslacher, A. Neveu, Nonperturbative methods and extended-hadron models in field theory. I. Semiclassical functional methods. Phys. Rev. D **10**, 4114 (1974)
223. R.F. Dashen, B. Hasslacher, A. Neveu, Nonperturbative methods and extended-hadron models in field theory. II. Two-dimensional models and extended hadrons. Phys. Rev. D **10**, 4130 (1974)
224. R.F. Dashen, B. Hasslacher, A. Neveu, Nonperturbative methods and extended-hadron models in field theory. III. Four-dimensional non-Abelian models. Phys. Rev. D **10**, 4138 (1974)
225. X. Chen-Lin, D. Medina-Rincon, K. Zarembo, Quantum String test of nonconformal holography. arXiv:1702.07954
226. R. Forman, Functional determinants and geometry. Invent. math. **88**, 447 (1987)
227. R. Forman, Functional determinants and geometry (Erratum). Invent. math. **108**, 453 (1992)
228. G.V. Dunne, K. Kirsten, Functional determinants for radial operators. J. Phys. A **39**, 11915 (2006). arXiv:hep-th/0607066

229. V. Forini, A.A. Tseytlin, E. Vescovi, Perturbative computation of string one-loop corrections to Wilson loop minimal surfaces in $AdS_5 \times S^5$. JHEP **1703**, 003 (2017). arXiv:1702.02164
230. A.M. Polyakov, Conformal fixed points of unidentified gauge theories, Mod. Phys. Lett. A **19**, 1649 (2004). arXiv:hep-th/0405106, [1159(2004)]
231. S. Giombi, R. Ricci, R. Roiban, A. Tseytlin, C. Vergu, Quantum $AdS_5 \times S^5$ superstring in the AdS light-cone gauge. JHEP **1003**, 003 (2010). arXiv:0912.5105
232. S. Giombi, R. Ricci, R. Roiban, A. Tseytlin, Two-loop $AdS_5 \times S^5$ superstring: testing asymptotic Bethe ansatz and finite size corrections. J. Phys. A **44**, 045402 (2011). arXiv:1010.4594
233. R. Metsaev, A.A. Tseytlin, Superstring action in $AdS_5 \times S^5$. Kappa symmetry light cone gauge. Phys. Rev. D **63**, 046002 (2001). arXiv:hep-th/0007036
234. R. Metsaev, C.B. Thorn, A.A. Tseytlin, Light cone superstring in AdS space-time. Nucl. Phys. B **596**, 151 (2001). arXiv:hep-th/0009171
235. M. Kruczenski, A Note on twist two operators in $\mathcal{N} = 4$ SYM and Wilson loops in Minkowski signature. JHEP **0212**, 024 (2002). arXiv:hep-th/0210115
236. D. Uvarov, $AdS_4 \times \mathbb{CP}^3$ superstring in the light-cone gauge. Nucl. Phys. B **826**, 294 (2010). arXiv:0906.4699
237. D. Uvarov, Light-cone gauge Hamiltonian for $AdS_4 \times \mathbb{CP}^3$ superstring. Mod. Phys. Lett. A **25**, 1251 (2010). arXiv:0912.1044
238. R. Roiban, A.A. Tseytlin, Spinning superstrings at two loops: Strong-coupling corrections to dimensions of large-twist SYM operators. Phys. Rev. D **77**, 066006 (2008). arXiv:0712.2479
239. L. Bianchi, M.S. Bianchi, Quantum dispersion relations for the $AdS_4 \times \mathbb{C}P^3$ GKP string. JHEP **1511**, 031 (2015). arXiv:1505.00783
240. B. Basso, A. Rej, Bethe ansaetze for GKP strings. Nucl. Phys. B **879**, 162 (2014). arXiv:1306.1741
241. I. Montvay, G. Muenster, *Quantum Fields on a Lattice* (Cambridge University Press, 1994)
242. T. DeGrand, C.E. Detar, *Lattice Methods for Quantum Chromodynamics* (World Scientific, 2006)
243. C. Gattringer, C. Lang, *Quantum Chromodynamics on the Lattice: An Introductory Presentation* (Springer, Berlin Heidelberg, 2009)
244. R.W. McKeown, R. Roiban, The quantum $AdS_5 \times S^5$ superstring at finite coupling. arXiv:1308.4875
245. N. Beisert, B. Eden, M. Staudacher, Transcendentality and crossing. J. Stat. Mech. **0701**, P01021 (2007). arXiv:hep-th/0610251
246. M. Hanada, What lattice theorists can do for quantum gravity. arXiv:1604.05421
247. A.A. Tseytlin, On semiclassical approximation and spinning string vertex operators in $AdS_5 \times S^5$. Nucl. Phys. B **664**, 247 (2003). arXiv:hep-th/0304139
248. R. Roiban, A.A. Tseytlin, On semiclassical computation of 3-point functions of closed string vertex operators in $AdS_5 \times S^5$. Phys. Rev. D **82**, 106011 (2010). arXiv:1008.4921
249. A. Joseph, Review of lattice supersymmetry and gauge-gravity duality. Int. J. Mod. Phys. A **30**, 1530054 (2015). arXiv:1509.01440
250. G. Bergner, S. Catterall, Supersymmetry on the lattice. arXiv:1603.04478
251. D.B. Kaplan, M. Unsal, A Euclidean lattice construction of supersymmetric Yang-Mills theories with sixteen supercharges. JHEP **0509**, 042 (2005). arXiv:hep-lat/0503039
252. M. Unsal, Twisted supersymmetric gauge theories and orbifold lattices. JHEP **0610**, 089 (2006). arXiv:hep-th/0603046
253. S. Catterall, From twisted supersymmetry to orbifold lattices. JHEP **0801**, 048 (2008). arXiv:0712.2532
254. P.H. Damgaard, S. Matsuura, Geometry of orbifolded supersymmetric lattice gauge theories. Phys. Lett. B **661**, 52 (2008). arXiv:0801.2936
255. S. Catterall, D.B. Kaplan, M. Unsal, Exact lattice supersymmetry. Phys. Rept. **484**, 71 (2009). arXiv:0903.4881
256. D. Schaich, T. DeGrand, Parallel software for lattice $\mathcal{N} = 4$ supersymmetric Yang-Mills theory. Comput. Phys. Commun. **190**, 200 (2015). arXiv:1410.6971

257. S. Catterall, D. Schaich, P.H. Damgaard, T. DeGrand, J. Giedt, $\mathcal{N} = 4$ Supersymmetry on a space-time lattice. Phys. Rev. D **90**, 065013 (2014). arXiv:1405.0644
258. S. Catterall, J. Giedt, D. Schaich, P.H. Damgaard, T. DeGrand, Results from lattice simulations of $\mathcal{N} = 4$ supersymmetric Yang-Mills. PoS LATTICE2014, 267 (2014). arXiv:1411.0166 (In: Proceedings, 32nd International Symposium on Lattice Field Theory (Lattice 2014), p. 267)
259. D. Schaich, Aspects of lattice $\mathcal{N} = 4$ supersymmetric Yang-Mills, PoS LATTICE2015, 242 (2015). arXiv:1512.01137 (In: Proceedings, 33rd International Symposium on Lattice Field Theory (Lattice 2015), p. 242)
260. T. Ishii, G. Ishiki, S. Shimasaki, A. Tsuchiya, $\mathcal{N} = 4$ super Yang-Mills from the plane wave matrix model. Phys. Rev. D **78**, 106001 (2008). arXiv:0807.2352
261. G. Ishiki, S.-W. Kim, J. Nishimura, A. Tsuchiya, Deconfinement phase transition in $\mathcal{N} = 4$ super Yang-Mills theory on $\mathbb{R} \times S^3$ from supersymmetric matrix quantum mechanics. Phys. Rev. Lett. **102**, 111601 (2009). arXiv:0810.2884
262. G. Ishiki, S.-W. Kim, J. Nishimura, A. Tsuchiya, Testing a novel large-N reduction for $\mathcal{N} = 4$ super Yang-Mills theory on $\mathbb{R} \times S^3$. JHEP **0909**, 029 (2009). arXiv:0907.1488
263. M. Hanada, S. Matsuura, F. Sugino, Two-dimensional lattice for four-dimensional $\mathcal{N} = 4$ supersymmetric Yang-Mills. Prog. Theor. Phys. **126**, 597 (2011). arXiv:1004.5513
264. M. Honda, G. Ishiki, J. Nishimura, A. Tsuchiya, Testing the AdS/CFT correspondence by Monte Carlo calculation of BPS and non-BPS Wilson loops in 4d $\mathcal{N} = 4$ super-Yang-Mills theory. PoS LATTICE2011, 244 (2011). arXiv:1112.4274 (In: Proceedings, 29th International Symposium on Lattice field theory (Lattice 2011), p. 244)
265. M. Honda, G. Ishiki, S.-W. Kim, J. Nishimura, A. Tsuchiya, Direct test of the AdS/CFT correspondence by Monte Carlo studies of $\mathcal{N} = 4$ super Yang-Mills theory. JHEP **1311**, 200 (2013). arXiv:1308.3525
266. M. Hanada, Y. Hyakutake, G. Ishiki, J. Nishimura, Holographic description of quantum black hole on a computer. Science **344**, 882 (2014). arXiv:1311.5607
267. V. Forini, V.G.M. Puletti, L. Griguolo, D. Seminara, E. Vescovi, Remarks on the geometrical properties of semiclassically quantized strings. J. Phys. A **48**, 475401 (2015). arXiv:1507.01883
268. V. Forini, V.G.M. Puletti, M. Pawellek, E. Vescovi, One-loop spectroscopy of semiclassically quantized strings: bosonic sector. J. Phys. A **48**, 085401 (2015). arXiv:1409.8674
269. V. Forini, V. Giangreco M. Puletti, L. Griguolo, D. Seminara, E. Vescovi, Precision calculation of 1/4-BPS Wilson loops in $AdS_5 \times S^5$. JHEP **1602**, 105 (2016). arXiv:1512.00841
270. L. Bianchi, M.S. Bianchi, A. Bres, V. Forini, E. Vescovi, Two-loop cusp anomaly in ABJM at strong coupling. JHEP **1410**, 13 (2014). arXiv:1407.4788
271. V. Forini, L. Bianchi, M.S. Bianchi, B. Leder, E. Vescovi, Lattice and string worldsheet in AdS/CFT: a numerical study. PoS LATTICE2015, 244 (2015). arXiv:1601.04670 (In: Proceedings, 33rd International Symposium on Lattice Field Theory (Lattice 2015), p. 244)
272. L. Bianchi, M.S. Bianchi, V. Forini, B. Leder, E. Vescovi, Green-Schwarz superstring on the lattice. JHEP **1607**, 014 (2016). arXiv:1605.01726
273. J. Aguilera-Damia, A. Faraggi, L.A. Pando Zayas, V. Rathee, G.A. Silva, D. Trancanelli, E. Vescovi, in preparation
274. V. Forini, L. Bianchi, B. Leder, P. Toepfer, E. Vescovi, Strings on the lattice and AdS/CFT. PoS LATTICE2016, 206 (2016). arXiv:1702.02005 (In: Proceedings, 34th International Symposium on Lattice Field Theory (Lattice 2016, Southampton, UK, July 24–30, 2016), p. 206)
275. V. Forini, L. Bianchi, B. Leder, P. Toepfer, E. Vescovi, in preparation

Chapter 2
Superstring Actions in $AdS_5 \times S^5$ and $AdS_4 \times \mathbb{CP}^3$ Spaces

Following the earlier construction of the covariant action for supersymmetric particles [1], the Green-Schwarz (GS) superstring action in flat space was proposed in [2] and interpreted as a coset sigma-model of Wess-Zumino type by Henneaux and Mezincescu in [3]. The action displays a local fermionic symmetry, called *κ-symmetry*, which generalizes the one exhibited by massive and massless superparticles [4, 5]. Superstrings of the type IIB can be formulated on a generic supergravity background with preservation of this gauge invariance [6].

Along the lines of the approach in [3], R.R. Metsaev and A.A. Tseytlin constructed the covariant κ-symmetric action for type IIB superstring on $AdS_5 \times S^5$ in [7, 8]. Given its central relevance in the main instance (1.1) of AdS/CFT, we devote Sect. 2.1 to review the sigma-model based on the supercoset $\frac{PSU(2,2|4)}{SO(1,4)\times SO(5)}$.

The importance of constructing superstring theory for the lower-dimensional duality (1.5) prompted the formulation of the $\frac{OSp(4|6)}{SO(1,3)\times U(3)}$ supercoset action on the $AdS_4 \times \mathbb{CP}^3$ background [9, 10]. However, some subtleties related to the κ-symmetry transformations of this sigma-model will lead us in Sect. 2.2 to consider the alternative formulation of the action proposed by Uvarov [11, 12].

2.1 Supercoset Construction of the String Action in $AdS_5 \times S^5$

This $AdS_5 \times S^5$ background is a maximally supersymmetric solution [13] of the type IIB supergravity equations of motion in ten-dimensions, together with the flat Minkowski space $\mathbb{R}^{1,9}$ and the "plane-wave" background [14]. The presence of the self-dual Ramond-Ramond (RR) five-form flux supporting this "vacuum" geometry precludes the use of the Neveu-Schwarz-Ramond (NSR) approach [15, 16] to construct the action. The Green-Schwarz formalism [2, 17] has proven to be a viable

E. Vescovi, *Perturbative and Non-perturbative Approaches to String Sigma-Models in AdS/CFT*, Springer Theses, DOI 10.1007/978-3-319-63420-3_2

method when the RR fields are not vanishing, with the additional advantage of realizing supersymmetry manifestly in the ten-dimensional ambient space. For any type II supergravity background, the formal expression of the superstring action exists [6], but it is not very practical to explicitly find the action in terms of the coordinate fields.[1] For this reason, one is led to devise an alternative approach to write the complete action on this space.

A more advantageous route to the superstring action in the $AdS_5 \times S^5$ background is tailored to the peculiar structure of the superisometry group $PSU(2, 2|4)$, namely the supersymmetric extension of the $SU(2, 2) \times SU(4)$ group which is locally isomorphic to the bosonic isometries $SO(2, 4) \times SO(6)$ of this product manifold.

The generalization to the curved $AdS_5 \times S^5$ background has been developed in [7, 8], prompted by the then-recent conjecture of the AdS/CFT correspondence [18], and it is also conceptually very close to the construction of the GS action for strings moving in $\mathbb{R}^{1,9}$ [3]. Taking inspiration from the flat-space counterpart, the superstring action can be formulated as a type of Wess-Zumino-Witten (WZW) non-linear sigma-model in two-dimensions with the supercoset $PSU(2, 2|4)/(SO(1, 4) \times SO(5))$ as target space. We recall that the bosonic reduction of this coset is precisely a representation of the $AdS_5 \times S^5$ space, as the quotient of $SO(2, 4) \times SO(6) \subset PSU(2, 2|4)$ over $SO(1, 4) \times SO(5)$, where the 10 bosonic degrees of freedom of the superstring propagate. The additional 32 fermionic degrees of freedom, which would parametrize the two Majorana-Weyl fermions of 10d type IIB supergravity, are provided by the corresponding anticommuting generators of $PSU(2, 2|4)$.

The next section reviews the basic facts [19] about sigma-model actions on general coset spaces, which will provide a natural way to include the couplings to the background RR fields for the $AdS_5 \times S^5$ case in question. This is also a general strategy for constructing sigma-models in other integrable AdS/CFT systems.

Let us mention that the supercoset general construction naturally applies to the string theory on $AdS_4 \times \mathbb{CP}^3$ [9, 10], $AdS_3 \times S^3 \times S^3 \times S^1$ and $AdS_3 \times S^3 \times T^4$ [20] and also to the η-model [21, 22] and λ-model [23, 24], which are integrable deformations of the Metsaev-Tseytlin supercoset action named after the corresponding deformation parameters. However, they will not be covered in what follows.

2.1.1 String Sigma-Model for Coset Spaces and κ-Symmetry

The construction of the Lagrangian makes use of a general description valid for any homogeneous manifold expressed as coset space G/H, where G is the isometry group of the Killing vectors on such manifold and H is the stabilizer subgroup. The resulting sigma-model is based on the action of the supercoset $\tilde{G}/H$, where $\tilde{G}$ is the supergroup embedding G as its even part. We denote the (super)algebras associated to G, H and $\tilde{G}$ by $\mathfrak{g}$, $\mathfrak{h}$ and $\tilde{\mathfrak{g}}$ respectively. In most of the relevant examples

[1] Indeed, given a bosonic background, one has to determine the exact expressions for the supervielbeins, which is a difficult problem to solve in non-trivial cases.

of AdS backgrounds, we restrict to a Lie superalgebra $\tilde{\mathfrak{g}}$ admitting an order-four *automorphism*, which is a linear map $\Omega : \tilde{\mathfrak{g}} \to \tilde{\mathfrak{g}}$ with

$$\Omega([a, b]) = [\Omega(a), \Omega(b)], \qquad a, b \in \tilde{\mathfrak{g}}, \qquad \Omega^4 = \text{id} \tag{2.1}$$

that decomposes $\tilde{\mathfrak{g}}$ into a direct sum of four graded subspaces

$$\tilde{\mathfrak{g}} = \tilde{\mathfrak{g}}^{(0)} \oplus \tilde{\mathfrak{g}}^{(1)} \oplus \tilde{\mathfrak{g}}^{(2)} \oplus \tilde{\mathfrak{g}}^{(3)}, \tag{2.2}$$

where each of them is an eigenspace of Ω

$$\Omega(\tilde{\mathfrak{g}}^{(k)}) = i^k \tilde{\mathfrak{g}}^{(k)}. \tag{2.3}$$

The grading is compatible with the supercommutator $[\tilde{\mathfrak{g}}^{(k)}, \tilde{\mathfrak{g}}^{(m)}] \subset \tilde{\mathfrak{g}}^{(k+m \bmod 4)}$. Note that $\tilde{\mathfrak{g}}^{(0)}$ is a subalgebra and the set of stationary points of Ω. In addition to the $\mathbb{Z}_2$-grading implicit in the definition of superalgebra $\mathfrak{g}$—for which $\tilde{\mathfrak{g}}^{(0)}$, $\tilde{\mathfrak{g}}^{(2)}$ are even and $\tilde{\mathfrak{g}}^{(1)}$, $\tilde{\mathfrak{g}}^{(3)}$ are odd subspaces—the automorphism endows $\mathfrak{g}$ with the structure of a $\mathbb{Z}_4$-graded algebra [25].

The description of the superstring action is given in terms of a *coset representative* $g(\tau, \sigma) \in \tilde{G}$ defined on 2d worldsheet, spanned by the τ and σ coordinates. We such g to build the $\tilde{\mathfrak{g}}$-valued one-form current

$$A \equiv -g^{-1}dg = -(g^{-1}\partial_\tau g)d\tau - (g^{-1}\partial_\sigma g)d\sigma \tag{2.4}$$

and split it as $A = A^{(0)} + A^{(1)} + A^{(2)} + A^{(3)}$ according to the $\mathbb{Z}_4$-decomposition. By construction the current is *flat*, namely it has vanishing two-form curvature $F \equiv dA - A \wedge A = 0$. The current exhibits other important properties:

- it is invariant under the left group action $g \to hg$ of a constant $h \in \tilde{G}$,
- under the local right action $g \to gh$ with $h(\tau, \sigma) \in H$, it undergoes the "gauge" transformation $A \to h^{-1}Ah - h^{-1}dh$, which in components splits into

$$A^{(0)} \to h^{-1}A^{(0)}h - h^{-1}dh, \qquad A^{(k)} \to h^{-1}A^{(k)}h, \quad k = 1, 2, 3. \tag{2.5}$$

Having fixed the coset representatives g, the supercoset sigma-model action with target superspace $\tilde{G}/H$ is given by

$$S = -\frac{T}{2}\int d\tau d\sigma\, \mathcal{L}, \qquad \mathcal{L} = \sqrt{-g}g^{\alpha\beta}\text{str}\left(A^{(2)}_\alpha A^{(2)}_\beta\right) + \kappa\,\epsilon^{\alpha\beta}\text{str}\left(A^{(1)}_\alpha A^{(3)}_\beta\right), \tag{2.6}$$

where the dimensionless quantity T is the string tension, $\alpha, \beta = 0, 1$ are the worldsheet indices, $g_{\alpha\beta}$ is the worldsheet metric with $g \equiv \det g_{\alpha\beta}$, the antisymmetric symbol $\epsilon^{\alpha\beta}$ is defined by $\epsilon^{01} = 1$ and we dropped the wedge operator between the one-forms. The Lagrangian density is built by means of the supertrace operator (str) acting on a suitable matrix representation of the $\mathbb{Z}^4$-components of the current $A^{(k)}$ in the algebra $\tilde{\mathfrak{g}}$.

- The first term in $\mathcal{L}$ is the usual kinetic term of a sigma-model. In addition to the proper quadratic kinetic term, it also contains interactions, hence (2.6) has to be regarded as a *non-linear* sigma-model.
- The second addend is a Wess-Zumino (WZ) type term. Although not directly visible in this form, it can be written indeed as the integral of a closed three-form over a 3d space M_3 with the 2d worldsheet $\Sigma = \partial M_3$ as boundary:

$$\frac{1}{2}\int_{\Sigma} \mathrm{str}\left(A^{(1)}A^{(3)}\right) = \int_{M_3} \mathrm{str}\left(A^{(2)}A^{(3)}A^{(3)} - A^{(2)}A^{(1)}A^{(1)}\right). \tag{2.7}$$

The value of the real coefficient κ can be fixed by the requirement that the action possesses the (κ-symmetry which we discuss below.

Physical motivations for the proposal (2.6) come from the special case when fermionic supermatrix elements of A are set to zero: one can show that S reduces to the standard Polaykov action for bosonic strings in the background G/H. It is also important to observe that the action, although it depends on $g \in \tilde{G}$, it only explicitly depends on the equivalence class of coset elements in $\tilde{G}/H$. This is a consequence of the supertrace in (2.6) being insensible to the similarity transformations (2.5) of the components $A^{(k)}$ with $k = 1, 2, 3$. Different parametrizations of the coset element lead to equivalent Lagrangians that differ in the explicit dependence on the coset degrees of freedom. This observation will motivate a preferred coset parametrization, accompanied by an appropriate gauge-fixing of the bosonic and fermionic symmetries, to drastically simplify the sigma model.

The group of global symmetry of $\mathcal{L}$, acting on the coset space by multiplication from the left, is $\tilde{G}$. In fact, given a coset representative $g \in \tilde{G}$, the action of a constant element $g' \in \tilde{G}$ is the map

$$g(\tau,\sigma) \;\rightarrow\; g^{'}g(\tau,\sigma) \equiv g^{''}(\tau,\sigma)h(\tau,\sigma)\,, \tag{2.8}$$

where the image of g is rewritten as the *right* action of a suitable local element $h \in H$ on the new coset representative $g^{''}$. Then the global $\tilde{G}$-invariance of the Lagrangian $\mathcal{L}$ is a consequence of the local H-symmetry of the action that gauges away $h(\tau, \sigma)$.

As already anticipated, the symmetry group of the action can be enhanced to include κ-symmetry, a local fermionic symmetry first discovered for massive and massless supersymmetric particles [4, 5] and then found in the GS superstring in flat space [2]. It plays a crucial role for the consistency of the supercoset model, ensuring that there is the correct number of physical fermionic degrees of freedom. It is realized as a certain right local action of $g^{'}(\tau,\sigma) \equiv \exp\varepsilon(\tau,\sigma) \in \tilde{G}$, with parameter $\varepsilon \in \tilde{\mathfrak{g}}$, on the coset element

$$g(\tau,\sigma) \;\rightarrow\; g(\tau,\sigma)g^{'}(\tau,\sigma) \equiv g^{''}(\tau,\sigma)h(\tau,\sigma), \tag{2.9}$$

where we needed a compensating element $h(\tau, \sigma) \in H$ as done in (2.8) and g'' is a new coset element. The fundamental difference with the case of global symmetry analyzed above is the fact that invariance of the action under (2.9) is guaranteed only for some parametrizations of the infinitesimal fermionic parameter $\varepsilon \equiv \varepsilon^{(1)} + \varepsilon^{(3)} \in \tilde{\mathfrak{g}}$. Given the variations of the four components of the current, one can prove that the infinitesimal variation of the Lagrangian amounts to

$$\delta_\varepsilon \mathcal{L} = \delta_\varepsilon(\sqrt{-g}g^{\alpha\beta})\text{str}\left(A_\alpha^{(2)} A_\beta^{(2)}\right) - 4\text{str}\left(P_+^{\alpha\beta}\left[A_\beta^{(1)}, A_\alpha^{(2)}\right]\varepsilon^{(1)} + P_-^{\alpha\beta}\left[A_\beta^{(3)}, A_\alpha^{(2)}\right]\varepsilon^{(3)}\right), \tag{2.10}$$

where the two projector operators are

$$P_\pm^{\alpha\beta} = \frac{\sqrt{-g}g^{\alpha\beta} \pm \kappa\, \epsilon^{\alpha\beta}}{2}. \tag{2.11}$$

Finding those ε that make the variation vanish needs an ansatz appropriate to the particular supercoset under investigation. Here we just want to point out that a necessary requirement for the invariance under κ-symmetry of the supercoset sigma-model for $AdS_5 \times S^5$ is that the projectors (2.11) are orthogonal to each other $g_{\beta\gamma} P_+^{\alpha\beta} P_-^{\gamma\delta} = 0$, which is true only if the real parameter in the WZ term is either $\kappa = 1$ or $\kappa = -1$.

Another crucial question for the consistency of the model is to understand the number of independent parameters of ε, which equals the numbers of fermionic degrees of freedom that can be gauged away by a κ-symmetry transformation. One can show that it is always possible to reduce the 32 real degrees of freedom in the coset element by a factor of one half. Therefore the gauge-fixed coset model involves 16 real physical fermions, which indeed coincides with the number of fermionic degrees of freedom in the κ-symmetry gauge-fixed GS action in ten dimensions.

2.1.2 *Classical Integrability of the Supercoset Model*

Classical integrability is a bonus symmetry[2] in all supercoset sigma-models with $\mathbb{Z}_4$-grading. This property[3] is equivalent to the statement that the Euler-Lagrange equations of motion following from (2.6) can be cast into the zero-curvature condition

$$dL + L \wedge L = 0 \tag{2.12}$$

for the one-parameter family of *Lax connection* (or *Lax pair*) one-forms $L(\tau, \sigma, z)$, functions of the fields of the theory and spanned by the *spectral parameter* $z \in \mathbb{C}$.

---

[2]This does not obviously mean that integrability is not spoiled by quantum effects [26, 27].

[3]Most of the discussion here is tailored to Sect. 2.1.1. A primer on classical and quantum field theories with emphasis on modern AdS/CFT applications can be found in [28].

It is worth appreciating that (2.12) is a strong condition on the classical dynamics of the model because it has to be satisfied for any value of z. There is, however, a certain level of arbitrariness in constructing L, as reflected by the observation that that gauge transformations

$$L \rightarrow h^{-1}Lh - h^{-1}dh \tag{2.13}$$

leaves the vanishing of the "field strength" built out of L unaffected. Both the local parameter h and the Lax connection are generically square matrices taking values in the some non-abelian algebra.

The existence of a Lax connection is sufficient condition to write the integrals of motion of the action. One can construct the *monodromy matrix*

$$T(z) \equiv \mathcal{P}\exp\left(\int_0^{2\pi} d\sigma L_\sigma(\tau,\sigma,z)\right) \tag{2.14}$$

as the path-ordered exponential of L parallel-transported along a closed path encircling the spacelike direction of the worldsheet cylinder, where for definiteness we assumed that all the quantities are periodic in $\sigma \in [0, 2\pi)$. The flatness condition (2.12) guarantees that (2.14) does not depend on the timeslice at fixed τ of such loop. In other words, the monodromy matrix encapsulates the time-independent information of the model. Practically, the Taylor expansion of T in the *continuous* spectral parameter delivers the *infinite* number of local conserved charges.

The Lax pair formulation of integrability at our disposal is very convenient, but it requires some effort to be set up for the supercoset sigma models with $\mathbb{Z}_4$-grading, as pedagogically shown in [19]. One can start with an ansatz for the Lax connection in terms of the components $A^{(k)}$ of the one-form current (2.4) and constrain it by imposing the flatness condition (2.12). At the end the string equations of motion lead to the zero-curvature condition for

$$L_\alpha = A_\alpha^{(0)} + \frac{1}{2}\left(z^2 + \frac{1}{z^2}\right)A_\alpha^{(2)} - \frac{1}{2\kappa}\left(z^2 - \frac{1}{z^2}\right)\gamma_{\alpha\beta}\epsilon^{\beta\gamma}A_\gamma^{(2)} + zA_\alpha^{(1)} + \frac{1}{z}A_\alpha^{(3)}. \tag{2.15}$$

An intriguing byproduct of integrability is that the zero-curvature condition for the Lax connection (2.15) implies $\kappa = \pm 1$ in (2.6). This is exactly the same condition that is found by imposing the κ-symmetry invariance of the sigma-model.

2.1.3 The $AdS_5 \times S^5$ String Action in the AdS Light-Cone Gauge

The formulation of a supercoset non-linear sigma-model on the curved $AdS_5 \times S^5$ background space is a special case of (2.6). We recall that the bosonic part of the supercoset where the string moves

$$AdS_5 \times S^5 = \frac{SO(2,4)}{SO(1,4)} \times \frac{SO(6)}{SO(5)} \,. \tag{2.16}$$

is the quotient between the isometry group $G = SO(2,4) \times SO(6)$ and isotropy group $H = SO(1,4) \times SO(5)$ of $AdS_5 \times S^5$. The group G is enlarged to $\tilde{G} = PSU(2,2|4)$ to endow the model with fermionic degrees of freedom.

The action constructed in [7] is the unique generalization of the flat-space GS action [3] that meets the following conditions:

- the bosonic reduction of the action is the usual Polyakov action in $AdS_5 \times S^5$,
- it is invariant under global $PSU(2,2|4)$ transformations and (local) κ-symmetry,
- it reduces to the type IIB GS action in flat space (after an appropriate rescaling of fermionic variables) in the limit of infinite AdS_5 and S^5 radius,
- it has the general form of the GS action corresponding to the $AdS_5 \times S^5$ supergravity solution.

The form of the action (2.6) is convenient to make the symmetries of the model apparent, but it is not amenable to direct sigma-model computations. The explicit expansion in terms of the bosonic and fermionic degrees of freedom is generally highly non-linear and dependent on the particular embedding of the coset element g into $PSU(2,2|4)$ written in terms of a set of generators of the superalgebra $\mathfrak{psu}(2,2|4)$.

Secondly, substantial simplifications occur when κ-symmetry is gauge-fixed. This is achieved in the flat-space GS action by selecting a light-cone gauge [2, 29] for which the action becomes at most quadratic in fermions. At variance with flat space, two possible ways of imposing the light-cone gauge in $AdS_5 \times S^5$ are available, depending on the choice of a light-cone direction in the background geometry: we can pick either a null geodesic wrapping a great circle of S^5 or one lying entirely in AdS_5. The former is equivalent to expand near the pp-wave geometry [30–38] and it is related to the ferromagnetic vacuum of the spin-chain picture for local operators in $\mathcal{N} = 4$ SYM. However, it is also known to lead to a complicated non-polynomial form of the action that is not suitable for direct calculations in sigma-model perturbation theory. On the other hand, the latter option was considered in [8] and is known to produce a gauge-fixed form of the action with at most quartic powers of the physical fermions. Since our eventual aim in the following chapters will be to quantize the theory perturbatively around some particular vacua, in this section we will proceed with the form of the AdS light-cone gauge that selects a massless geodesic entirely in AdS_5.

The AdS light-cone gauge is conveniently described in the Poincaré patch of AdS_5 (see 3.5 below), where the radial AdS coordinate is $z = e^{\phi}$ and we define the light-cone directions $x^{\pm}$ running in the AdS boundary and their transversal complex coordinates $x, \bar{x}$:

$$x^{\pm} \equiv \frac{x^3 \pm x^0}{\sqrt{2}}\,, \qquad x = \frac{x^1 + i x^2}{\sqrt{2}}\,, \tag{2.17}$$
$$\bar{x} = \frac{x^1 - i x^2}{\sqrt{2}}\,, \qquad x^a = (x^+, x^-, x, \bar{x})\,, \quad a = 1, 2, 3, 4\,,$$

The appropriate *light-cone basis* of $\mathfrak{psu}(2,2|4)$[4] corresponds to a set of generators that respect the splitting of its even part into $\mathfrak{so}(2,4) \oplus \mathfrak{so}(6)$. If we identity $\mathfrak{so}(2,4)$ with the superconformal algebra in four dimensions, then this algebra comprises the momenta P^a, the angular momenta J^{ab}, the conformal boosts K^{μ} and the dilatation generator D, with the indices $a, b = +, -, x, \bar{x}$ being in the light-cone frame notation of (2.17). In addition to this, we need to include the $\mathfrak{so}(6) \sim \mathfrak{su}(4)$ rotations $J^i{}_j$ $(i, j = 1, \ldots 4)$. The odd part of the superalgebra is spanned by 32 generators spinors $\{Q^{\pm i}, Q^{\pm}_i, S^{\pm i}, S^{\pm}_i\}$ labeled by upper (lower) index $i = 1, \ldots 4$ transforming in the (anti-)fundamental of the $SU(4)$ R-symmetry. The super-Poincaré Q's and superconformal S's account for the 32 Killing spinors preserved by the (maximally supersymmetric) supergravity background of $AdS_5 \times S^5$.

A natural choice for the coset representative is given by[5]

$$g = e^{x^a P_a + \theta \cdot Q}\, e^{\eta \cdot S}\, e^{y^i{}_j J^j{}_i}\, e^{\phi D}\,, \tag{2.18}$$

in terms of linear combinations $y^i{}_j = \frac{i}{2} y^{A'} (\gamma^{A'})^i{}_j$ of the S^5 coordinates $y^{A'}$ involving the $SO(5)$ Dirac matrices $\gamma^{A'}$ $(A' = 1, \ldots 5)$, in addition to the combinations $\theta \cdot Q \equiv \theta^{-i} Q^+_i + \theta^-_i Q^{+i} + \theta^+_i Q^{-i} + \theta^{+i} Q^-_i$ and $\eta \cdot S \equiv \eta^{-i} S^+_i + \eta^-_i S^{+i} + \eta^+_i S^{-i} + \eta^{+i} S^-_i$. The 16 θ- and 16 η-variables are the Grassmann-odd partners of the 10 bosonic coordinates $x^{\pm}, x, \bar{x}, \phi, y_{AA'}$ and all the supercoordinates are in correspondence with a generator of the superalgebra. Raising and lowering $SU(4)$ indices in the supercharges correspond to take the operation of complex conjugation.

We impose the *κ-symmetry light-cone gauge* by setting to zero the half of the fermions that have positive charges under the generator J^{+-}, namely

$$\theta^+_i = \theta^{+i} = \eta^+_i = \eta^{+i} = 0\,. \tag{2.19}$$

The non-vanishing fermions $\theta^-_i, \theta^{-i}, \eta^-_i, \eta^{-i}$ (we drop the minus label from now on) are the 16 physical degrees for freedom and their number matches the fermionic content of a gauge-fixed GS action in ten dimensions. The induced coordinate parametrization of the one-form current (2.18) can be read off by projecting the $\mathfrak{psu}(2,2|4)$-invariant current A on the light-cone basis[6]:

[4]Commutation relation of the $\mathfrak{psu}(2,2|4)$ superalgebra and further details of derivation of the action are given in the seminal paper [8].

[5]This form of the expression defining g goes under the name of *Killing parametrization* of the superspace spanned by the supercoodinates $(x, \phi, y, \theta, \eta)$. The action can be equivalently put in the *Wess-Zumino parametrization* discussed in (2.33) below.

[6]We changed the sign of the current compared to (2.4).

$$A = g^{-1}dg \equiv A_P^a P_a + A_K^a K_a + A_D D + \frac{1}{2}A^{ab}J_{ab} + A^i{}_j J^j{}_i + A_Q^{-i} Q_i^+ + A_{Qi}^- Q^{+i} + A_Q^{+i} Q_i^- + A_{Qi}^+ Q^{-i} + A_S^{-i} S_i^+ + A_{Si}^- S^{+i} + A_S^{+i} S_i^- + A_{Si}^+ S^{-i} , \tag{2.20}$$

where the non-vanishing *Cartan one-forms* are

$$\begin{aligned}
&A_P^+ = e^\phi dx^+ , \quad A_P^- = e^\phi (dx^- - \frac{i}{2}\tilde{\theta}^i d\tilde{\theta}_i - \frac{i}{2}\tilde{\theta}_i d\tilde{\theta}^i) , \\
&A_P^x = e^\phi dx , \quad A_P^{\bar{x}} = e^\phi d\bar{x} , \quad A_D = d\phi , \\
&A_K^- = e^{-\phi}\big[\frac{1}{4}(\tilde{\eta}^2)^2 dx^+ + \frac{i}{2}\tilde{\eta}^i d\tilde{\eta}_i + \frac{i}{2}\tilde{\eta}_i d\tilde{\eta}^i\big] , \\
&A^i{}_j = (dUU^{-1})^i{}_j + i(\tilde{\eta}^i\tilde{\eta}_j - \frac{1}{4}\tilde{\eta}^2\delta^i_j)dx^+ , \\
&A_Q^{-i} = e^{\phi/2}(\tilde{d\theta}^i + i\tilde{\eta}^i dx) , \quad A_{Qi}^- = e^{\phi/2}(\tilde{d\theta}_i - i\tilde{\eta}_i d\bar{x}) , \\
&A_Q^{+i} = -ie^{\phi/2}\tilde{\eta}^i dx^+ , \quad A_{Qi}^+ = ie^{\phi/2}\tilde{\eta}_i dx^+ , \\
&A_S^{-i} = e^{-\phi/2}(\tilde{d\eta}^i + \frac{i}{2}\tilde{\eta}^2\tilde{\eta}^i dx^+) , \quad A_{Si}^- = e^{-\phi/2}(\tilde{d\eta}_i - \frac{i}{2}\tilde{\eta}^2\tilde{\eta}_i dx^+) .
\end{aligned} \tag{2.21}$$

We introduced the shorthand $\tilde{\eta}^2 \equiv \tilde{\eta}^i\tilde{\eta}_i$, the fermions

$$\tilde{\theta}^i \equiv U^i{}_j\theta^j , \quad \tilde{\theta}_i \equiv \theta_j(U^{-1})^j{}_i , \tag{2.22}$$

$$\tilde{d\theta}^i \equiv U^i{}_j d\theta^j , \quad \tilde{d\theta}_i \equiv d\theta_j(U^{-1})^j{}_i , \tag{2.23}$$

and similar ones for η, all obtained through the local unitary matrix

$$U \equiv \cos\frac{|y|}{2} + i\gamma^{A'}n^{A'}\sin\frac{|y|}{2} , \quad |y| \equiv \sqrt{y^{A'}y^{A'}} , \quad n^{A'} \equiv \frac{y^{A'}}{|y|} . \tag{2.24}$$

After a field-dependent rescaling $\eta^i \to \sqrt{2}e^\phi\eta^i$, $\eta_i \to \sqrt{2}e^\phi\eta_i$ to have fermions of homogeneous conformal dimension and the sign flip $x^a \to -x^a$ to change the sign of the kinetic term, one eventually reaches a form of the Lagrangian (2.6) that reads

$$S = T\int d\tau d\sigma\, \mathcal{L} , \tag{2.25}$$

$$\begin{aligned}
\mathcal{L} = &\sqrt{-g}g^{\mu\nu}\Big[e^{2\phi}(\partial_\mu x^+\partial_\nu x^- + \partial_\mu x\partial_\nu\bar{x}) + \frac{1}{2}\partial_\mu\phi\partial_\nu\phi + \frac{1}{2}G_{\mathcal{AB}}(y)D_\mu y^{\mathcal{A}}D_\nu y^{\mathcal{B}}\Big] \\
&+ \frac{i}{2}\sqrt{-g}g^{\mu\nu}e^{2\phi}\partial_\mu x^+\Big[\theta^i\partial_\nu\theta_i + \theta_i\partial_\nu\theta^i + \eta^i\partial_\nu\eta_i + \eta_i\partial_\nu\eta^i + ie^{2\phi}\partial_\nu x^+(\eta^2)^2\Big] \\
&- \Big\{\epsilon^{\mu\nu}e^{2\phi}\partial_\mu x^+\eta^i\Big[C'_{ij}\cos|y| + i(C'\gamma^{A'})_{ij}n^{A'}\sin|y|\Big](\partial_\nu\theta^j - i\sqrt{2}e^\phi\eta^j\partial_\nu x) + \text{h.c.}\Big\} .
\end{aligned} \tag{2.26}$$

A few explanations of the objects in the Lagrangian follow in order. The S^5 metric has a factorized form in terms of the vielbien

$$G_{\mathcal{AB}} = e_{\mathcal{A}}^{A'} e_{\mathcal{B}}^{A'} , \qquad e_{\mathcal{A}}^{A'} = \frac{\sin |y|}{|y|}(\delta_{\mathcal{A}}^{A'} - n_{\mathcal{A}} n^{A'}) + n_{\mathcal{A}} n^{A'} , \tag{2.27}$$

and we made a distinction between curved $\mathcal{A}, \mathcal{B}, \ldots$, and flat indices $A', B', \ldots$, both with range from 1 to 5, while $\mu, \nu = 0, 1$ denote the worldsheet indices. C' is the constant charge conjugation matrix of the $SO(5)$ Dirac matrix algebra generated by $\gamma^{A'}$. The differential $D_\mu y^{\mathcal{A}}$ is defined by

$$D_\mu y^{\mathcal{A}} = \partial_\mu y^{\mathcal{A}} - 2i\eta_i (V^{\mathcal{A}})^i{}_j \eta^j e^{2\phi} \partial_\mu x^+ , \tag{2.28}$$

where $(V^{\mathcal{A}})^i{}_j$ are the components of the $su(4)$-valued Killing vectors of S^5

$$\begin{aligned}(V^{\mathcal{A}})^i{}_j \partial_{y^{\mathcal{A}}} &= \frac{1}{4}(\gamma^{A'B'})^i{}_j [y^{A'} \partial_{y^{B'}} - y^{B'} \partial_{y^{A'}}] \\ &\quad + \frac{i}{2}(\gamma^{A'})^i{}_j \Big[|y| \cot |y| (\delta^{A'\mathcal{A}} - n^{A'} n^{\mathcal{A}}) + n^{A'} n^{\mathcal{A}} \Big] \partial_{y^{\mathcal{A}}} ,\end{aligned} \tag{2.29}$$

with $\delta^{A'\mathcal{A}}$ being the Kronecker delta symbol and $y^{\mathcal{A}} = \delta^{\mathcal{A}}_{A'} y^{A'}$, $n^{\mathcal{A}} = \delta^{\mathcal{A}}_{A'} n^{A'}$, $n^{\mathcal{A}} = n_{\mathcal{A}}$.

In (2.26) the kinetic term is proportional to the Weyl-invariant combination $\sqrt{-g} g^{\mu\nu}$, while the Wess-Zumino term contains the antisymmetric symbol ϵ with $\epsilon^{01} = 1$. The formula shows also that in the light-cone κ-symmetry gauge-fixed action, together with the choice of supercoodinates implicit in the supercoset parametrization (2.18), no fermionic interactions with powers higher than four appear, as the action is quadratic in one half of them (θ) and quartic in the other one (η).

The gauge-fixed action can be expressed in a related form, in which the matrix (2.24) can be either incorporated in the definition of a covariant derivative for fermions (Wess-Zumino parametrization) or eliminated through a change of coordinates in S^5 that manifestly realizes the $SO(6)$ symmetry of its coordinates. We proceed to present only the latter strategy, as it will lead to the from of the action needed in Chap. 7.

The Lagrangian can be put into a manifestly $SU(4)$-invariant form by combining the $y^{A'}$ and the radial coordinate ϕ into an $SO(6)$ vector z^M $(M = 1, \ldots 6)$

$$z^{A'} = e^{-\phi} \sin |y| \, n^{A'} , \qquad z^6 = e^{-\phi} \cos |y| , \quad z \equiv \sqrt{z^M z^M} = e^{-\phi} , \tag{2.30}$$

in terms of which the metric appears in the "4+6 parametrization"

$$ds^2_{AdS_5 \times S^5} = \frac{dx^a dx_a + dz^M dz_M}{z^2} . \tag{2.31}$$

If we start again with (2.26) and use the identifications $(\gamma^{A'})^i{}_j = i(\rho^{A'})^{il}\rho^6_{lj}$ and $C'_{ij} = \rho^6_{ij}$ with the ρ-matrices in (F.2), we arrive to the final form of the AdS light-cone gauge-fixed action

$$S = T \int d\tau d\sigma\, \mathcal{L} \tag{2.32}$$

$$\mathcal{L} = \sqrt{-g}g^{\mu\nu}z^{-2}\Big[\partial_\mu x^+\partial_\nu x^- + \partial_\mu x\partial_\nu \bar{x} + \frac{1}{2}(\partial_\mu z^M + i\eta_i(\rho^{MN})^i{}_j z^N\eta^j z^{-2}\partial_\mu x^+) \tag{2.33}$$

$$\times(\partial_\nu z^M + i\eta_i(\rho^{MP})^i{}_j z^P\eta^j z^{-2}\partial_\nu x^+)\Big] + \frac{i}{2}\sqrt{-g}g^{\mu\nu}z^{-2}\partial_\mu x^+\Big[\theta^i\partial_\nu\theta_i + \theta_i\partial_\nu\theta^i + \eta^i\partial_\nu\eta_i$$

$$+\eta_i\partial_\nu\eta^i + iz^{-2}\partial_\nu x^+(\eta^2)^2\Big] - \Big[\epsilon^{\mu\nu}z^{-3}\partial_\mu x^+\eta^i\rho^M_{ij}z^M\Big(\partial_\nu\theta^j - i\sqrt{2}z^{-1}\eta^j\partial_\nu x\Big) + \text{h.c.}\Big]\,.$$

2.2 The $AdS_4 \times \mathbb{CP}^3$ String Action in the AdS Light-Cone Gauge

The $AdS_4 \times \mathbb{CP}^3$ space is another prominent example of AdS background in the context of integrable systems in the gauge/gravity duality. The background, supported by with RR four-form flux through AdS_4 and a RR two-form flux through a $\mathbb{CP}_1 \subset \mathbb{CP}_3$, arises as a ten-dimensional solution of the IIA supergravity equations. The crucial difference with $AdS_5 \times S^5$ is the absence of maximal supersymmetry, since $AdS_4 \times \mathbb{CP}^3$ preserves only 24 out of 32 supersymmetries [39]. The Green-Schwarz approach to find the superstring action suffers from an obstruction: although the formal expression of the GS action is known for any type IIA supergravity background, it is was explicitly written up to quartic terms only [40], so the complete form remains unknown for the $AdS_4 \times \mathbb{CP}^3$ space.

The supercoset approach outlined in Sect. 2.1.1 provides a pragmatic strategy that was pursued in [9, 10]. In fact, $AdS_4 \times \mathbb{CP}^3$ is an homogeneous space

$$AdS_4 \times \mathbb{CP}^3 = \frac{SO(2,3)}{SO(1,3)} \times \frac{SO(6)}{U(3)} \tag{2.34}$$

written as the quotient of $G = USp(2,3) \times SO(6)$, locally isomorphic to $SO(2,3) \times SO(6)$ and $H = SO(1,3) \times U(3)$. Together with the observation that G is the bosonic subgroup of the orthosymplectic group $\tilde{G} = OSp(4|6)$, the coset structure of the space (2.34) hinted at the formulation of a sigma-model on the supercoset $OSp(4|6)/(SO(1,3) \times U(3))$. The construction of the sigma-model parallels the case of the $AdS_5 \times S^5$ superstring in many aspects. It is again possible to define a $\mathbb{Z}_4$-grading of $\mathfrak{osp}(4|6)$ to decompose the zero-curvature current one-form (2.4) and plug its components into (2.6) after an appropriate choice of the coset representative g. The supercoset straightforwardly inherits the classical integrability—as a mean to write the superstring equations of motion in a flat Lax connection (2.12)—found in [41] for the case of $AdS_5 \times S^5$. The standard kinetic term is supplemented with a Wess-Zumino term so that it was possible to establish κ-symmetry transformations similar to the usual one in $AdS_5 \times S^5$ [7].

On the other hand, there is a problem that did not exist in the other case since here the background is not maximally supersymmetric. The superspace consists of 24 fermionic directions in correspondence with an equal number of supercharges preserved by the background space. The other 8 fermionic coordinates, required to match the total number of 32 degrees of freedom of the GS string, were argued to be gauged away by half of the 16 parameters of κ-symmetry. In [10, 42] the supercoset sigma-model was interpreted as equivalent to a *partially κ-symmetry gauge-fixed* GS action in relation to these missing 8 coordinates associated to the odd $\mathfrak{osp}(4|6)$-generators broken by the background.

For this interpretation to be correct, for a *generic* configuration where the string motion occurs in both AdS_4 and $\mathbb{CP}^3$ the supercoset enjoys a (residual) κ-symmetry of rank 8, namely capable of eliminating precisely 8 of the 24 fermionic degrees of freedom in the supercoset to yield a *totally κ-symmetry gauge-fixed* action with only 16 physical fermionic degrees of freedom.

This consideration is not true for a string embedded purely in the AdS_4 sector[7] of the background, in which case the rank of the (residual) κ-symmetry is enhanced to 12 and the supercoset action with κ-symmetry totally gauge-fixed would have only 12 fermionic degrees of freedom [10, 42]. The supercoset formulation is not equivalent to the GS action for these "singular" configurations: the implicit κ-symmetry gauge choice puts the 8 broken fermions to zero in the supercoset action, and this turns out to be incompatible in some string configurations.

The null cusp classical string studied in Chap. 6 is an example of these singular backgrounds, being embedded only in the AdS_4 part, and cannot be properly described[8] by the correct number of physical fermions within the supercoset formalism. In other words, the trouble of the semiclassically quantization of this string would be a sudden change in the number of fermionic degrees of freedom (from 12 to 8) as soon as the classical solution is perturbed in $\mathbb{CP}^3$. Since the work in of Chap. 6 will be the only instance of strings in the $AdS_4 \times \mathbb{CP}^3$ geometry, we will rather concentrate[9] on an alternative non-coset form of the $AdS_4 \times \mathbb{CP}^3$ action—developed by Uvarov in [11, 12]—that is capable of capturing the dynamics of all bosonic string configurations.

---

[7]The same situation occurs when a string forms a worldsheet instanton in $\mathbb{CP}^3$ [43].

[8]Despite this obstruction, one explicit calculation suggests that the supercoset action can be still used for one-loop calculations for classical solutions that are point-like in $\mathbb{CP}^3$ [44]. This is possible because a "regularization" parameter (non-vanishing angular momentum J) is kept throughout the calculation and the observable (one-loop string energy) admits a smooth limit when the classical string becomes a singular configuration ($J \to 0$).

[9] Let us mention that an alternative κ-symmetry gauge-fixing was considered in [45], based on the the complete $AdS_4 \times \mathbb{CP}^3$ superspace with 32 fermionic directions [42] that allows to cover regions of the space that are not reachable by the supercoset sigma model of [9, 10]. A different ("superconformal") realization of the κ-symmetry gauge-fixing was presented in [46].

The starting point is the supersymmetric membrane action [47] based on the super-coset $OSp(4|8)/(SO(1,3) \times SO(7))$ in the maximally supersymmetric background $AdS_4 \times S^7$. The string action is eventually obtained performing a double dimensional reduction to $AdS_4 \times \mathbb{CP}^3$ and choosing a κ-symmetry light-cone gauge for which both light-like directions lie in AdS_4. The result is an action that is a close counterpart of the sigma-model action for type IIB superstrings in the $AdS_5 \times S^5$ background [8, 48].

In the construction of [11, 12], the space $AdS_4 \times S^7$ is seen as the bosonic coset

$$AdS_4 \times S^7 = \frac{SO(2,3)}{SO(1,3)} \times \frac{SO(8)}{SO(7)} \tag{2.35}$$

which admits the supersymmetric extension $OSp(4|8)/(SO(1,3) \times SO(7))$. The latter includes 32 fermionic coordinates (called θ and η) associated to the each of the (respectively super-Poincaré and superconformal) Grassmann-odd generators in the algebra $\mathfrak{osp}(4|8)$. An analogue of the AdS κ-symmetry light-cone gauge sets to zero half of the fermionic coordinates, where the "half" corresponds to those 16 fermionic generators carrying negative charge under the $SO(1,1)$ generator M^{+-} from the Lorentz group of the 3d boundary of AdS_4.

The output is an action including physical fermions up to the fourth power[10] that is able to capture the string dynamics in any submanifold of $AdS_4 \times \mathbb{CP}^3$. As for classical integrability, the standard Lax pair construction of [10, 41] does not clearly carry over to this *non-coset* model. The zero-curvature Lax pair was built for any string configuration in the full $AdS_4 \times \mathbb{CP}^3$ superspace up to quadratic order in the fermionic degrees of freedom [49].[11]

Here we will be mostly interested in presenting the final gauge-fixed string action. The starting point is the metric of $AdS_4 \times \mathbb{CP}^3$

$$ds^2_{AdS_4 \times \mathbb{CP}^3} = R^2_{\mathbb{CP}^3} \left(\frac{1}{4} ds^2_{AdS_4} + ds^2_{\mathbb{CP}^3} \right), \tag{2.36}$$

where we factored out the $\mathbb{CP}^3$ radius $R_{\mathbb{CP}^3}$, which can be then set to 1 for simplicity. The factor of $(1/2)^2$ accounts for the relative size of the radius of $\mathbb{CP}^3$ being twice the one of AdS_4. We parametrize AdS_4 in Poincaré patch

$$ds^2_{AdS_4} = \frac{dw^2 + dx^+ dx^- + dx^1 dx^1}{w^2}, \qquad x^\pm \equiv x^2 \pm x^0, \tag{2.37}$$

[10] This is a feature shared with the κ-symmetry light-cone gauge-fixed action in $AdS_5 \times S^5$ action (2.26).

[11] In [50] is was shown that the string is classically integrable up to quadratic order in fermions *before* fixing κ-symmetry, for the full $AdS_4 \times \mathbb{CP}^3$ and other AdS backgrounds relevant in the AdS/CFT correspondence.

where $w \equiv e^{2\varphi}$ is the radial coordinate in AdS_4 and we pick two light-cone directions $x^{\pm}$ in the three-dimensional boundary of AdS_4. For the moment we do not specify the $\mathbb{CP}^3$ coordinates z^M $(M = 1, \dots 6)$:

$$ds^2_{\mathbb{CP}^3} = g_{MN}\, dz^M dz^N \; . \tag{2.38}$$

In addition to the embedding coordinates of $AdS_4 \times \mathbb{CP}^3$, the model has 16 physical complex fermions: the 3+3 η_a and θ_a (and their 3+3 conjugates $\bar{\eta}^a$ and $\bar{\theta}^a$) transform in the fundamental (anti-fundamental) representation of $SU(3)$ $(a = 1, 2, 3)$ and stem from those 24 supersymmetries of type IIA supergravity unbroken by the $AdS_4 \times \mathbb{CP}^3$ background, while the broken 8 supersymmetries bring the remaining 1+1 fermions η_4, θ_4 and their 1+1 conjugates η^4, θ^4.

The κ-symmetry light-cone gauge-fixed Lagrangian [11, 12] reads

$$\begin{aligned}
S =& \frac{T}{2} \int d\tau\, d\sigma\, \mathcal{L}\,, \\
\mathcal{L} =& g^{ij} \Big[\frac{e^{-4\varphi}}{4} \left(\partial_i x^+ \partial_j x^- + \partial_i x^1 \partial_j x^1\right) + \partial_i \varphi \partial_j \varphi + g_{MN} \partial_i z^M \partial_j z^N \\
&+ e^{-4\varphi} \left(\partial_i x^+ \varpi_j + \partial_i x^+ \partial_j z^M h_M + e^{-4\varphi} B \partial_i x^+ \partial_j x^+\right) \Big] \\
&- 2\, \epsilon^{ij} e^{-4\varphi} \left(\omega_i \partial_j x^+ + e^{-2\varphi} C \partial_i x^1 \partial_j x^+ + \partial_i x^+ \partial_j z^M \ell_M\right) .
\end{aligned} \tag{2.39}$$

Some brief explanations of the objects present in the Lagrangian are in order. The string tension T was discussed around (1.9) and g_{ij} $(i, j = 0, 1)$ is the usual auxiliary metric field in the Polyakov part of the action. The action exhibits highly non-linear bosonic interactions with fermionic fields in the coefficients

$$\varpi_i = i \left(\partial_i \theta_a \bar{\theta}^a - \theta_a \partial_i \bar{\theta}^a + \partial_i \theta_4 \bar{\theta}^4 - \theta_4 \partial_i \bar{\theta}^4 + \partial_i \eta_a \bar{\eta}^a - \eta_a \partial_i \bar{\eta}^a + \partial_i \eta_4 \bar{\eta}^4 - \eta_4 \partial_i \bar{\eta}^4\right), \tag{2.40}$$

$$\omega_i = \hat{\eta}_a \hat{\partial}_i \bar{\theta}^a + \hat{\partial}_i \theta_a \hat{\bar{\eta}}^a + \frac{1}{2} \left(\partial_i \theta_4 \bar{\eta}^4 - \partial_i \eta_4 \bar{\theta}^4 + \eta_4 \partial_i \bar{\theta}^4 - \theta_4 \partial_i \bar{\eta}^4\right) , \tag{2.41}$$

$$B = 8 \left[(\hat{\eta}_a \hat{\bar{\eta}}^a)^2 + \epsilon_{abc} \hat{\bar{\eta}}^a \hat{\bar{\eta}}^b \hat{\bar{\eta}}^c \bar{\eta}^4 + \epsilon^{abc} \hat{\eta}_a \hat{\eta}_b \hat{\eta}_c \eta_4 + 2\eta_4 \bar{\eta}^4 \left(\hat{\eta}_a \hat{\bar{\eta}}^a - \theta_4 \bar{\theta}^4\right) \right] , \tag{2.42}$$

$$C = 2\, \hat{\eta}_a \hat{\bar{\eta}}^a + \theta_4 \bar{\theta}^4 + \eta_4 \bar{\eta}^4 \,, \tag{2.43}$$

$$h_M = 2 \left[\Omega^a_M \epsilon_{abc} \hat{\bar{\eta}}^b \hat{\bar{\eta}}^c - \Omega_{aM} \epsilon^{abc} \hat{\eta}_b \hat{\eta}_c + 2 \left(\Omega_{aM} \hat{\bar{\eta}}^a \bar{\eta}^4 - \Omega^a_M \hat{\eta}_a \eta_4\right) + 2 \left(\theta_4 \bar{\theta}^4 + \eta_4 \bar{\eta}^4\right) \tilde{\Omega}_a{}^a{}_M \right] , \tag{2.44}$$

$$\ell_M = 2i \left[\Omega_{aM} \hat{\bar{\eta}}^a \bar{\theta}^4 + \Omega^a_M \hat{\eta}_a \theta_4 + \left(\theta_4 \bar{\eta}^4 - \eta_4 \bar{\theta}^4\right) \tilde{\Omega}_a{}^a{}_M \right] . \tag{2.45}$$

At variance with [11, 12], we operated a field-dependent rescaling of the fermionic fields

$$\theta_a \to \sqrt{2}\, \theta_a \qquad \theta_4 \to \sqrt{2}\, e^{-\varphi} \theta_4 \qquad \eta_a \to \sqrt{2}\, e^{-2\varphi} \eta_a \qquad \eta_4 \to \sqrt{2}\, e^{-\varphi} \eta_4 \tag{2.46}$$

and similarly for their complex conjugates, inspired by the analogue one below (2.24) for $AdS_5 \times S^5$. The η_a and θ_a appear in fully contracted combinations in (2.40)–(2.45). The manifest symmetry of the action is therefore only the $SU(3)$ subgroup of the $SU(4)$ global symmetry of $\mathbb{CP}^3 = SU(4)/U(3)$ that rotates the unbroken fermions into themselves.[12]

The Ω^a_M and Ω_{aM} are the complex vielbein of $\mathbb{CP}^3$ which satisfy $ds^2_{\mathbb{CP}^3} \equiv \Omega^a_M \Omega_{aN}\, dz^M\, dz^N$, namely the components of the Cartan one-forms of $SU(4)/U(3)$, $\Omega^a = \Omega^a_M\, dz^M$ and $\Omega_a = \Omega_{aM}\, dz^M$. In the Uvarov's supercoset construction [11], $\tilde{\Omega}_a{}^a$ is related to a one-form corresponding to the reduction direction coordinate in S^7 and its explicit expression can be found below in terms of the $\mathbb{CP}^3$ coordinates. The Ω^a_M and $\tilde{\Omega}_a{}^a$ appear in [11] in a "dressed" supercoset representative for $OSp(4|6)/(SO(1,3) \times U(3)))$ where the dressing incorporates the information on the broken supersymmetries and the $U(1)$ fiber direction.

Hatted variables in the Lagrangian are related to unhatted ones through a local rotation depending on the $\mathbb{CP}^3$ coordinates

$$\hat{\eta}_a \equiv T_a{}^b\, \eta_b + T_{ab}\, \bar{\eta}^b\,, \qquad\qquad \hat{\bar{\eta}}^a \equiv T^a{}_b\, \bar{\eta}^b + T^{ab}\, \eta_b \tag{2.47}$$

and similarly for the θ fermions, in the same spirit of the unitary in U matrices (2.24) above.

We can now parametrize $\mathbb{CP}^3$ with with complex variables z^a (and their conjugates $\bar{z}_a$) transforming in the fundamental (anti-fundamental) of the symmetry group $SU(3)$ [51]. Then (2.38) can be expanded as

$$ds^2_{\mathbb{CP}^3} = g_{ab}\, dz^a\, dz^b + g^{ab}\, d\bar{z}_a\, d\bar{z}_b + 2\, g_a{}^b\, dz^a\, d\bar{z}_b\,, \tag{2.48}$$

with coefficients[13]

$$\begin{aligned} g_{ab} &= \frac{1}{4|z|^4}\left(|z|^2 - \sin^2|z| + \sin^4|z|\right)\bar{z}_a\,\bar{z}_b\,,\\ g^{ab} &= \frac{1}{4|z|^4}\left(|z|^2 - \sin^2|z| + \sin^4|z|\right)z^a\, z^b\,,\\ g_a{}^b &= \frac{\sin^2|z|}{2|z|^2}\,\delta^b_a + \frac{1}{4|z|^4}\left(|z|^2 - \sin^2|z| - \sin^4|z|\right)\bar{z}_a\, z^b\,, \end{aligned} \tag{2.49}$$

where $|z|^2 \equiv z^a\,\bar{z}_a$ for short. We read off the one-forms

$$\Omega^a = \Omega^a{}_{,b}\, dz^b + \Omega^{a,b}\, d\bar{z}_b\,, \qquad \Omega_a = \Omega_{a,b}\, dz^b + \Omega_a{}^{;b}\, d\bar{z}_b\,, \qquad \tilde{\Omega}^a_a = \tilde{\Omega}^a_{a\,,b}\, dz^b + \tilde{\Omega}^{a,b}_a\, d\bar{z}_b\,, \tag{2.50}$$

---

[12]The broken symmetry will remain visible, for instance, at the level of semiclassical fluctuations around the light-like cusp in Sect. 6.2.

[13]They are related to the conventional Fubini-Study metric of $\mathbb{CP}^3$, see [12].

from the vielbein of this metric

$$\Omega_a = d\bar{z}_a \frac{\sin|z|}{|z|} + \bar{z}_a \frac{\sin|z|(1-\cos|z|)}{2|z|^3}(dz^c\bar{z}_c - z^c d\bar{z}_c) + \bar{z}_a\left(\frac{1}{|z|} - \frac{\sin|z|}{|z|^2}\right) d|z| \,, \tag{2.51}$$

$$\Omega^a = dz^a \frac{\sin|z|}{|z|} + z^a \frac{\sin|z|(1-\cos|z|)}{2|z|^3}(z^c d\bar{z}_c - dz^c\bar{z}_c) + z^a\left(\frac{1}{|z|} - \frac{\sin|z|}{|z|^2}\right) d|z| \tag{2.52}$$

and

$$\tilde{\Omega}_a{}^a = i\,\frac{\sin^2|z|}{|z|^2}\left(dz^a\,\bar{z}_a - z^a\,d\bar{z}_a\right) . \tag{2.53}$$

The matrices (2.47) can be usefully incapsulated into a 6×6 unitary matrix $T_{\hat{a}}{}^{\hat{b}}$ [51]

$$T_{\hat{a}}{}^{\hat{b}} \equiv \begin{pmatrix} T_a{}^b & T_{ab} \\ T^{ab} & T^a{}_b \end{pmatrix} = \begin{pmatrix} \delta_a^b\,\cos|z| + \bar{z}_a\,z^b\,\frac{1-\cos|z|}{|z|^2} & i\,\epsilon_{acb}\,z^c\,\frac{\sin|z|}{|z|} \\ -i\,\epsilon^{acb}\,\bar{z}_c\,\frac{\sin|z|}{|z|} & \delta_b^a\,\cos|z| + z^a\,\bar{z}_b\,\frac{1-\cos|z|}{|z|^2} \end{pmatrix} . \tag{2.54}$$

In Sect. 6.1 we will be interested in Wick-rotating the gauge-fixed Lagrangian (2.39) to Euclidean AdS_4 and perform a diagrammatical computation at two loops. One of the main motivations behind this analysis will be also to put the action in this gauge to a stringent test at the quantum level, where the $\mathbb{CP}^3$ geometry indirectly manifests in more complicated structures (2.40)–(2.45) compared to the compact form of the $AdS_5 \times S^5$ supercoset action (2.33). Let us also remark that the action (2.39) can be rewritten in a more compact form that resembles the Wess-Zumino type parametrization of [8, 48] by the introduction of a covariant derivative for the terms quadratic in fermions.[14]

References

1. L. Brink, J.H. Schwarz, Quantum superspace. Phys. Lett. B **100**, 310 (1981)
2. M.B. Green, J.H. Schwarz, Covariant description of superstrings. Phys. Lett. B **136**, 367 (1984)
3. M. Henneaux, L. Mezincescu, A sigma model interpretation of green-schwarz covariant superstring action. Phys. Lett. B **152**, 340 (1985)
4. J.A. de Azcarraga, J. Lukierski, Supersymmetric particles with internal symmetries and central charges. Phys. Lett. B **113**, 170 (1982)
5. W. Siegel, Hidden local supersymmetry in the supersymmetric particle action. Phys. Lett. B **128**, 397 (1983)
6. M.T. Grisaru, P.S. Howe, L. Mezincescu, B. Nilsson, P.K. Townsend, $\mathcal{N} = 2$ Superstrings in a supergravity background. Phys. Lett. B **162**, 116 (1985)
7. R. Metsaev, A.A. Tseytlin, Type IIB superstring action in $AdS_5 \times S^5$ background. Nucl. Phys. B **533**, 109 (1998). arxiv:hep-th/9805028
8. R. Metsaev, A.A. Tseytlin, Superstring action in $AdS_5 \times S^5$. Kappa symmetry light cone gauge. Phys. Rev. D **63**, 046002 (2001). arxiv:hep-th/0007036
9. J.B. Stefanski, Green-Schwarz action for Type IIA strings on $AdS_4 \times \mathbb{CP}^3$. Nucl. Phys. **B808**, 80 (2009). arxiv:0806.4948

[14]This was first illustrated in appendix A of [52] and reported with further details in [53].

10. G. Arutyunov, S. Frolov, Superstrings on $AdS_4 \times \mathbb{CP}^3$ as a coset sigma-model. JHEP **0809**, 129 (2008). arxiv:0806.4940
11. D. Uvarov, $AdS_4 \times \mathbb{CP}^3$ superstring in the light-cone gauge. Nucl. Phys. B **826**, 294 (2010). arxiv:0906.4699
12. D. Uvarov, Light-cone gauge Hamiltonian for $AdS_4 \times \mathbb{CP}^3$ superstring. Mod. Phys. Lett. A **25**, 1251 (2010). arxiv:0912.1044
13. J.H. Schwarz, Covariant field equations of chiral $\mathcal{N} = 2$ $D = 10$ supergravity. Nucl. Phys. B **226**, 269 (1983)
14. M. Blau, J.M. Figueroa-O'Farrill, C. Hull, G. Papadopoulos, A New maximally supersymmetric background of IIB superstring theory. JHEP **0201**, 047 (2002). arxiv:hep-th/0110242
15. A. Neveu, J.H. Schwarz, Factorizable dual model of pions. Nucl. Phys. B **31**, 86 (1971)
16. P. Ramond, Dual theory for free fermions. Phys. Rev. D **3**, 2415 (1971)
17. M.B. Green, J.H. Schwarz, Properties of the covariant formulation of superstring theories. Nucl. Phys. B **243**, 285 (1984)
18. J.M. Maldacena, The large N limit of superconformal field theories and supergravity. Adv. Theor. Math. Phys. **2**, 231 (1998). arxiv:hep-th/9711200
19. G. Arutyunov, S. Frolov, Foundations of the $AdS_5 \times S^5$ superstring. part I. J. Phys. A **42**, 254003 (2009). arxiv:0901.4937
20. A. Babichenko, B. Stefanski Jr., K. Zarembo, Integrability and the AdS_3/CFT_2 correspondence. JHEP **1003**, 058 (2010). arxiv:0912.1723
21. F. Delduc, M. Magro, B. Vicedo, An integrable deformation of the $AdS_5 \times S^5$ superstring action. Phys. Rev. Lett. **112**, 051601 (2014). arxiv:1309.5850
22. F. Delduc, M. Magro, B. Vicedo, Derivation of the action and symmetries of the q-deformed $AdS_5 \times S^5$ superstring. JHEP **1410**, 132 (2014). arxiv:1406.6286
23. T.J. Hollowood, J.L. Miramontes, D.M. Schmidtt, Integrable deformations of strings on symmetric spaces. JHEP **1411**, 009 (2014). arxiv:1407.2840
24. T.J. Hollowood, J.L. Miramontes, D.M. Schmidtt, An integrable deformation of the $AdS_5 \times S^5$ superstring. J. Phys. A **47**, 495402 (2014). arxiv:1409.1538
25. N. Berkovits, M. Bershadsky, T. Hauer, S. Zhukov, B. Zwiebach, Superstring theory on $AdS_2 \times S^2$ as a coset supermanifold. Nucl. Phys. B **567**, 61 (2000). arxiv:hep-th/9907200
26. E. Abdalla, M. Forger, M. Gomes, On the origin of anomalies in the quantum non-local charge for the generalized non-linear sigma models. Nucl. Phys. B **210**, 181 (1982)
27. E. Abdalla, M. Forger, A. Lima, Santos, non-local charges for non-linear sigma models on grassmann manifolds. Nucl. Phys. B **256**, 145 (1985)
28. N. Beisert, Integrability in QFT and AdS/CFT, Lecture notes, http://edu.itp.phys.ethz.ch/hs14/14HSInt/IntAdSCFT14Notes.pdf
29. M.B. Green, J.H. Schwarz, Supersymmetrical string theories. Phys. Lett. B **109**, 444 (1982)
30. R.R. Metsaev, Type IIB Green-Schwarz superstring in plane wave Ramond-Ramond background. Nucl. Phys. B **625**, 70 (2002). arxiv:hep-th/0112044
31. R.R. Metsaev, A.A. Tseytlin, Exactly solvable model of superstring in Ramond-Ramond plane wave background. Phys. Rev. D **65**, 126004 (2002). arxiv:hep-th/0202109
32. D.E. Berenstein, J.M. Maldacena, H.S. Nastase, Strings in flat space and pp waves from $\mathcal{N} = 4$ superYang-Mills. JHEP **0204**, 013 (2002). arxiv:hep-th/0202021
33. S.S. Gubser, I.R. Klebanov, A.M. Polyakov, A semi-classical limit of the gauge/string correspondence. Nucl. Phys. B **636**, 99 (2002). arxiv:hep-th/0204051
34. S. Frolov, A.A. Tseytlin, Semiclassical quantization of rotating superstring in $AdS_5 \times S^5$. JHEP **0206**, 007 (2002). arxiv:hep-th/0204226
35. J. Callan, G. Curtis, H.K. Lee, T. McLoughlin, J.H. Schwarz, I. Swanson et al., Quantizing string theory in $AdS_5 \times S^5$: beyond the pp wave. Nucl. Phys. **B673**, 3 (2003). arxiv:hep-th/0307032
36. C.G. Callan Jr., T. McLoughlin, I. Swanson, Holography beyond the Penrose limit. Nucl. Phys. B **694**, 115 (2004). arxiv:hep-th/0404007
37. G. Arutyunov, S. Frolov, Integrable Hamiltonian for classical strings on $AdS_5 \times S^5$. JHEP **0502**, 059 (2005). arxiv:hep-th/0411089

38. G. Arutyunov, S. Frolov, J. Plefka, M. Zamaklar, The off-shell symmetry algebra of the light-cone $AdS_5 \times S^5$ superstring. J. Phys. A **40**, 3583 (2007). arxiv:hep-th/0609157
39. B.E.W. Nilsson, C.N. Pope, Hopf fibration of eleven-dimensional supergravity. Class. Quant. Grav. **1**, 499 (1984)
40. M. Cvetic, H. Lu, C.N. Pope, K.S. Stelle, T duality in the Green-Schwarz formalism, and the massless/massive IIA duality map. Nucl. Phys. B **573**, 149 (2000). arxiv:hep-th/9907202
41. I. Bena, J. Polchinski, R. Roiban, Hidden symmetries of the $AdS_5 \times S^5$ superstring. Phys. Rev. D **69**, 046002 (2004). arxiv:hep-th/0305116
42. J. Gomis, D. Sorokin, L. Wulff, The Complete $AdS_4 \times \mathbb{CP}^3$ superspace for the type IIA superstring and D-branes. JHEP **0903**, 015 (2009). arxiv:0811.1566
43. A. Cagnazzo, D. Sorokin, L. Wulff, String instanton in $AdS_4 \times \mathbb{CP}^3$. JHEP **1005**, 009 (2010). arxiv:0911.5228
44. T. McLoughlin, R. Roiban, Spinning strings at one-loop in $AdS_4 \times \mathbb{CP}^3$. JHEP **0812**, 101 (2008). arxiv:0807.3965
45. P.A. Grassi, D. Sorokin, L. Wulff, Simplifying superstring and D-brane actions in $AdS_4 \times \mathbb{CP}^3$ superbackground. JHEP **0908**, 060 (2009). arxiv:0903.5407
46. K. Zarembo, Worldsheet spectrum in AdS_4/CFT_3 correspondence. JHEP **0904**, 135 (2009). arxiv:0903.1747
47. B. de Wit, K. Peeters, J. Plefka, A. Sevrin, The M theory two-brane in $AdS_4 \times S^7$ and $AdS_7 \times S^4$. Phys. Lett. B **443**, 153 (1998). arxiv:hep-th/9808052
48. R. Metsaev, C.B. Thorn, A.A. Tseytlin, Light cone superstring in AdS space-time. Nucl. Phys. B **596**, 151 (2001). arxiv:hep-th/0009171
49. D. Sorokin, L. Wulff, Evidence for the classical integrability of the complete $AdS_4 \times \mathbb{CP}^3$ superstring. JHEP **1011**, 143 (2010). arxiv:1009.3498
50. L. Wulff, Superisometries and integrability of superstrings. JHEP **2014**, 115 (2014). arxiv:1402.3122
51. D. Uvarov, $AdS_4 \times \mathbb{CP}^3$ superstring and $D = 3$ $\mathcal{N} = 6$ superconformal symmetry. Phys. Rev. D **79**, 106007 (2009). arxiv:0811.2813
52. L. Bianchi, M.S. Bianchi, A. Bres, V. Forini, E. Vescovi, Two-loop cusp anomaly in ABJM at strong coupling. JHEP **1410**, 13 (2014). arxiv:1407.4788
53. L. Bianchi, Perturbation theory for string sigma models, Ph.D. thesis. arxiv:1604.01676

Chapter 3
Geometric Properties of Semiclassically Quantized Strings

The geometric properties of string worldsheets embedded in a higher-dimensional space-time, and of linearized perturbations above them, have been object of various studies since the seminal observation on the relevance of quantizing string models [1]. In the framework of the AdS/CFT correspondence, a particularly important setting where these analyses have been performed is the non-linear sigma-model on the curved $AdS_5 \times S^5$. The aim of the chapter is to provide a minimal, practical manual applicable to any classical solution of the string sigma-model, generalizing previous analyses [2, 3][1] that were focussing on the bosonic sector. The formulas that we present require only the knowledge of the generic properties of the classical configuration and basic information about the spacetime background, and they output the bosonic and fermionic operators for fluctuations over it.

The chapter is modelled upon [6] as well as on Appendix B in [7], albeit using the index conventions in Appendix D of this thesis. The objective shall be to write the fluctuation Lagrangian in terms of intrinsic and extrinsic geometric invariants of the classical worldsheet in $AdS_5 \times S^5$ space (Sect. 3.1). In Sect. 3.2 we begin by reviewing the statement that a classical string configuration is a surface of minimal area. In Sect. 3.3 we treat the bosonic sector with the expansion of the Polyakov action in Riemann normal coordinates and discuss the appropriate gauge-fixing, while in Sect. 3.4 we show how to reduce the quadratic Green-Schwarz action for the 10d type IIB Majorana-Weyl fermions to the action for two-dimensional spinors.

While we cannot regularize the determinants of the differential operators entering the free action for an *arbitrary* classical solution—this will be done in the specific examples of Chaps. 4–6—in Sect. 3.5 we extract only their divergent parts. With this we will review the mechanism responsible for the cancellation of the one-loop conformal anomaly in the $AdS_5 \times S^5$ action (see comments in Sect. 1.4.1).

---

[1] See also [4, 5] and references therein, where this analysis has been exploited for the description of QCD strings or stability effects for membrane solutions.

E. Vescovi, *Perturbative and Non-perturbative Approaches to String Sigma-Models in AdS/CFT*, Springer Theses, DOI 10.1007/978-3-319-63420-3_3

The novelty in this chapter is the treatment of the fermionic operator (due to complications related to the flux term) and the general expressions for bosonic and fermionic masses (as in all cases previously analyzed simplifications occurred due to the "flatness" of the normal bundle associated to the classical surface).

Although our main focus is $AdS_5 \times S^5$, almost all statements are independent on the dimensionality d and the specific spacetime (provided that $M = 1, \ldots d$ and similarly for other target-space indices in Appendix D), so in principle it is directly generalizable to other important AdS/CFT backgrounds ($AdS_4 \times \mathbb{CP}^3$, $AdS_3 \times S^3 \times M_4$, $AdS_3 \times S^2 \times M_5$, $AdS_2 \times S^2 \times M_6$, with $M_4 = T^4$, $S^3 \times S^1$ and $M_5 = S^3 \times T^2$) by exploiting some general properties shared by their geometries (see the "separability" condition for the 10d Riemann tensor in (3.51) below).

3.1 Geometry of the $AdS_5 \times S^5$ Space

The $AdS_5 \times S^5$ space is a direct product of the five dimensional anti-de Sitter space and the five-dimensional sphere. The AdS space is the isometric embedding $-X_0^2 + \sum_{i=1}^4 X_i^2 - X_5^2 = -R^2$ into the flat space $\mathbb{R}^{2,4}$ with metric $ds^2_{\mathbb{R}^{2,4}} = -dX_0^2 + \sum_{i=1}^4 dX_i^2 - dX_5^2$. By construction, AdS_5 is an homogeneous space with isometry group $SO(2,4)$. On the other hand, the sphere S^5 is the homogeneous space $\sum_{i=1}^6 Y_i^2 = R^2$ with $SO(6)$ symmetry and embedded in Euclidean $\mathbb{R}^6$ with $ds^2_{\mathbb{R}^6} = \sum_{i=1}^6 dY_i^2$. Here, we were careful to keep the same radius of curvature R in the two subspaces as in (1.1). For simplicity, we set it equal to one from now on (Fig. 3.1).

The parametrization of $AdS_5 \times S^5$ in *global coordinates*

$$\begin{aligned} X_1 + iX_2 &= \sinh\rho\,\cos\theta\, e^{i\phi_1}\,, & X_3 + iX_4 &= \sinh\rho\,\sin\theta\, e^{i\phi_2}\,, \\ X_5 + iX_0 &= \cosh\rho\, e^{it}\,, & Y_5 + iY_6 &= \cos\gamma\, e^{i\varphi_3}\,, \\ Y_1 + iY_2 &= \sin\gamma\,\cos\psi\, e^{i\varphi_1}\,, & Y_3 + iY_4 &= \sin\gamma\,\sin\psi e^{i\varphi_1} \end{aligned} \tag{3.1}$$

gives the explicit expression for its metric $ds^2_{AdS_5\times S^5} = ds^2_{AdS^5} + ds^2_{S^5}$

$$ds^2_{AdS_5} = -\cosh^2\rho\, dt^2 + d\rho^2 + \sinh^2\rho\,(d\theta^2 + \cos^2\theta\, d\phi_1^2 + \sin^2\theta\, d\phi_2^2)\,, \tag{3.2}$$

$$ds^2_{S^5} = d\gamma^2 + \cos^2\gamma\, d\varphi_3^2 + \sin^2\gamma\,(d\psi^2 + \cos^2\psi\, d\varphi_1^2 + \sin^2\psi\, d\varphi_2^2)\,. \tag{3.3}$$

Above, θ, ϕ_1 and ϕ_2 parametrize a three-sphere, $\rho > 0$ is the radial coordinate stretching from the AdS center $\rho = 0$ all the way to the boundary with geometry $S^1 \times S^3$ located at spatial infinity $\rho = \infty$. The periodicity of the *global time* is restricted to $t \in [0, 2\pi)$, but in AdS/CFT applications it is usually decompactified to consider the *universal cover* of AdS_5 with $t \in \mathbb{R}$, by removing the identification

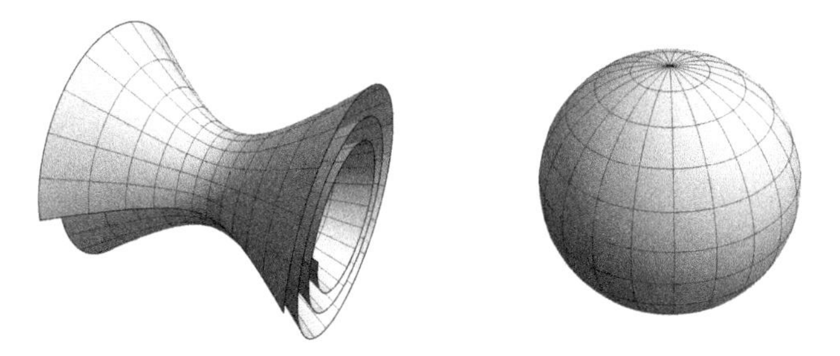

Fig. 3.1 Sketch of the universal cover of the AdS_5 space, wrapping the hyperboloid an infinite number of times, and the sphere S^5 [10]

$t \sim t + 2\pi$, in order to avoid closed time-like curves and to make contact with the time coordinate of the boundary gauge theory.[2]

In addition to the global parametrization (3.2), the AdS part can be written in different forms that will turn out to be useful later in this thesis. The analysis of strings surfaces ending on the AdS boundary, e.g. in Chap. 5, is conveniently done in the *Poincaré patch*

$$\begin{aligned} X_0 &= \frac{x_0}{z} = \cosh\rho \sin t\,, & X_i &= \frac{x_i}{z} = n_i \sinh\rho\,, \qquad i = 1, 2, 3, \\ X_4 &= \frac{-1 + z^2 - x_0^2 + \sum_{1=1}^{3} x_i^2}{2z} = n_4 \sinh\rho\,, & & \sum_{i=1}^{4} n_i^2 = 1\,, \\ X_5 &= \frac{1 + z^2 - x_0^2 + \sum_{1=1}^{3} x_i^2}{2z} = \cosh\rho \cos t\,, \end{aligned} \tag{3.4}$$

which brings the conformally-flat metric

$$ds^2_{AdS_5} = \frac{-dx_0^2 + \sum_{m=1}^{3} dx_m^2 + dz^2}{z^2}\,. \tag{3.5}$$

The x's coordinates are identified with the Cartesian coordinates of the four-dimensional boundary of AdS located at $z = 0$. In Chap. 7 we make yet use of another form of the full $AdS_5 \times S^5$ metric by combining the five angles (3.3) and the coordinate z into the sextuplet z^M $(M = 1, \ldots 6)$

$$ds^2_{AdS_5 \times S^5} = \frac{-dx_0^2 + \sum_{m=1}^{3} dx_m^2 + \sum_{M=1}^{6} dz^M dz^M}{z^2}\,, \qquad z^2 = \sum_{M=1}^{6} z_M^2\,. \tag{3.6}$$

[2]The AdS/CFT correspondence prescribes a *conformal compactification* of $AdS_5 \times S^5$ and the dual $\mathbb{R}^{1,3}$ space, as explained in the review [8]. We thank Hagen Münkler for pointing out to us the reference [9] where such construction for $\mathbb{R}^{p,q}$ is rigorously spelt out.

in which $\sum_{M=1}^{6} dz^M dz^M \equiv dz^2 + z^2 ds^2_{S^5}$.

3.2 The Minimal-Surface Equations

We start by recalling some basic facts about classical string theory. Fermionic fields always vanish on a classical solution and the bosonic string action can be either in the *Nambu-Goto* or *Polyakov* form. First we shall deal with classical backgrounds which extremize the former action functional (here defined after dropping the overall string tension T)

$$S_{NG} = \int d\tau d\sigma \sqrt{h}\,, \tag{3.7}$$

where $h_{ij} \equiv G_{MN}\partial_i X^M \partial_j X^N$ is the induced metric, namely the pull-back of the target space metric G_{MN} on the 2d worldsheet Σ spanned by the pair of coordinates (τ, σ). It is important to stress that here h is the absolute value of the determinant of h_{ij}.

As explained in Sect. 1.4, in semiclassical approximation we are interested in finding the surface of minimal area X^M which solves the Euler-Lagrangian equations

$$\Box_h X^M + h^{ij}\,\Gamma^M_{NP}\,\partial_i X^N \partial_j X^P = 0\,, \tag{3.8}$$

here written in terms of the covariant Laplacian on worldsheet scalars $\Box_h = \frac{1}{\sqrt{h}}\,\partial_i(\sqrt{h}\,h^{ij}\,\partial_j)$ and the Christoffel connection Γ^M_{NP} of the target-space metric G_{MN}. The covariant Laplacian is expanded in terms of h_{ij} and its Christoffel connection Λ^i_{jk} to yield

$$h^{ij}(\partial_i\partial_j X^M - \Lambda^k_{ij}\partial_k X^M + \Gamma^M_{NP}\partial_i X^N \partial_j X^P) \equiv h^{ij} K^M_{ij} = 0\,, \tag{3.9}$$

where the *second fundamental form* (or *extrinsic curvature*) K^M_{ij} of the worldsheet in the embedding space has been introduced. One can see that the equations of motion (3.9) state that the all the "curvatures" K^M vanish[3]

$$K^M \equiv h^{ij} K^M_{ij} = 0\,. \tag{3.10}$$

By construction the extrinsic curvature is orthogonal to the two vectors $t^M_i \equiv \partial_i X^M$ tangent to the worldsheet

[3]Note that the "shape" of a minimal surface is deeply influenced by the signature of the 10d G_{MN} and 2d metric h_{ij}. This also implies that a rigorous mathematical treatment of the differential equations (3.10)—e.g. existence and uniqueness theorems of its solutions—is substantially different in a Riemannian or pseudo-Riemannian ambient space. We acknowledge an elucidating discussion with Thomas Klose about this point.

$$G_{MN} t_i^M K_{jk}^N = 0. \tag{3.11}$$

Physically this means that only 8 of the 10 equations in (3.10) are independent and they govern the 8 transverse degrees of freedom. The 2 longitudinal ones are gauge degrees of freedom associated to the *2d diffeomorphism invariance* $(\tau, \sigma) \to (\tau', \sigma') = (\tau'(\tau, \sigma), \sigma'(\tau, \sigma))$ of (3.8) (and of its equivalent formulation (3.12) below).

We arrive at the same equations of motion by starting from the Polyakov form of the string sigma-model (omitting now a factor of $\frac{T}{2}$)

$$S_P = \int d\tau d\sigma \sqrt{g} g^{ij} G_{MN} \partial_i X^M \partial_j X^N \tag{3.12}$$

where the g_{ij} is an independent two-dimensional metric field and g is the absolute value of its determinant. In this case the dynamical equations for the embedding coordinates X^M acquire an extra term

$$\Box_g X^M + g^{ij} \Gamma^M_{NP} \partial_i X^N \partial_j X^P = g^{ij} K^M_{ij} + g^{ij} (\Lambda^k_{ij} - \tilde{\Gamma}^k_{ij}) \partial_k X^M = 0, \tag{3.13}$$

where $\Box_g$ denotes the covariant scalar Laplacian and $\tilde{\Gamma}^i_{\ jk}$ the Christoffel symbol of the auxiliary metric g_{ij}, whereas $\Lambda^i_{\ jk}$ is still the one of the induced metric. The (algebraic) equations of motion $\frac{\delta S_P}{\delta g_{ij}} = 0$ for the auxiliary metric g_{ij} (*Virasoro constraints*) set it proportional to the induced one

$$g_{ij} = e^{\varphi} h_{ij} \, . \tag{3.14}$$

The local factor φ is not uniquely fixed because the combination $\sqrt{g} g^{ij}$ in the Polyakov action is invariant under *Weyl transformations* ($\phi = \phi(\tau, \sigma)$ is an arbitrary scalar function)

$$X^M \to X^M \, , \qquad g_{ij} \to e^{\phi} g_{ij} \, . \tag{3.15}$$

The last addend in (3.13) vanishes due to (3.14) and the equations of motion again reduce to (3.10).

The problem of finding a surface of least area stretched across a given closed contour is as old as the calculus of variations [11]. The equations (3.10) are typically too complicated to allow for a direct solution. One can instead begin with an ansatz for X^M—using certain unknown functions of (τ, σ) and, possibly, some adjustable parameters—that respect the expected symmetries of the solution. Out of this infinite family of surfaces, the one of least extension optimises the action functional. The advantage of this method lies in the ability of decomposing the 2d variational problem into a set of ordinary differential equations for the auxiliary functions, which can be then solved separately. A rather non-trivial example of this strategy was pursued for finding the worldsheet dual to a non-BPS cusped Wilson loop in Appendix C.2 of [12].

Another ingenious way to construct new open string solutions from known ones stems from conformal invariance. In the following argument the S^5 subspace does not play any role, so in the metric (3.5) we indicate a minimal surface by $X^M(\tau,\sigma) = (x^\mu(\tau,\sigma), z(\tau,\sigma))$ for short and its boundary $X^M(\tau,0) = (x^\mu(\tau,0),0)$ with $\mu = 0,\ldots 3$. Suppose we are interested in the solution of the variational problem for $X'^M(\tau,\sigma) = (x'^\mu(\tau,\sigma), z'(\tau,\sigma))$ with a given boundary shape $X'^M(\tau,0) = (x'^\mu(\tau,0),0)$ being the image of $X^M(\tau,0)$ under an inversion at the unit circle $x^\mu(\tau,0) \to x'^\mu(\tau,0) = \frac{x^\mu(\tau,0)}{x^\mu(\tau,0)x_\mu(\tau,0)}$. Conformal mappings of $\mathbb{R}^4$ can be promoted to isometries in the bulk of AdS_5

$$x^\mu \to \frac{x^\mu}{x^\mu x_\mu + z^2}\,, \qquad\qquad z \to \frac{z}{x^\mu x_\mu + z^2}\,. \tag{3.16}$$

Therefore, we can easily construct the new surface X'^M as the image of the minimal one X^M through this map which, being an AdS isometry, guarantees that X'^M is of least area as well. The argument was employed, for instance, for the worldsheet dual to a Wilson loop made of two circular arcs [13].

3.3 Bosonic Fluctuations

In this section we shall discuss the semiclassical expansion of the bosonic action around a classical background by truncating it at quadratic order in the bosonic fluctuations. The starting point can be either the Polyakov form with the independent metric g_{ij} or the Nambu-Goto form, as they should yield equivalent expressions for the one-loop partition function. However, we prefer the Polyakov formulation because the definition of the path-integral measure and the cancellation of the one-loop conformal anomaly are better understood in this form [3].

After reviewing previous analysis [2, 3] based on background field method for non-linear sigma-models and the virtues of the expansion in geodesic normal coordinates [14], we will write the relevant contributions in terms of intrinsic and extrinsic geometric invariants of the classical solution.

3.3.1 *The Bosonic Lagrangian*

We will discuss the bosonic sector starting from (3.12) for the fields $\tilde{X}^M$

$$S_P = \int d\tau d\sigma \sqrt{g} g^{ij} \partial_i \tilde{X}^M \partial_j \tilde{X}^N G_{MN}(\tilde{X})\,, \tag{3.17}$$

while keeping the notation X^M for the solution of the equations of motion (3.10). A well-known subtlety of the expansion of a non-linear sigma-model around a classical background X^M [14] is that writing it as a power series in terms of fluctuations $\delta X^M = \tilde{X}^M - X^M$ does not lead to a manifestly covariant expression for the series coefficients. As a matter of fact, the difference δX^M does not transform simply under reparametrization. The easiest way to obtain a manifestly covariant form for the coefficients is to take advantage of the *Riemann normal coordinates* of the embedding space, i.e. expressing δX^m as a power series in the spacetime vectors that are tangent to the geodesic connecting X^m with $X^m + \delta X^m$ [14]. One considers a geodesic line $X^M(t)$ with t parametrizing the arc length such that

$$X^M(0) = X^M\,, \qquad\qquad X^M(1) = \tilde{X}^M \equiv X^M + \delta X^M\,. \tag{3.18}$$

Solving then the geodesic equation for $X^M(t)$

$$\ddot{X}^M(t) + \Gamma^M_{NP}\dot{X}^N(t)\dot{X}^P(t) = 0 \tag{3.19}$$

in terms of the tangent vector $\zeta^M \equiv \dot{X}^M(0)$ to the geodesic in $t = 0$, one finds

$$X^M(t) = X^M + t\zeta^M - \frac{1}{2}t^2\Gamma^M_{NP}\zeta^N\zeta^P + O(t^3)\,, \tag{3.20}$$

Setting $t = 1$, this means[4]

$$\delta X^M = \tilde{X}^M - X^M = \zeta^M - \frac{1}{2}\Gamma^M_{NP}\zeta^N\zeta^P + O(\zeta^3)\,. \tag{3.21}$$

Where $\Gamma^M_{NP} \equiv \Gamma^M_{NP}(X^M)$ is evaluated on the classical surface. The difference δX^M is now the desired local power series in the vector ζ^M, which can then be conveniently used as a fundamental variable. Plugging this expression into (3.17), we find the fluctuation action (see [2] for example)

$$S = S_B^{(0)}(X) + \int d\tau d\sigma \sqrt{g} g^{ij}\left(\nabla_i\zeta^M\nabla_j\zeta^N G_{MN} - R_{RMSN}\zeta^R\zeta^S\partial_i X^M\partial_j X^N\right) + O(\zeta^3). \tag{3.22}$$

Here, $\nabla_i\zeta^M \equiv \partial_i\zeta^M + \Gamma^M_{NP}\,\partial_i X^N\,\zeta^p$ is the pullback of the covariant derivative onto the classical surface. In the quadratic truncation of the Polyakov action above, the term $S_B^{(0)}(X)$ denotes the *classical action*, while the integral describes the *quadratic fluctuations* and it will be denoted with $S_B^{(2)} = \int d\tau d\sigma \mathcal{L}_B^{(2)}$ in the following. In order to have a canonically normalized kinetic term it is convenient to introduce a set of vielbein E^A_M and its inverse E^M_A for the target metric as shown in

[4] If we reintroduce the string tension $T/2$ in the Polyakov action, then we need to rescale $\delta X^M \to \sqrt{\frac{2}{T}}\delta X^M$ and the effective expansion parameter $\sqrt{\frac{2}{T}}\zeta^M$ is small in semiclassical approximation $T \to \infty$.

Appendix D. In terms of the redefined fluctuations fields

$$\xi^A = E^A_M \zeta^M , \tag{3.23}$$

the quadratic fluctuation Lagrangian becomes [2, 3]

$$\mathcal{L}^{(2)}_B = \sqrt{g}\left(g^{ij} D_i \xi^A D_j \xi_A - M_{A,B} \xi^A \xi^B\right), \tag{3.24}$$

where the *mass matrix* [2, 3]

$$M_{AB} = R_{AC,BD} t^{aC} t^D_a \tag{3.25}$$

is defined through two vectors (D.4) tangent to the worldsheet and the covariant derivative in the tangent-frame indices now reads

$$D_i \xi^A = \partial_i X^A + \Omega^A{}_{BM} \xi^B \partial_i X^M , \tag{3.26}$$

with the spin-connection $\Omega^A{}_{BM}$ replaced by the usual Christoffel symbols. To better understand the geometrical structure of the Lagrangian (3.24), we introduce 8 orthonormal vector fields $N^A_{\underline{a}}$ orthogonal to the worldsheet satisfying (D.5) and project the field ζ^A onto directions tangent ($x^{\tilde{a}}$) and orthogonal ($y^{\underline{i}}$) to the surface

$$\xi^A \equiv x^{\tilde{a}} t^A_{\tilde{a}} + y^{\bar{i}} N^A_{\bar{i}} . \tag{3.27}$$

It is known from the general theory of submanifolds [15] that this decomposition carries over to the covariant derivatives

$$t^{\tilde{a}}_A D_i \xi^A = \mathcal{D}_i x^{\tilde{a}} - K^{\tilde{a}}_{Ai} N^A_{\bar{j}} y^{\bar{j}} , \qquad N^{\underline{i}}_A D_j \xi^A = \mathcal{D}_j y^{\underline{i}} + x^{\tilde{a}} N^{\underline{i}}_A K^A_{\tilde{a}j} . \tag{3.28}$$

Here, $\mathcal{D}_i$ is the covariant derivative on the worldsheet and it acts differently on $x^{\tilde{a}}$ and $y^{\underline{i}}$

$$\mathcal{D}_i x^{\tilde{a}} \equiv \partial_i x^{\tilde{a}} + \omega^{\tilde{a}}_{\tilde{b}i} x^{\tilde{b}} , \qquad \mathcal{D}_i y^{\underline{i}} \equiv \partial_i y^{\underline{i}} - A^{\underline{i}}{}_{\underline{k}i} y^{\underline{k}}, \tag{3.29}$$

since $x^{\tilde{a}}$ lives in the tangent bundle of the worldsheet, while $y^{\underline{i}}$ is a section of the normal bundle. The field $A^{\underline{i}}{}_{\underline{j}k}$ is a two-component *gauge connection* ($k = 1, 2$) taking values in the $SO(8)$ normal bundle ($\underline{i}, \underline{j} = 2, \ldots 9$) induced by the classical solution:

$$A^{\underline{i}}{}_{\underline{j}k} \equiv N^B_{\underline{j}} D_k N^{\underline{i}}_B = N^B_{\underline{j}} (\partial_k N^{\underline{i}}_B - N^{\underline{i}}_C \Omega^C{}_{Bk}). \tag{3.30}$$

As usual the action of $\mathcal{D}_i$ on tensors with indices in both bundles is obtained combining the two actions in (3.29). The tensor $K^a_{Aj} = E_{Am} e^{ai} K^M_{ij}$ in (3.28) is the extrinsic curvature (3.9) of the embedding expressed in a mixed basis.

In the following we will make use of the *Gauss-Codazzi equation*

$$R_{ACBD} t^A_{\tilde{a}} t^C_{\tilde{c}} t^B_{\tilde{b}} t^D_{\tilde{d}} = {}^{(2)}R_{\tilde{a}\tilde{c}\tilde{b}\tilde{d}} + \eta_{AB} K^A_{\tilde{c}\tilde{b}} K^B_{\tilde{d}\tilde{a}} - \eta_{AB} K^A_{\tilde{c}\tilde{d}} K^B_{\tilde{b}\tilde{a}} \,, \tag{3.31}$$

an integrability condition relating the worldsheet curvature ${}^{(2)}R_{\tilde{a}\tilde{c}\tilde{b}\tilde{d}}$ to the extrinsic and background geometry as characterized by the extrinsic curvature $K^A_{\tilde{a}\tilde{b}}$ and the space-time Riemann tensor $R^A{}_{CBD}$. Another useful constraint on the covariant derivative of the extrinsic curvature is provided by the *Codazzi-Mainardi equation*

$$\mathcal{D}_j K^i_{kl} - \mathcal{D}_k K^i_{jl} = R_{MNRS} t^M_j t^N_k t^S_l N^{Ri} \,, \qquad\qquad K^i_{jk} \equiv K^A_{jk} N^i_A \,. \tag{3.32}$$

Taking into account (3.31), (3.32) and the equations of motion (3.10) for the background, the quadratic Lagrangian (3.24) finally appears to be

$$\begin{aligned} \mathcal{L}^{(2)}_B = \sqrt{g}\big[(g^{ij}\mathcal{D}_i x^{\tilde{a}} \mathcal{D}_j x_{\tilde{a}} - {}^{(2)}R_{\tilde{a}\tilde{b}} x^{\tilde{a}} x^{\tilde{b}}) + g^{ij} \mathcal{D}_i y^{\underline{i}} \mathcal{D}_j y_{\underline{i}} - \\ - 2h^{ij}(\mathcal{D}_i x^a K_{i,aj} y^i - \mathcal{D}_i y^{\underline{i}} x^{\tilde{a}} K_{\underline{i},\tilde{a}j}) - 2m_{\tilde{a}\underline{i}} x^{\tilde{a}} y^{\underline{i}} - m_{\underline{i}\underline{j}} y^{\underline{i}} y^{\underline{j}}\big] \,. \end{aligned} \tag{3.33}$$

Above, the matrices appearing in the mass terms are given by

$$m_{\tilde{a}\underline{i}} = -g^{kj} \nabla_k K_{i,aj} \quad \text{and} \quad m_{\underline{i}\underline{j}} = R_{AC,BD} t^{cC} t^D_c N^A_{\underline{i}} N^B_{\underline{j}} - g^{kl} g^{mn} K_{\underline{i},km} K_{\underline{j},jn} \,. \tag{3.34}$$

So far we have treated the independent metric g_{ij} as a non-dynamical field. We should recall that in the Polyakov formulation it actually fluctuates

$$\tilde{g}_{ij} = g_{ij} + \mathfrak{g}_{ij} \tag{3.35}$$

around a classical background g_{ij} (3.14). In particular this means that all the g_{ij} appearing in the previous analysis must be replaced with $\tilde{g}_{ij}$. The quadratic part of the Polyakov action involving its fluctuations $\mathfrak{g}_{ij}$ reads

$$\begin{aligned} S_{\rm g} &= \int d\tau d\sigma \mathcal{L}_{\rm g} \,, \\ \mathcal{L}_{\rm g} &= e^{\varphi} \sqrt{h} \big[\tfrac{1}{4} (h^{li} h^{kj} + h^{lj} h^{ki} - h^{ij} h^{kl}) \mathfrak{g}_{ij} \mathfrak{g}_{kl} - (\mathcal{D}^i x^j + \mathcal{D}^j x^i - h^{ij} \mathcal{D}_{\tilde{i}} x^{\tilde{i}} - 2 y_{\underline{i}} K^{\underline{i}ij}) \mathfrak{g}_{ij} \big], \end{aligned} \tag{3.36}$$

where $x^i = e^i_{\tilde{a}} x^{\tilde{a}}$. To deal with the quadratic fluctuation Lagrangians (3.33) and (3.36), we have different possibilities (see [6] for a more exhaustive list of gauge fixings). In the *conformal gauge* for the metric fluctuations

$$\mathfrak{g}_{ij} = e^{\varphi} h_{ij} \delta\varphi \tag{3.37}$$

the action $S_{\rm g}$ vanishes while the *ghost action* [16, 17] associated to the gauge-fixing (3.37) is

$$S_{\text{ghost}} = \int d\tau d\sigma \mathcal{L}_{\text{ghost}} , \qquad \mathcal{L}_{\text{ghost}} = \frac{1}{\sqrt{2}} b^{ij} \left(\mathcal{D}_i c_j + \mathcal{D}_j c_i - h_{ij} \mathcal{D}_k c^k \right) . \tag{3.38}$$

The parameter c and the symmetric traceless tensor b^{ij} are Grassmann-odd ghost fields. The full ghost contribution is encoded into a functional determinant, obtained by integrating over c and b^{ij}. Concretely, the computation of the ghost determinant means here to solve the following eigenvalue problem in the background geometry (n labels eigenvalues and eigenvectors)

$$\begin{cases} \frac{1}{\sqrt{2}} (\mathcal{D}_i c_{(n)j} + \mathcal{D}_j c_{(n)i} - h_{ij} \mathcal{D}_k c^k_{(n)}) = \lambda_n b_{(n)ij} \\ - \sqrt{2} \mathcal{D}_i b^{ij}_{(n)} = \lambda_n c^{\nu}_{(n)} . \end{cases} \tag{3.39}$$

The system of equations leads to the eigenvalue equation for the *ghost operator*

$$- (\Box_h \delta^i_j + R^i{}_j) c^j_{(n)} = \lambda_n^2 c^i_{(n)} \tag{3.40}$$

and the value of its determinant

$$\triangle_{\text{ghost}} = \prod_n \lambda_n = \text{Det}^{1/2} (- \Box_h \delta^i_j - R^j{}_i) . \tag{3.41}$$

We can now decouple the longitudinal fluctuation $x^{\tilde{a}}$ from the transverse ones $y^{\underline{i}}$. We start from the action (3.33), and derive the equations of motion for the fluctuation parallel to the worldsheet:

$$\Box_h x_{\tilde{i}} + R_{\tilde{i}\tilde{j}} x^{\tilde{j}} = \mathcal{D}^{\tilde{j}} B_{\tilde{j}\tilde{i}} , \qquad B_{\tilde{i}\tilde{j}} = 2 y_{\underline{i}} K^{\underline{i}}_{\tilde{i}\tilde{j}} , \tag{3.42}$$

where we introduced the traceless tensor $B_{\tilde{i}\tilde{j}}$. This equation can be equivalently written as follows

$$(P_1)_{\tilde{i}\tilde{j}} \equiv \mathcal{D}_{\tilde{j}} x_{\tilde{i}} + \mathcal{D}_{\tilde{i}} x_{\tilde{j}} - h_{\tilde{i}\tilde{j}} \mathcal{D}_{\tilde{k}} x^{\tilde{k}} = B^{\parallel}_{\tilde{i}\tilde{j}} \tag{3.43}$$

where we have introduced a local projector P_1 mapping vectors to traceless symmetric tensors. We can conveniently decompose the traceless symmetric tensor

$$B_{\tilde{i}\tilde{j}} = B^{\parallel}_{\tilde{i}\tilde{j}} + B^{\perp}_{\tilde{i}\tilde{j}} , \qquad B^{\parallel}_{\tilde{i}\tilde{j}} \in \text{range}(P_1) , \qquad B^{\perp}_{\tilde{i}\tilde{j}} \in \text{range}(P_1)^{\perp} = \text{Ker}(P_1^{\dagger}) . \tag{3.44}$$

If we are only interested in worldsheet with the topology of the sphere or the disc, then $B^{\perp}_{\tilde{i}\tilde{j}} = 0$ and $B_{\tilde{i}\tilde{j}} = B^{\parallel}_{\tilde{i}\tilde{j}}$.[5] Considering now a solution $\bar{x}^{\tilde{i}}$ of (3.43) and

[5]This corresponds to the absence of non-trivial *Beltrami differentials* at genus 0.

performing the shift $x^{\tilde{i}} \to \bar{x}^{\tilde{i}} + x^{\tilde{i}}$ in the path-integral, all mixed terms xy in (3.33) disappear and we are left therefore with the quadratic Lagrangian

$$\mathcal{L}_B^{(2)} \equiv \mathcal{L}_{\text{long}}^{(2)} + \mathcal{L}_{\text{transv}}^{(2)} \tag{3.45}$$

where

$$\mathcal{L}_{\text{long}}^{(2)} = \sqrt{h}\left(h^{ij}\mathcal{D}_i x^{\tilde{a}}\mathcal{D}_j x_{\tilde{a}} - {}^{(2)}R_{\tilde{a}\tilde{b}}x^{\tilde{a}}x^{\tilde{b}}\right) \tag{3.46}$$

and

$$\begin{aligned}
\mathcal{L}_{\text{transv}}^{(2)} &= \sqrt{h}(h^{ij}\mathcal{D}_i y^{\underline{i}}\mathcal{D}_j y_{\underline{i}} - \mathcal{M}_{\underline{ij}}^{(B)} y^{\underline{i}} y^{\underline{j}}), \\
\mathcal{M}_{\underline{ij}}^{(B)} &= R_{AC,BD}t^{cC}t_c^{D}N_{\underline{i}}^{A}N_{\underline{j}}^{B} + K_{\underline{i},ij}K_{\underline{j}}^{ij}.
\end{aligned} \tag{3.47}$$

After the above redefinition, the operator controlling the fluctuations $x^{\tilde{a}}$ parallel to the worldsheet in (3.46) coincides with the one for ghosts in (3.41), but we remark that this does *not* mean in general that the corresponding determinants will simply cancel. For instance in the case of open strings different boundary conditions should be imposed for the two determinants. Moreover the treatment of the ghost operator requires additional care since it might contain zero modes associated to the Killing vectors of the worldsheet metric h_{ij}.

Another option to get rid of the longitudinal fluctuations is to choose the *static gauge* by setting $x^{\tilde{i}} = 0$. Then (3.33) reduces to a Lagrangian for the transverse fluctuations only, with a mass matrix given not by $\mathcal{M}_{\underline{ij}}$ but by $m_{\underline{ij}}$. However, we must recall that the transverse fluctuations are still coupled to the metric fluctuations (see Eq. (3.36)):

$$\mathcal{L}_{\text{g}} = e^{\varphi}\sqrt{h}\left(\tfrac{1}{2}h^{li}h^{kj}\bar{\mathfrak{g}}_{ij}\bar{\mathfrak{g}}_{kl} + 2y_{\underline{i}}K^{\underline{i}ij}\bar{\mathfrak{g}}_{ij}\right). \tag{3.48}$$

If we again eliminate the metric fluctuation $\bar{\mathfrak{g}}_{ij}$ through its equation of motion, we get back to the mass matrix $\mathcal{M}_{\underline{ij}}^{(B)}$ and to $\mathcal{L}_{\text{transv}}^{(2)}$ in (3.47).

We also remark that starting from the Nambu-Goto action, where no dynamical worldsheet metric is present, in the static gauge $x^{\tilde{i}} = 0$ we would directly obtain $\mathcal{L}_{\text{transv}}^{(2)}$.

In view of the analysis above, we will focus our attention on transverse fluctuations and closely examine the structure of their Lagrangian (3.47). These modes are in general coupled between themselves and we can distinguish two different sources of coupling: the $SO(8)$ gauge connection $A^{\underline{i}}{}_{\underline{j}k}$ induced on the normal bundle and the mass matrix $\mathcal{M}_{\underline{ij}}^{(B)}$.

3.3.2 The Normal Bundle

Let us begin with the geometric structure of the normal bundle. The *normal-bundle curvature* of the gauge connection (3.30)

$$F^{\underline{i}}{}_{\underline{j}kl} \equiv \partial_k A^{\underline{i}}{}_{\underline{j}l} - \partial_l A^{\underline{i}}{}_{\underline{j}k} + A^{\underline{i}}{}_{\underline{k}k} A^{\underline{k}}{}_{\underline{j}l} - A^{\underline{i}}{}_{\underline{k}l} A^{\underline{k}}{}_{\underline{j}k} \tag{3.49}$$

can be easily evaluated in terms of the Riemann curvature of the target space and of the extrinsic curvature through the *Ricci equation*

$$F^{\underline{i}}{}_{\underline{j}kl} = -R_{ABCD} t_k^A t_l^B D^{\underline{i}C} N_{\underline{j}}^D - h^{mn}(K^{\underline{i}}_{mk} K_{\underline{j}n} - K^{\underline{i}}_{ml} K_{\underline{j},mk}). \tag{3.50}$$

If we make the choice of $AdS^5 \times S^5$ explicit, we see that the contribution of the Riemann tensor in this background to the curvature of the normal bundle

$$R_{ABCD} = -(\hat{P}_{AC}\hat{P}_{BD} - \hat{P}_{AD}\hat{P}_{BC}) + (\bar{P}_{AC}\bar{P}_{BD} - \bar{P}_{AD}\bar{P}_{BC}) \tag{3.51}$$

vanishes identically as in flat space. We called $\hat{P}_{AB}$ the projector onto AdS_5 and $\bar{P}_{AB}$ the one onto S^5, with $\hat{P}_{AB} + \bar{P}_{AB} = \eta_{AB}$. We remain with the expression for the normal curvature

$$F^{\underline{i}}{}_{\underline{j}ij} = (h^{kl}\epsilon^{mn} K^{\underline{i}}_{km} K_{\underline{j}ln})\epsilon_{ij} \equiv \sqrt{h}\epsilon_{ij}\mathcal{F}^{\underline{i}}{}_{\underline{j}}\,, \tag{3.52}$$

which holds also in flat space. The normal bundle is *flat* when $F^{\underline{i}}{}_{\underline{j}ij} = 0$. In that case, we can always choose the 8 normal vectors in (D.5) so that $A^{\underline{i}}{}_{\underline{j}k} = 0$ and the covariant derivative acting on the transverse fluctuations (3.29) reduces to the standard partial derivative. This occurs, for instance, when the minimal surface is confined in a three-dimensional subspace of the target space: the extrinsic curvature is in fact not vanishing just in one normal direction [3, 18–23].

For a generic worldsheet, which solves the equations of motion in the $AdS_5 \times S^5$ background, the extrinsic curvarture defines, at most, two independent vector fields normal to the worldsheet.[6] In fact the matrix $\mathcal{F}^{\underline{i}}{}_{\underline{j}}$ can be always put in the form

$$\mathcal{F}^{\underline{i}}{}_{\underline{j}} = (k^{\underline{i}} h_{\underline{j}} - k_{\underline{j}} h^{\underline{i}}), \tag{3.53}$$

where $(h^{\underline{i}} \cdot t_{\tilde{a}}) = (k^{\underline{i}} \cdot t_{\tilde{a}}) = 0$ and we can choose $(k \cdot h) = 0$ without loss of generality. Here and in the following, we make use of the inner product over flat 10d indices

$$(V \cdot W) \equiv \eta_{AB}\, V^A\, W^B\,. \tag{3.54}$$

[6] This follows from the fact that we have just two independent components of the extrinsic curvature (*e.g.* K^M_{11} and K^M_{12}). Alternatively, we can argue this result, in a covariant way, from the following matrix relation $\mathcal{F}^4 = \frac{1}{2}\mathrm{tr}(\mathcal{F}^2)\mathcal{F}^2$ satisfied by $\mathcal{F}$.

3.3.3 Mass Matrix and Sum Rules

The next step is to examine more closely the mass matrix (3.47) for the transverse bosonic degrees of freedom. Since we have not used the properties of $AdS_5 \times S^5$ yet, in a generic ambient space it has the form

$$\mathcal{M}^{(B)}_{\underline{i}\underline{j}} = R_{AC,BD} t^{\tilde{a}C} t^{D}_{\tilde{a}} N^{A}_{\underline{i}} N^{B}_{\underline{j}} + K_{\underline{i},ij} K^{ij}_{\underline{j}} . \tag{3.55}$$

There are few general properties of $\mathcal{M}^{(B)}_{\underline{i}\underline{j}}$ that can be easily read from (3.55), since the embedding equations for a sub-manifold do not provide a direct constraint on the contraction $R_{AC,BD} t^{\tilde{a}C} t^{D}_{\tilde{a}} N^{A}_{i} N^{B}_{j}$. However, the trace $\mathrm{tr}(\mathcal{M}^{(B)})$ admits a quite simple and compact expression in terms of geometric quantities. If we use the completeness relation (D.5) and the Gauss equation (3.31) we can rewrite the trace of (3.55) as follows

$$\mathrm{tr}(\mathcal{M}^{(B)}) = R_{MN} t^{cM} t^{N}_{c} - R_{AM,BN} t^{cM} t^{N}_{c} t^{A}_{d} t^{dB} + \mathrm{tr}(K^2) = R_{AB}\, t^{\tilde{a}A} t^{B}_{\tilde{a}} - {}^{(2)}R \,, \tag{3.56}$$

where we used the traced of the square of the extrinsic curvature

$$\mathrm{tr}(K^2) = h^{ij} h^{kl} \eta_{AB} K^{A}_{kj} K^{B}_{li} \,, \tag{3.57}$$

the trace of the Ricci tensor R_{MN} and the two-dimensional (intrinsic) scalar curvature ${}^{(2)}R$ of the two-dimensional curvature.

We now explicitly use the geometric properties of $AdS_5 \times S^5$ background, referring the reader to [6] for a detailed derivation. We shall use hats and bars over vectors to denote their projection on AdS_5 and S^5 respectively

$$\hat{V}^{A} = \hat{P}^{A}{}_{B} V^{B} \,, \qquad \bar{V}^{A} = \bar{P}^{A}{}_{B} V^{B} . \tag{3.58}$$

The complete mass matrix is given by

$$\begin{aligned} \mathcal{M}^{(B)}_{\underline{i}\underline{j}} &= -m^2_{AdS_5} (\hat{N}_{\underline{i}} \cdot \hat{N}_{\underline{j}}) - m^2_{S^5} (\bar{N}_{\underline{i}} \cdot \bar{N}_{\underline{j}}) + K_{\underline{i}ij} K^{ij}_{\underline{i}} \,, \\ m^2_{AdS_5} &\equiv h^{\tilde{a}\tilde{b}} (\hat{t}_{\tilde{a}} \cdot \hat{t}_{\tilde{b}}) = -\frac{1}{2} \left({}^{(2)}R + \mathrm{tr}(K^2)\right) + 1 \,, \\ m^2_{S_5} &\equiv -h^{\tilde{a}\tilde{b}} (\bar{t}_{\tilde{a}} \cdot \bar{t}_{\tilde{b}}) = -\frac{1}{2} \left({}^{(2)}R + \mathrm{tr}(K^2)\right) - 1 \end{aligned} \tag{3.59}$$

and the trace of the mass matrix $\mathrm{tr}(\mathcal{M}^{(B)})$ simplifies to

$$\mathrm{tr}(\mathcal{M}^{(B)}) = 3\, {}^{(2)}R + 4\, \mathrm{tr}(K^2) . \tag{3.60}$$

The structure of the mass matrix (3.59) can be further constrained assuming particular properties of the classical background. The simplest geometrical configuration is when the minimal surface is confined in one of the two spaces: AdS_5 if $\bar{t}_{\tilde{a}} = 0$ or S^5 if $\hat{t}_{\tilde{a}} = 0$. Let us focus on the first possibility, as the second one can be discussed in complete analogy.

If the classical worldsheet lies entirely in AdS_5, then the mass matrix reduces to

$$\mathcal{M}^{(B)}_{\underline{i}\underline{j}} = -2(\hat{N}_{\underline{i}} \cdot \hat{N}_{\underline{j}}) + K_{\underline{i}kl} K^{kl}_{\underline{j}}, \tag{3.61}$$

where we have used that $m^2_{S^5} = 0$ and $m^2_{AdS_5} = 2$. The extrinsic curvature $K_{\underline{i}kl}$ is different from zero only for orthogonal directions lying in AdS_5. Therefore we have 5 massless scalar ($\mathfrak{m}_i = m_{S^5} = 0$ with $i = 1, \ldots 5$), one for each direction of S^5. We can always choose a sixth direction (lying in AdS_5) orthogonal to the worldsheet and to the two normal directions defined by $K_{\underline{i}kl}$. The mass $\mathfrak{m}_6$ of this sixth scalar is $\mathfrak{m}_6^2 = -2$.

Finally, we have to select the last two orthogonal directions ($i = 8, 9$) and we choose the only two orthonormal eigenvectors of $K_{\underline{i}kl} K^{kl}_{\underline{j}}$ with non vanishing eigenvalues. They always exist if the normal bundle is not flat. Then the two masses are given by

$$\mathfrak{m}_7^2 = \lambda_1 - 2\,, \qquad \mathfrak{m}_8^2 = \lambda_2 - 2. \tag{3.62}$$

Here λ_1 and λ_2 are the two non-vanishing eigenvalues of $K_{\underline{i}kl} K^{kl}_{\underline{j}}$ and they are determined in terms of the geometric quantity of the surface through the relations:

$$\lambda_1 + \lambda_2 = \mathrm{tr}(K^2) = -{}^{(2)}R - 2\,, \qquad \lambda_1\lambda_2 = \frac{1}{2}[(\mathrm{tr}(K^2))^2 - \mathrm{tr}(K^4)] = \frac{1}{2}\mathrm{tr}(\mathcal{F}^2), \tag{3.63}$$

where $\mathrm{tr}(K^4) \equiv K^{\underline{i}}_{jk} K^{jk}_{\underline{j}} K^{\underline{j}}_{mn} K^{mn}_{\underline{i}}$. If the normal bundle is flat, $\mathcal{F}^{\underline{i}}_{\underline{j}} = 0$ and one of the two eigenvalues vanishes, *e.g.* $\lambda_1 = 0$. Then the two masses collapse to the known result [3, 24]

$$\mathfrak{m}_7^2 = -2\,, \qquad \mathfrak{m}_8^2 = -{}^{(2)}R - 4. \tag{3.64}$$

Let us turn our attention to the general case where the worldsheet extends both in AdS_5 and S^5 spaces and the mass matrix has the general form (3.59). The first step is to choose two of the fluctuations ($i = 8, 9$) along the two orthogonal directions (h and k in (3.53)) with non-vanishing extrinsic curvature. These two directions are defined up to a rotation in the (h, k)-plane. We fix this freedom by choosing h and k to be the only two orthonormal eigenvectors of $K_{\underline{i}kl} K^{kl}_{\underline{j}}$ with non vanishing eigenvalues. Then the only non vanishing component of the field strength in the normal bundle is $\mathcal{F}^8{}_9$.

The bosonic masses can be analysed in details if the field strength $\mathcal{F}^8{}_9$ is essentially *abelian*, namely if the only component of the connection different from zero is given by $A^8{}_9$. In this case one can show that the Codazzi-Mainardi equation (3.32) implies

$$(\bar{t}_{\tilde{a}} \cdot N_{\underline{i}}) = (\hat{t}_{\tilde{a}} \cdot N_{\underline{i}}) = 0 \quad \text{for } i \neq 8, 9. \tag{3.65}$$

We find that the remaining six normal directions are orthogonal both to $\bar{t}_{\tilde{a}}$ and to $\hat{t}_{\tilde{a}}$, implying that some of these vectors completely lie in AdS_5, while the others in S^5. Generically we expect to find three of them in AdS_5 and three in S^5 (a different partition of the six vectors between the two subspaces may occur when some of $\hat{t}_{\tilde{a}}$ or of $\bar{t}_{\tilde{a}}$ vanishes). Because of the orthogonality relations (3.65), the mass matrix $\mathcal{M}^{(B)}$ takes the form[7]

$$\mathcal{M}^{(B)}_{\underline{i}\underline{j}} = \begin{pmatrix} -m^2_{AdS_5} & 0 & 0 & 0 & 0 & 0 & 0 & 0 \\ 0 & -m^2_{AdS_5} & 0 & 0 & 0 & 0 & 0 & 0 \\ 0 & 0 & -m^2_{AdS_5} & 0 & 0 & 0 & 0 & 0 \\ 0 & 0 & 0 & -m^2_{S^5} & 0 & 0 & 0 & 0 \\ 0 & 0 & 0 & 0 & -m^2_{S^5} & 0 & 0 & 0 \\ 0 & 0 & 0 & 0 & 0 & -m^2_{S^5} & 0 & 0 \\ 0 & 0 & 0 & 0 & 0 & 0 & m_{88} & m_{89} \\ 0 & 0 & 0 & 0 & 0 & 0 & m_{89} & m_{99} \end{pmatrix} \tag{3.66}$$

with $m^2_{AdS_5}$ and $m^2_{S^5}$ given in (3.59). With the help of the trace condition (3.56)

$$m_{88} + m_{99} = \mathrm{tr}(K^2)\,, \tag{3.67}$$

the matrix subblock for the directions 8 and 9 can be cast into to the form

$$\begin{pmatrix} m_{88} & m_{89} \\ m_{98} & m_{99} \end{pmatrix} = \begin{pmatrix} -m^2_{AdS_5} + 2(\bar{N}_8 \cdot \bar{N}_9) + \lambda_1 & 2(\bar{N}_8 \cdot \bar{N}_9) \\ 2(\bar{N}_8 \cdot \bar{N}_9) & m^2_{AdS_5} - 2(\bar{N}_8 \cdot \bar{N}_8) + \lambda_2 \end{pmatrix} \tag{3.68}$$

where λ_1 and λ_2 again obey (3.63).

In the general case, when $\mathcal{F}^8{}_9$ is not generated by only taking $A^8{}_9$ different from zero, the structure of the mass matrix may become more intricate. However we can always choose at least two orthogonal directions, one in S^5 and one in AdS_5, which are orthogonal to the minimal surface and to the extrinsic curvature. The masses of these two fluctuations are then given by (3.59).

3.4 Fermionic Fluctuations

The full covariant Green-Schwarz string action in $AdS_5 \times S^5$ has a complicated non-linear structure [25, 26], but to analyze the relevant fermionic contributions it is sufficient to consider only its quadratic part

[7] The matrix elements are labelled by $\underline{i}, \underline{j} = 2, \ldots 9$ in our index conventions.

$$S_F^{(2)} = \int d\tau d\sigma\, \mathcal{L}_F^{(2)}\,, \qquad \mathcal{L}_F^{(2)} = i\left(\sqrt{h}h^{ij}\,\delta^{IJ} - \epsilon^{ij}s^{IJ}\right)\bar{\Psi}^I\,\rho_i\,D_j^{JK}\,\Psi^K\,. \tag{3.69}$$

Above, Ψ^I ($I = 1, 2$) are two ten-dimensional Majorana-Weyl spinors with the same chirality, $s^{IJ} = \text{diag}(1, -1)$, ρ_i are the worldsheet projections of the ten-dimensional Dirac matrices

$$\rho_i = E_{AM}\,\partial_i\,X^M\,\Gamma^A \tag{3.70}$$

and D_i^{JK} is the two-dimensional pullback $\partial_i X^M\, D_M^{JK}$ of the ten-dimensional covariant derivative, sum of an ordinary spinor covariant derivative and an additional "Pauli-like" coupling to the Ramond-Ramond flux background, $D_M^{JK} = \mathfrak{D}_M\,\delta^{JK} - \frac{1}{8\cdot 5!}F_{M_1\ldots M_5}\Gamma^{M_1\ldots M_5}\Gamma_M\,\epsilon^{JK}$.

The pullback of D_M^{JK}

$$D_i^{JK} = \mathfrak{D}_i^{JK} + \mathcal{F}_i^{JK} \tag{3.71}$$

can be written as the sum of a "kinetic" part

$$\mathfrak{D}_i^{JK} = \delta^{JK}\,(\partial_i + \Omega_i)\,, \qquad \Omega_i = \frac{1}{4}\,\partial_i X^M \Omega_M^{AB}\Gamma_{AB} = \frac{1}{4}\,\Omega_i^{AB}\Gamma_{AB} \tag{3.72}$$

and of a flux term $\mathcal{F}_i^{JK}$[8]

$$\mathcal{F}_i^{JK} = -\frac{i}{2}\epsilon^{JK}\Gamma_\star\rho_i\,, \qquad \Gamma_\star = i\,\Gamma_{01234}\,. \tag{3.75}$$

This splitting suggests to define (3.69) as the sum of

$$\mathcal{L}_{\text{kin}}^{(2)} = i\left(\sqrt{h}h^{ij}\,\delta^{IJ} - \epsilon^{ij}s^{IJ}\right)\bar{\Psi}^I\,\rho_i\,\mathfrak{D}_j^{JK}\,\Psi^K \tag{3.76}$$

$$\mathcal{L}_{\text{flux}}^{(2)} = i\left(\sqrt{h}h^{ij}\,\delta^{IJ} - \epsilon^{ij}s^{IJ}\right)\bar{\Psi}^I\,\rho_i\,\mathcal{F}_j^{JK}\,\Psi^K\,. \tag{3.77}$$

Looking for a general formalism for fluctuations, there is a natural choice for the κ-symmetry gauge-fixing that is viable in type IIB string action, where both

[8] An alternative form for the flux [25] is

$$\mathcal{F}_i^{JK} = -\frac{i}{2}\epsilon^{JK}\,\tilde{\rho}_i\,, \qquad \tilde{\rho}_i = \Gamma_A\hat{t}_i^A + i\Gamma_A\bar{t}_i^A \tag{3.73}$$

with which the corresponding part of the gauge-fixed Lagrangian reads

$$\mathcal{L}_{\text{flux}}^{(2)} = -\epsilon^{ij}\,\bar{\Psi}\,\rho_i\,\tilde{\rho}_j\,\Psi\,. \tag{3.74}$$

Its equivalence with (3.75) and (3.86) below is manifest in the 5 + 5 basis of [25], see also discussion in [3].

Majorana-Weyl fermions in the Green-Schwarz action have the same chirality, namely[9]

$$\Psi^1 = \Psi^2 \equiv \Psi \,. \tag{3.78}$$

This choice has the advantage of having a trivial gauge-fixing determinant [3]. Since $s^{11} = -s^{22} = 1$, in the kinetic part of the fermionic action only the term proportional to h^{ij} will survive after the κ-symmetry gauge-fixing, while in the flux part only the term proportional to ϵ^{ij}.

3.4.1 The Kinetic Term

Let us first focus on the "kinetic" part of the action (3.76) containing the ordinary spinor covariant derivative

$$\mathcal{L}^{(2)}_{\rm kin} = 2i\sqrt{h}h^{ij}\,\bar{\Psi}\Gamma^A\,e_{Ai}\,(\partial_j + \Omega_j)\,\Psi \,. \tag{3.79}$$

We begin with the observation that the Dirac algebra is decomposed in two subsets: the components along the worldsheet and those orthogonal to the worldsheet, which in the ten-dimensional case means

$$\rho_{\tilde{a}} = t_{A\tilde{a}}\Gamma^A \,, \qquad\qquad \rho_{\underline{i}} = N_{A\underline{i}}\Gamma^A \,. \tag{3.80}$$

As used earlier in [18–22, 33, 34] and made explicit in this context in [3], since a two-dimensional Clifford algebra holds by construction for the $\{\rho_i, \rho_j\} = 2\,h_{ij}$, it is always possible to find a *local* $SO(1,9)$ rotation S that transforms ρ_i into two-dimensional Dirac matrices contracted with zweibein

$$\rho_{\tilde{a}} = S\Gamma^a S^{-1} e_{a\tilde{a}} \,, \qquad\qquad \rho_{\underline{i}} = S\Gamma^a S^{-1} e_{a\underline{i}} \,, \tag{3.81}$$

where

$$\Gamma^1 = i\,\sigma_2 \otimes \mathbb{I}_{16} \,, \qquad \Gamma^2 = \sigma_1 \otimes \mathbb{I}_{16} \,, \qquad \Gamma^a = \sigma_3 \otimes \Sigma^a \,, \qquad a = 3, \ldots 10 \,, \tag{3.82}$$

σ_a are Pauli matrices ($a = 1, 2, 3$) and Σ^a are 16×16 Dirac matrices in 8 Euclidean dimensions. Defining now

[9] A widely used alternative to (3.78) is the light-cone gauge-fixing $\Gamma^+\Psi^I = 0$, where the light-cone might lie entirely in S^5 [27, 28] or being shared between AdS_5 and S^5 [29] (and [30] for further references therein). One of the obvious advantages of the "covariant" gauge-fixing (3.78) is preservation of global bosonic symmetries of the action. A more general choice is $\Psi_1 = k\,\Psi_2 \equiv k\,\Psi$ where k is a real parameter whose dependence is expected to cancel in the effective action, see discussion in [31]. Yet another κ-symmetry gauge-fixing, albeit equivalent to (3.78) [3], has been used for studying stringy fluctuations in $AdS_5 \times S^5$ in [32].

$$\Theta = S^{-1}\Psi \quad \text{and} \quad \hat{\Omega}_i = S^{-1}\Omega_i S + S^{-1}\partial_i S\,, \tag{3.83}$$

one ends up with the following rotated expression

$$\mathcal{L}^{(2)}_{\text{kin}} = 2i\sqrt{h}h^{ij}\,\bar{\Theta}\Gamma^A\,e_{Ai}\,(\partial_j + \hat{\Omega}_j)\,\Theta. \tag{3.84}$$

We remark here that the present analysis is only valid at classical level: the local rotation produces quantum mechanically a non-trivial Jacobian in the path-integral measure [3], whose contribution is crucial to recover the correct structure of the divergent terms in Sect. 3.5. A tedious computation brings the fermion kinetic part of the rotated Lagrangian in the form

$$\mathcal{L}^{(2)}_{\text{kin}} = 2i\,\sqrt{h}h^{ij}\,\bar{\Theta}\,\Gamma^A\,e_{Ai}\left(\partial_j + \tfrac{1}{4}\Gamma_{AB}\omega_j^{AB} - \tfrac{1}{4}A_j^{\underline{ij}}\Gamma_{\underline{ij}}\right)\Theta\,. \tag{3.85}$$

Namely, in the rotated basis (3.81)–(3.82) the Green-Schwarz kinetic operator (3.76) results in a standard two-dimensional Dirac fermion action on a curved two-dimensional background with geometry defined by the induced metric [3]. The spinor covariant derivative can be written as a 2d ordinary spinor covariant derivative plus one additional term that consists in the normal bundle gauge connection (3.30), with respect to which the 16 two-dimensional spinors making up the 32-component Majorana-Weyl spinor Θ transform in the spinor representation of $SO(8)$. This interacting term vanishes (*i.e.* a set of normal vectors exists such that $A_k^{\underline{ij}} = 0$) when the field-strength associated to the normal connection vanishes, see discussion below (3.52). As mentioned there, this is always the case, for example, for embeddings of the string worldsheet in AdS_3, where indeed the normal direction is just one and the normal bundle is then trivial. For a more general embedding extending in both $AdS_5 \times S^5$ subspaces, the presence or not of this interaction term has to be checked case by case.[10]

3.4.2 The Flux Term

We now analyze the flux term (3.77), which after the κ-symmetry gauge-fixing (3.78) reads

$$\mathcal{L}^{(2)}_{\text{flux}} = -\epsilon^{ij}\,\bar{\Psi}\,\rho_i\,\Gamma_*\rho_j\,\Psi\,, \qquad \Gamma_* = i\Gamma_{01234}\,. \tag{3.86}$$

In order to understand the geometrical meaning of the terms in (3.86), we again decompose the gamma matrices as in (3.81)–(3.82) and rotate the spinor as in (3.83). Defining the antisymmetric product of the Dirac matrices projected onto the worldsheet as

[10]We will encounter an example of string background with non-flat connection in (5.29), see the induced fermionic Lagrangian in (5.53) and comments below. This was first noticed in (5.35) of [6].

$$\rho_3 \equiv \frac{1}{2\sqrt{h}} \epsilon^{ij} \rho_{ij} \,, \qquad \rho^{ij} = -\frac{1}{\sqrt{h}} \epsilon^{ij} \rho_3 \,, \tag{3.87}$$

one can rearrange the flux term in the following way

$$\mathcal{L}^{(2)}_{\text{flux}} = \bar{\Theta}\, \tilde{\Gamma}_* \Big[\sqrt{h}\, \sigma_3 \, (m^2_{AdS_5} + m^2_{S^5}) + 2\, \epsilon^{ij} e^A_j \, \Gamma_{AB} \, \nabla_k K^{Bk}_i \Big] \Theta \,, \quad B = 8, 9 \,, \tag{3.88}$$

in terms of the bosonic masses (3.59), $\sigma_3 = S^{-1} \rho_3 \, S$, and $\tilde{\Gamma}_* = S^{-1} \Gamma_* S$. In general, the rotation of Γ_* is written as

$$\tilde{\Gamma}_* = i \hat{\epsilon}^{ABCDE} (\hat{t}^{\tilde{a}}_A \Gamma_{\tilde{a}} + \hat{N}^{\underline{i}}_A \Gamma_{\underline{i}} + N^{\underline{r}}_A \Gamma_{\underline{r}}) ... (\hat{t}^{\tilde{e}}_E \Gamma_{\tilde{e}} + \hat{N}^{\underline{j}}_E \Gamma_{\underline{j}} + N^{\underline{s}}_E \Gamma_{\underline{s}}) \,. \tag{3.89}$$

Here the hat in $\hat{\epsilon}^{ABCDE}$ is to signal that A, B, C, D, E take values $0, 1, \ldots 4$, as clear from (3.86). We also split $\underline{i}, \underline{j} = 2, \ldots 7$ and $\underline{r}, \underline{s} = 8, 9$. A clever choice among the basis vectors, made possible by the string motion, drastically simplifies the expression for $\hat{\Gamma}_*$.

3.4.3 *Mass Matrix and Sum Rule*

In analogy to the bosonic mass rule (3.60), now we derive the one for the fermion "mass matrix". The total fermionic action is the sum of (3.79) and (3.86)

$$\mathcal{L}^{(2)}_{\text{kin}} + \mathcal{L}^{(2)}_{\text{flux}} \equiv 2\sqrt{h} \bar{\Psi} \mathcal{O}_F \Psi \,. \tag{3.90}$$

For the self-adjoint Dirac operator $\mathcal{O}_F$ we can define the "mass matrix" $\mathcal{M}^{(F)}$ by posing

$$\mathcal{O}_F \equiv i \rho^i \mathcal{D}_i - \mathcal{M}^{(F)} \,, \qquad \mathcal{M}^{(F)} = \frac{1}{2} \sqrt{h} \epsilon^{ij} \rho_i \Gamma_* \rho_j \,. \tag{3.91}$$

When ρ_i commutes with $\mathcal{M}^{(F)}$ the invariant $\text{tr}\left(\rho^i \mathcal{M}^{(F)} \rho_i \mathcal{M}^{(F)}\right)$ in (3.109) reduces to the more familiar $\text{tr}\left((M^{(F)})^2\right)$, analysed in [3] or in [24] for example. However, for a general string solution, like the classical configuration in (Sect. 5.4), the two matrices do not commute, leading to the sum rule explained below. Notice that, due to the cyclicity of the trace, we are computing the relevant trace before performing any local $SO(1, 9)$ rotations, which turns out a convenient strategy in this case. Let us begin with splitting the trace in threes summands

$$\mathrm{tr}\left(\rho^i \mathcal{M}^{(F)} \rho_i \mathcal{M}^{(F)}\right) = \mathrm{tr}\left(\frac{h^{ij}}{4h}\epsilon^{kl}\epsilon^{mn}\rho_i\rho_k\Gamma_*\rho_l\rho_j\rho_m\Gamma_*\rho_n\right) \tag{3.92}$$

$$= \mathrm{tr}\left(h^{ij}4h\epsilon^{kl}\epsilon^{mn}\rho_i\rho_k\Gamma_*\left(h_{lj}\rho_m + h_{jm}\rho_l - h_{lm}\rho_j\right)\Gamma_*\rho_n\right)$$
$$\equiv I_1 + I_2 + I_3 \tag{3.93}$$

and compute the three terms above separately:

$$I_1 \equiv \frac{h^{ij}}{4h}\epsilon^{kl}\epsilon^{mn}h_{lj}\mathrm{tr}\left(\rho_i\rho_k\Gamma_*\rho_m\Gamma_*\rho_n\right) = -\frac{1}{2}\left(m^2_{AdS_5} + m^2_{S^5}\right)\mathrm{tr}\left(\rho_3^2\right), \tag{3.94}$$

$$I_2 = \frac{h^{ij}}{4h}\epsilon^{kl}\epsilon^{nm}h_{jn}\mathrm{tr}\left(\rho_i\rho_k\Gamma_*\rho_l\Gamma_*\rho_m\right) = -\frac{1}{2}\left(m^2_{AdS_5} + m^2_{S^5}\right)\mathrm{tr}(\mathbb{I}), \tag{3.95}$$

$$I_3 = -\frac{h^{ij}}{4h}\epsilon^{kl}\epsilon^{nm}h_{ln}\mathrm{tr}\left(\rho_i\rho_k\Gamma_*\rho_j\Gamma_*\rho_m\right) = -\frac{1}{4}\left(m^2_{AdS_5} + m^2_{S^5}\right)\mathrm{tr}(\mathbb{I} - \rho_3^2). \tag{3.96}$$

Recalling that $\rho_3 = \frac{1}{2\sqrt{h}}\epsilon^{ij}\rho_{ij}$ squares to the identity matrix, we obtain the fermionic analogue of (3.60)

$$\mathrm{tr}\left(\rho^i \mathcal{M}^{(F)} \rho_i \mathcal{M}^{(F)}\right) = -\left(m^2_{AdS_5} + m^2_{S^5}\right)\mathrm{tr}(\mathbb{I}) = \left({}^{(2)}R + \mathrm{tr}(K^2)\right)\mathrm{tr}(\mathbb{I}) \tag{3.97}$$

by means of the expressions (3.59).

3.5 Quantum Divergences

Combining the gauge-fixed Lagrangians (3.47) and (3.90), we arrive at the expressions

$$\mathcal{L}^{(2)}_{\mathrm{transv}} \equiv \sqrt{h}\, y_{\underline{i}}(\mathcal{O}_B)^{\underline{i}}{}_{\underline{j}}\, y^{\underline{j}}, \qquad \mathcal{L}^{(2)}_{\mathrm{kin}} + \mathcal{L}^{(2)}_{\mathrm{flux}} \equiv 2\sqrt{h}\bar{\Psi}\mathcal{O}_F\Psi\,. \tag{3.98}$$

We stripped off a factor $\sqrt{h}$ in the definition of the 2d operators $\mathcal{O}$'s to make their eigenspectra independent of the parametrization of the worldsheet. In a semiclassical approximation of the path-integral, the leading quantum corrections to a classical solution are given by the ratio of determinants

$$Z_{\mathrm{string}} \approx e^{-S_B^{(0)}}\, Z^{(1)}, \qquad Z^{(1)} \equiv \frac{\mathrm{Det}^{1/2}\mathcal{O}_F}{\mathrm{Det}^{1/2}\mathcal{O}_B}\,. \tag{3.99}$$

In the formula the ghost determinant of (3.41) in the numerator cancels against the one for the longitudinal modes (3.46) in the denominator, implicitly assuming that they are evaluated under the same boundary conditions (see comments below (3.46)).

3.5.1 Regularization of the Classical Action

A first trouble with (3.99) is that the exponential would lead to a divergent result when the string stretches to the AdS boundary. In fact, the minimal surface has an infinite extension due to the near-boundary behaviour ($z \to 0$) of the AdS metric (3.4). The blow-up of the metric can be screened by a translation of the string boundary slightly into the bulk of the AdS space: if we cut off the part of the string submanifold with radial Poincaré coordinate z smaller than a regulator ϵ, the classical action behaves as

$$S_B^{(0)} = \frac{\ell}{\epsilon} + o\left(\epsilon^{-1}\right) \tag{3.100}$$

where the coefficient of the pole is the length of the worldsheet boundary.

Different ways to regularize the classical area have been proposed: a Legendre transform [35, 36] or a subtraction of the "boundary part" of the Euler characteristic of the surface [3]. They all result into a sort of "minimal subtraction scheme" that removes only the $1/\epsilon$ pole [37] and defines the *regularized area* to be plugged into (3.99) as the finite term in the expansion (3.100)

$$S_B^{(0)} \quad \to \quad \lim_{\epsilon \to 0} \left(S_B^{(0)} - \frac{\ell}{\epsilon} \right) . \tag{3.101}$$

The classical area (3.100) needs further renormalization when the loop at the AdS boundary has cusp points. A subleading term proportional to $\log \epsilon$ appears in (3.100) and it is not cured by the subtraction rule (3.101). In fact, in the holographic interpretation (1.11) logarithmic divergences correspond to physical UV divergences of the dual Wilson loop [35].

3.5.2 One-Loop Divergences

The definition of determinant of differential operators is not unique in the literature. In Appendix B we illustrated techniques based on *zeta-function regularization* [38, 39] widely used in many area of mathematical physics [40]. Although convenient to extract the renormalized value of (3.99), this regularization would only capture the logarithmic divergence of the determinant.[11]

In the spirit of the seminal paper [3], we prefer a better way to estimate the divergences of a determinant. The scheme that we shall use here is known as *proper-time cutoff (PTC) regularization* [42] because it restricts the "time" t in (1.15) at some small parameter Λ^{-2}, where Λ is a large mass scale. In fact, the infinite part

[11] This corresponds to the a_2 coefficient in (3.102), compare in [41] the zeta-function regularization (5.51) with the proper-time cutoff scheme (5.74).

of the determinant of a 2d operator $\mathcal{O}$—which we assume free of zero modes and negative eigenvalues—arises only from the small-t region of integration in

$$(\log \mathrm{Det}\, \mathcal{O})_{\mathrm{PTC},\,\Lambda} \equiv -\int_{\Lambda^{-2}}^{\infty} \frac{dt}{t} K_{\mathcal{O}}(t) \tag{3.102}$$
$$= -a_0\Lambda^2 - 2a_1\Lambda - a_2 \log(\Lambda^2) + O(\Lambda^0)\,.$$

and can be estimated from the first *Seeley-DeWitt (SDW) coefficients* a_i, namely the expansion coefficients of the functional trace of the heat kernel in (1.15) for small t

$$K_{\mathcal{O}}(t) \equiv \int d\tau d\sigma \sqrt{h}\, \mathrm{tr}\, K_{\mathcal{O}}(\tau,\sigma;\tau,\sigma;t) \overset{t\to 0^+}{=} \sum_{k=0}^{\infty} t^{(k-2)/2}\, a_k\,. \tag{3.103}$$

This is the complete structure of divergences for a second-order operator $\mathcal{O}$ on a two-dimensional manifold, here the worldsheet Σ. A method of evaluating the heat trace asymptotics was proposed by DeWitt [43] and generalized by Gilkey [44]. Here we will borrow the SDW coefficients from Chap. 4 of [45].

To estimate the divergences of the one-loop corrections to the partition function (3.104), we begin by rewriting (3.99)

$$Z_{\mathrm{string}} \approx Z^{(1)}\, e^{-S_B^{(0)}}\,, \qquad Z^{(1)} = \frac{\mathrm{Det}^{1/4}\mathcal{O}_F^2}{\mathrm{Det}^{1/2}\mathcal{O}_B}\,. \tag{3.104}$$

in terms of the square of the Dirac operator, which is a second-order operator admitting a well-defined heat kernel. We then expand the bosonic and fermionic heat traces

$$\left(\log Z^{(1)}\right)_{\mathrm{PTC},\,\Lambda} = -\int_{\Lambda^{-2}}^{\infty} \frac{dt}{t}\left(\frac{1}{4}K_{\mathcal{O}_F^2}(t) - \frac{1}{2}K_{\mathcal{O}_B}(t)\right) \tag{3.105}$$
$$= -\left(\frac{1}{4}a_0^{(F)} - \frac{1}{2}a_0^{(B)}\right)\Lambda^2 - 2\left(\frac{1}{4}a_1^{(F)} - \frac{1}{2}a_1^{(B)}\right)\Lambda$$
$$-\left(\frac{1}{4}a_2^{(F)} - \frac{1}{2}a_2^{(B)}\right)\log(\Lambda^2) + O(\Lambda^0)\,,$$

where the subscript "B" refers to the Laplace-type operator $\mathcal{O}_B$ and "F" to the Dirac-type operator squared $\mathcal{O}_F^2$. For simplicity we study only the closed string case. Odd-numbered SDW coefficients a_{2n+1} vanish on boundaryless manifolds, hence the linear divergences in (3.102) are set to zero, and we will neglect boundary integrals over $\partial\Sigma$ in the even-numbered ones a_{2n}.

The quadratic divergence is proportional to the worldsheet area

$$a_0^{(B)} = 8\left(\frac{1}{4\pi}\int d\tau d\sigma\,\sqrt{h}\right), \tag{3.106}$$

$$a_0^{(F)} = 16\left(\frac{1}{4\pi}\int d\tau d\sigma\,\sqrt{h}\right) \tag{3.107}$$

and trivially cancels in (3.105) because of the 8 bosonic fluctuations and the 16 fermionic degrees of freedom of the 10d Majorana-Weyl spinor Ψ distributed in the eight 2d spinors (see beginning of Sect. 3.4.1).

The counting of logarithmic divergences is lengthier because they depend on the trace of the mass matrices[12]

$$a_2^{(B)} = \frac{1}{4\pi}\int d\tau d\sigma\,\sqrt{h}\,\mathrm{tr}\left(\frac{1}{6}{}^{(2)}R\,\mathbb{I}_8 + \mathcal{M}_B\right), \tag{3.108}$$

$$a_2^{(F)} = \frac{1}{4\pi}\int d\tau d\sigma\,\sqrt{h}\,\mathrm{tr}\left(-\frac{1}{12}{}^{(2)}R\,\mathbb{I}_{16} + \frac{1}{2}\rho^i\mathcal{M}^{(F)}\rho_i\mathcal{M}^{(F)}\right), \tag{3.109}$$

with the mass matrices given in (3.59) and (3.91). We have also used the fact that $\mathcal{O}_F$ is represented by 16×16 matrices instead of the initial 32×32 Dirac matrices, since it acts upon a Weyl spinor Ψ with definite 10d chirality. The Majorana condition on Ψ has been already taken into account in the power of $1/2$ of the fermionic determinant in (3.99).

Armed with (3.97) we can compare the bosonic and fermionic contribution to the total logarithmic divergences in (3.105) by computing explicitly $a_2^{(B)} - \frac{1}{2}a_2^{(F)}$. The part depending on the traced mass matrices is

$$\mathrm{tr}\left(\mathcal{M}_B - \frac{1}{4}\rho^i\mathcal{M}^{(F)}\rho_i\mathcal{M}^{(F)}\right) = -\,{}^{(2)}R\,, \tag{3.110}$$

while the remaining one is instead more subtle [3]. The naive use of the fermionic SDW coefficient would lead to

$$8\times\left(\frac{1}{6}{}^{(2)}R\right) - \frac{16}{2}\times\left(-\frac{1}{12}{}^{(2)}R\right) = 2\,{}^{(2)}R, \tag{3.111}$$

that combined with (3.110) would *not* produce the expected coefficient $3^{(2)}R$ [3]:

$$a_2^{(B)} - \frac{1}{2}a_2^{(F)} = \frac{1}{4\pi}\int d\tau d\sigma\,\sqrt{h}(-{}^{(2)}R + 2\,{}^{(2)}R) = \frac{1}{4\pi}\int d\tau d\sigma\,\sqrt{h}\,{}^{(2)}R\,. \tag{3.112}$$

The reason of the apparent disagreement is well known in 10d flat space and curved space [3, 20–22, 33, 34] and it is due to a difference between the fermi-

[12] At variance with the notation in [6], from $a_2^{(F)}$ we stripped off a minus sign and the factor of $1/2$ of the Majorana condition. We also included the integration in the SDW coefficients as it is common in literature.

onic kinetic term of the initial Green-Schwarz fermions (3.76) and the one of the decomposed standard 2d Dirac fermions (3.85). In fact, the local $SO(1,9)$ rotation S that transforms ρ_i into two-dimensional Dirac matrices contracted with zweibein gives rise to a non-trivial Jacobian in the path-integral measure, that contributes additionally to the logarithmic divergence. The effect is to change the coefficient of the relevant two-dimensional Dirac fermions by a factor 4. In our case it amounts to modify the fermionic contribution to (3.111) by a factor 4 and therefore

$$8 \times \left(\frac{1}{6}\,{}^{(2)}R\right) - 4 \times \frac{16}{2} \times \left(-\frac{1}{12}\,{}^{(2)}R\right) = 4\,{}^{(2)}R, \tag{3.113}$$

recovering, in combination with (3.110), the result of [3]

$$a_2^{(B)} - \frac{1}{2}a_2^{(F)} = \frac{1}{4\pi}\int d\tau d\sigma\,\sqrt{h}(-{}^{(2)}R + 4\,{}^{(2)}R) = \frac{1}{4\pi}\int d\tau d\sigma\,\sqrt{h}\,(3\,{}^{(2)}R)\,. \tag{3.114}$$

This is the contribution of the fluctuation determinants on a closed string to the one-loop conformal anomaly. The divergence is proportional to the surface integral of the scalar curvature, which equals $3\,\chi$, where χ is the Euler number of the boundaryless surface. In the open-string case, all the factors of ${}^{(2)}R$ are completed by appropriate boundary terms [46–48] to produce again a divergence proportional to the Euler number. In the argument of [3] for the cancellation of the total one-loop anomaly, it is argued that some factors in the path-integral measure—associated with conformal Killing vectors and/or Teichmüller moduli—would have the net effect of cancelling this divergence.

References

1. A.M. Polyakov, Quantum geometry of bosonic strings. Phys. Lett. B **103**, 207 (1981)
2. C.G. Callan Jr., L. Thorlacius, Sigma models and string theory. Proceedings, Particles, strings and supernovae **2**, 795–878 (1988). (in Providence)
3. N. Drukker, D.J. Gross, A.A. Tseytlin, Green-Schwarz string in $AdS_5 \times S^5$: semiclassical partition function. JHEP **0004**, 021 (2000). arXiv:hep-th/0001204
4. R. Capovilla, J. Guven, Geometry of deformations of relativistic membranes. Phys. Rev. D **51**, 6736 (1995). arXiv:gr-qc/9411060
5. K. Viswanathan, R. Parthasarathy, String theory in curved space-time. Phys. Rev. D **55**, 3800 (1997). arXiv:hep-th/9605007
6. V. Forini, V.G.M. Puletti, L. Griguolo, D. Seminara, E. Vescovi, Remarks on the geometrical properties of semiclassically quantized strings. J. Phys. **A48**, 475401 (2015). arXiv:1507.01883
7. A. Faraggi, W. Mueck, L.A. Pando, Zayas, one-loop effective action of the holographic antisymmetric wilson loop. Phys. Rev. D **85**, 106015 (2012). arXiv:1112.5028
8. O. Aharony, S.S. Gubser, J.M. Maldacena, H. Ooguri, Y. Oz, Large N field theories, string theory and gravity. Phys. Rept. **323**, 183 (2000). arXiv:hep-th/9905111
9. M. Schottenloher, *A Mathematical Introduction to Conformal Field Theory* (Springer, Berlin Heidelberg, 2008)

10. A.A. Tseytlin, Review of AdS/CFT integrability, chap. II.1: classical $AdS_5 \times S^5$ string solutions. Lett. Math. Phys. **99**, 103 (2012). arXiv:1012.3986
11. J.L. Lagrange, Essai d'une nouvelle methode pour determiner les maxima: et les minima des formules integrales indefinies. Miscellanea Taurinensia II **173** (1760–1761)
12. N. Drukker, S. Giombi, R. Ricci, D. Trancanelli, Supersymmetric Wilson loops on S^3. JHEP **0805**, 017 (2008). arXiv:0711.3226
13. H. Dorn, Wilson loops at strong coupling for curved contours with cusps. J. Phys. **A49**, 145402 (2016). arXiv:1509.00222
14. L. Alvarez-Gaume, D.Z. Freedman, S. Mukhi, The background field method and the ultraviolet structure of the supersymmetric nonlinear sigma model. Ann. Phys. **134**, 85 (1981)
15. S. Kobayashi, K. Nomizu, *Foundations of Differential Geometry*, vol. II (Interscience, New York, 1963)
16. M. Nakahara, *Geometry, Topology and Physics* (CRC Press, 2003)
17. J. Polchinski, *String Theory. Vol. 1: An Introduction to the Bosonic String* (Cambridge University Press, Cambridge, 2007)
18. A.R. Kavalov, I.K. Kostov, A.G. Sedrakian, Dirac and weyl fermion dynamics on two-dimensional surface. Phys. Lett. B **175**, 331 (1986)
19. A. Sedrakian, R. Stora, Dirac and weyl fermions coupled to two-dimensional surfaces: determinants. Phys. Lett. B **188**, 442 (1987)
20. F. Langouche, H. Leutwyler, Two-dimensional fermion determinants as Wess-Zumino actions. Phys. Lett. B **195**, 56 (1987)
21. F. Langouche, H. Leutwyler, Weyl fermions on strings embedded in three-dimensions. Phys. C **36**, 473 (1987)
22. F. Langouche, H. Leutwyler, Anomalies generated by extrinsic curvature. Phys. C **36**, 479 (1987)
23. D.R. Karakhanian, Induced Dirac operator and smooth manifold geometry
24. N. Drukker, V. Forini, Generalized quark-antiquark potential at weak and strong coupling. JHEP **1106**, 131 (2011). arXiv:1105.5144
25. R. Metsaev, A.A. Tseytlin, Type IIB superstring action in $AdS_5 \times S^5$ background. Nucl. Phys. B **533**, 109 (1998). arXiv:hep-th/9805028
26. R. Kallosh, J. Rahmfeld, A. Rajaraman, Near horizon superspace. JHEP **9809**, 002 (1998). arXiv:hep-th/9805217
27. R. Metsaev, A.A. Tseytlin, Superstring action in $AdS_5 \times S^5$. Kappa symmetry light cone gauge. Phys. Rev. D **63**, 046002 (2001). arXiv:hep-th/0007036
28. R. Metsaev, C.B. Thorn, A.A. Tseytlin, Light cone superstring in AdS space-time. Nucl. Phys. B **596**, 151 (2001). arXiv:hep-th/0009171
29. J. Callan, G. Curtis, H.K. Lee, T. McLoughlin, J.H. Schwarz, I. Swanson et al., Quantizing string theory in $AdS_5 \times S^5$: beyond the pp wave. Nucl. Phys. B **673**, 3 (2003). arXiv:hep-th/0307032
30. G. Arutyunov, S. Frolov, Foundations of the $AdS_5 \times S^5$ superstring. Part I J. Phys. A **42**, 254003 (2009). arXiv:0901.4937
31. R. Roiban, A. Tirziu, A.A. Tseytlin, Two-loop world-sheet corrections in $AdS_5 \times S^5$ superstring. JHEP **0707**, 056 (2007). arXiv:0704.3638
32. S. Forste, D. Ghoshal, S. Theisen, Stringy corrections to the Wilson loop in $\mathcal{N} = 4$ superYang-Mills theory. JHEP **9908**, 013 (1999). arXiv:hep-th/9903042
33. P.B. Wiegmann, Extrinsic geometry of superstrings. Nucl. Phys. B **323**, 330 (1989)
34. K. Lechner, M. Tonin, The cancellation of world sheet anomalies in the $D = 10$ Green-Schwarz heterotic string sigma model. Nucl. Phys. B **475**, 535 (1996). arXiv:hep-th/9603093
35. N. Drukker, D.J. Gross, H. Ooguri, Wilson loops and minimal surfaces. Phys. Rev. D **60**, 125006 (1999). arXiv:hep-th/9904191
36. N. Drukker, B. Fiol, All-genus calculation of Wilson loops using D-branes. JHEP **0502**, 010 (2005). arXiv:hep-th/0501109
37. J.M. Maldacena, Wilson loops in large N field theories. Phys. Rev. Lett. **80**, 4859 (1998). arXiv:hep-th/9803002

38. J.S. Dowker, R. Critchley, Effective lagrangian and energy momentum tensor in de sitter space. Phys. Rev. D **13**, 3224 (1976)
39. S.W. Hawking, Zeta function regularization of path integrals in curved space-time. Commun. Math. Phys. **55**, 133 (1977)
40. E. Elizalde, *Ten Physical Applications of Spectral Zeta Functions* (Springer, New York, 2012)
41. D. Fursaev, D. Vassilevich, *Operators, Geometry and Quanta: Methods of Spectral Geometry in Quantum Field Theory* (Springer, New York, 2011)
42. J. Schwinger, Casimir effect in source theory II. Lett. Math. Phys. **24**, 59 (1992)
43. B. DeWitt, *Dynamical Theory of Groups and Fields* (Gordon and Breach, 1965)
44. P.B. Gilkey, The spectral geometry of a riemannian manifold. J. Differ. Geom. **10**, 601 (1975)
45. P.B. Gilkey, *Invariance Theory, the Heat Equation and the Atiyah-Singer Index Theorem* (CRC Press, Boca Raton, 1995)
46. D. Friedan, Introduction to Polyakov's string theory, in *Les Houches Summer School in Theoretical Physics: Recent Advances in Field Theory and Statistical Mechanics Les Houches*, August 2-September 10 (France, 1982)
47. O. Alvarez, Theory of strings with boundaries: fluctuations, topology, and quantum geometry. Nucl. Phys. B **216**, 125 (1983)
48. H. Luckock, Quantum geometry of strings with boundaries. Ann. Phys. **194**, 113 (1989)

Chapter 4
"Exact" Semiclassical Quantization of Folded Spinning Strings

In the last chapter we determined the structure of the differential operators governing the semiclassical excitations around an arbitrary classical solution of the $AdS_5 \times S^5$ sigma-model. While the expressions for the masses of the worldsheet fields arise from the solution and the background in a relatively plain way, the one-loop partition function (3.98) is a formal expression subordinated to the evaluation of functional determinants, which is a task in general non-trivial. For the class of inhomogeneous solutions defined in Sect. 1.3, the complicated dependence of the differential operators on the worldsheet coordinates renders the inversion of such operators (hence the free Feynman propagators) very complicated and higher-loop calculations of Feynman diagrams around such non-trivial vacua virtually impossible. Among the large class of inhomogeneous configurations, only a restricted set of operators admits a spectral problem solvable in closed form.

This applies to certain solitonic solutions of the string sigma-model, in which folded string solutions with more than one spin, or of single-spin solutions [1, 2] in conformal gauge where bosonic fluctuations couple via the Virasoro constraints, the evaluation of the classical energy requires the diagonalization of non-trivial matrix-valued differential operators of the second order whose coefficients have a complicated coordinate-dependence. The same is true, for instance, for fluctuations over open string solutions dual to cusped Wilson loops that have an expectation value depending on the cusp angle and an internal R-symmetry angle [3, 4]. In general, the solution of the eigenspectrum is simplified once one or more of the parameters of the solution is set to zero (in the cases above) or a limit on them is taken in order to make the solution homogeneous (e.g. [2] as the first of many examples).

The main perspective of this chapter is to enrich the class of problems that can be solved analytically [3–6] with the exact solution of a fourth-order linear differential equations with doubly periodic coefficients that emerges as a natural generalization of the *Lamé differential equation*. Our analysis (Appendix C) applies to two non-trivial solitonic configurations of the $AdS_5 \times S^5$ string sigma-model.

E. Vescovi, *Perturbative and Non-perturbative Approaches to String Sigma-Models in AdS/CFT*, Springer Theses, DOI 10.1007/978-3-319-63420-3_4

The first example is the folded string positioned at the center of AdS_5 and rotating in S^5 with two large angular momenta (J_1, J_2) in the limit where it becomes a solution of the *Landau–Lifshitz (LL) effective action* of [7] (see also detailed review in [8, 9]). These string states are dual to the simplest non-trivial set of single-trace operators of $\mathcal{N} = 4$ SYM, i.e. the $SU(2)$ flavour sector of operators $\mathrm{tr}(X^{J_1} Z^{J_2})$ composed of only two of the three complex scalars of $\mathcal{N} = 4$ SYM and carrying two independent R-charges J_1 and J_2. The interest in these operators has roots in the discovery of [10] (see also Sect. 1.2) that the one-loop anomalous dimension of operators in the larger $SO(6)$ sector follows from solving the spectrum of an integrable $SO(6)$ spin-chain. A deeper understanding of how gauge and string theory picture are intertwined is possible in the $SU(2)$ sector. Here at weak-coupling one-loop level the dilatation operator takes the form of the Hamiltonian of the ferromagnetic Heisenberg model [11] after one identifies the Z's and X's scalars with the orientations "up" and "down" of the chain states.[1] The anomalous dimension of these operators follows from the diagonalization of the spin-chain integrable Hamiltonian with Bethe ansatz techniques in the thermodynamic limit of large number of spins $J \equiv J_1 + J_2 \gg 1$ [10, 14, 15] and agrees with the energy correction [16, 17] of a rotating string with the same quantum numbers in $AdS_5 \times S^5$.

The identification between gauge-theory operators and string states was made more complete in [7] and further clarified and developed in [8, 18, 19]. Here it was shown that the same effective LL action describes both the spin-chain and the string action in limit of $J \gg 1$ (large number of sites and fast rotating strings) with $\tilde{\lambda} \equiv \lambda/J^2$ fixed. The approximation singles out low-energy modes comprising spins that tend to precess over large distances (*long-wave limit*) and are well approximated by semiclassical coherent states, i.e. superpositions of Bethe ansatz eigenstates with approximately the same energy. The same limit in string theory is equivalent to a semiclassical expansion in the "inverse string tension" J^{-1} followed by an expansion of the classical action in small $\tilde{\lambda}$. A brief derivation of the LL effective action on the string theory side is in Sect. 4.1.1.

The agreement between the effective actions from the discrete Heisenberg Hamiltonian and the Polyakov action, as seen order by order in small $\tilde{\lambda}$, implies the matching of scaling dimensions and energies in the limit stated above. Calculations for the spectrum of the LL model linearized around the folded $SU(2)$ string solution have been made in [20] using operator methods via perturbative evaluation (in the parameter $J_2/J = J_2/(J_1 + J_2)$) of characteristic frequencies, with a sum over them cured via zeta-function regularization.

In Sect. 4.2.1 we will apply the tools developed by us in Appendix C to evaluate the exact one-loop effective action over the same solution, regularized by referring the determinants to the limit $J_2 = 0$, which ensures the expected vanishing of the partition function, and proceeding with a zeta-function-inspired regularization of the path-integral. Indeed, as any effective theory, the LL model does not lead naturally to a well-defined quantum theory. From the string viewpoint, it lacks fermionic degrees of freedom and it only includes the bosonic fluctuations that are "outside" the given

[1]See also [12, 13] for transparent and concise introductions to the subject.

$SU(2)$ sector, therefore it must be equipped with an appropriate regularization to ensure UV finiteness for the energy. The "semiclassically exact" analysis explained below provides an efficient and elegant tool to find in one step the needed spectral information. The result obtained in [20] using perturbation theory up to the n-th order means here an n-th order Taylor expansion.

The second example that we consider is the folded spinning string with non-vanishing AdS_3 spin and S^1 orbital momentum studied in [2], dual to the scalar sector of operators $\mathrm{tr}(D_+^S Z^J)$ made of an arbitrary number J of complex scalars and S of light-cone derivatives D_+. This configuration (Sect. 4.1.2) interpolates between the one-spin case $S = 0$ of the *Berenstein-Maldacena-Nastase (BMN) vacuum* [21] and the one at $J = 0$ of the *Gubser-Klebanov-Polyakov (GKP) string* [1].

The arbitrary two-spin configuration is a stationary soliton problem for which the classical equations of motion consist in a one-dimensional sinh-Gordon equation. The case of non-zero S does not correspond to the BMN operator with protected conformal dimension and the string energy is shifted from its classical value by a non-trivial function of the quantum numbers S and J. The problem of computing the one-loop quantum corrections in the string sigma-model was addressed in [2]. The formal expression of the energy in terms of determinants of the fluctuations operators with elliptic-function potentials was computed explicitly only in some regimes of the S and J parameters when the solution simplifies drastically. In a static gauge where fluctuations along the worldsheet directions are set to zero, fluctuations turn out to be governed by differential operators of Lamé type [22]. In this paper, they were solved exactly in an integral expression for an arbitrary value of the spin S in the non-boosted string $J = 0$.

In Sect. 4.2.2 we will see that the same techniques of Appendix C allow to solve the mixed-modes bosonic sector of fluctuations of the two-spin string. However, this mixing has its supersymmetric counterpart in a non-trivial fermionic mass matrix (as noticed already in Appendix D of [4]). The differential equations governing the fermionic spectrum do not unfortunately satisfy the conditions which allowed us to diagonalize the bosonic system, thus preventing a solution to the full quantum problem.

However, the tools in Appendix C still allow an analytic proof of equivalence between the full (including fermions) exact one-loop partition function for the one-spin ($J = 0$) folded string in conformal and static gauge,[2] which is a non-trivial statement verified only numerically in [22].

4.1 Fluctuation Spectrum for the Folded Strings

We summarize the main relations for the Landau–Lifshitz sigma-model in Sect. 4.1.1 and for the classical string with non-vanishing AdS_3-spin S and S^1-momentum J in Sect. 4.1.2. We shall conventionally refer to the former with the subscript "LL"

[2]Even in the single-spin case, bosonic fluctuations are still coupled in conformal gauge, while fermionic ones are decoupled.

and to the latter with "folded". We consider the bosonic fluctuations over these two classical string configurations and present the coupled systems of fluctuations that we will able to diagonalize in Sect. 4.2, using the tools presented in Appendix C.

4.1.1 Landau-Lifshitz Effective Action for the (J_1, J_2)-String

The starting point is the LL effective action for the $SU(2)$ sector [7, 19], which is obtained considering a string state whose motion with two large spins is restricted to the S^3 part of S^5. The collective "fast" coordinate β associated to the total angular momentum is gauged away and only transverse "slow" coordinates remain to describe the low-energy string motion. This is practically implemented by parametrizing the three-sphere coordinates as $X_1 + iX_2 = U_1 e^{i\beta}$, $X_3 + iX_4 = U_2 e^{i\beta}$, $U_a U_a^\star = 1$, fixing the gauge $t = \tau$, $p_\beta = \text{const} = J$, and rescaling the t coordinate via $\tilde{\lambda} = \lambda / J^2$, which plays the role of an effective parameter. To first order in the $\tilde{\lambda}$ expansion [20] one gets

$$S_{\text{LL}} = J \int d\tau \int_0^{2\pi} \frac{d\sigma}{2\pi} \mathcal{L}, \qquad \mathcal{L} = -iU_a^\star \partial_\tau U_a - \frac{\tilde{\lambda}}{2} |D_\sigma U_a|^2 + O(\tilde{\lambda}^2), \tag{4.1}$$
$$DU_a = dU_a - iCU_a, \qquad DU_a^\star = dU_a^\star + iCU_a^\star, \qquad C = -iU_a^\star dU_a .$$

We will consider a two-spin folded closed string positioned at the center of AdS, at fixed angle in S^5 and rotating along two orthogonal planes within a $S^3 \subset S^5$ with arbitrary frequencies w_1, w_2. The non-vanishing part of the background metric in $R_t \times S^3$ (t is the time direction of AdS_5)

$$ds^2 = -dt^2 + d\psi^2 + \cos^2\psi \, d\varphi_1^2 + \sin^2\psi \, d\varphi_2^2, \tag{4.2}$$

where one parametrizes the S^5 metric in terms of three complex coordinates X_a [19]

$$ds^2 = -dt^2 + dX_a dX_a^\star, \qquad X_a X_a^\star = 1, \qquad X_a = e^{i\beta} U_a, \qquad a = 1, 2, 3, \tag{4.3}$$
$$U_1 = \cos\psi \, e^{i\varphi}, \qquad U_2 = \sin\psi \, e^{-i\varphi}, \qquad \varphi = \frac{\varphi_1 - \varphi_2}{2}, \qquad \beta = \frac{\varphi_1 + \varphi_2}{2}. \tag{4.4}$$

Hence, the initial Lagrangian (4.1) becomes [20]

$$\mathcal{L} = \cos 2\psi \, \dot{\varphi} - \frac{\tilde{\lambda}}{2} \left(\psi'^2 + \sin^2 2\psi \varphi'^2 \right). \tag{4.5}$$

The equations of motion take the form of a one-dimensional sine-Gordon equation

$$\psi'' + 2\mathrm{w} \sin 2\psi = 0\,, \qquad \varphi = -w\,t\,, \qquad w = \frac{w_2 - w_1}{2} > 0\,, \qquad \mathrm{w} = \frac{w}{\tilde{\lambda}}\,, \tag{4.6}$$
$$\psi'^2 = 2\mathrm{w}\,(\cos 2\psi - \cos 2\psi_0)\,,$$

whose solution is 2π-periodic in σ and written in terms of Jacobi elliptic functions [20]

$$\sin\psi_{\rm cl}(\sigma) = k\,\mathrm{sn}(C\sigma, k^2), \qquad \cos\psi_{\rm cl}(\sigma) = \mathrm{dn}(C\sigma, k^2), \qquad k^2 = \sin^2\psi_0\,, \tag{4.7}$$
$$\sqrt{\mathrm{w}} = \frac{1}{\pi}\mathbb{K}(k^2), \qquad C = \frac{2}{\pi}\mathbb{K}(k^2) = 2\sqrt{\mathrm{w}}\,, \qquad \frac{\mathbb{E}(k^2)}{\mathbb{K}(k^2)} = 1 - \frac{J_2}{J}, \qquad 0 \le k < 1.$$

The two non-zero spins are the integrals of motion

$$J_1 \equiv w_1\sqrt{\lambda}\int_0^{2\pi}\frac{d\sigma}{2\pi}\cos^2\psi\,, \qquad J_2 \equiv w_2\sqrt{\lambda}\int_0^{2\pi}\frac{d\sigma}{2\pi}\sin^2\psi\,, \qquad \frac{J_1}{w_1} + \frac{J_2}{w_2} = \sqrt{\lambda}\,. \tag{4.8}$$

The next step is expanding the Lagrangian (4.5) around the classical solution (4.7) in the large semiclassical parameter J. Noticing that the total momentum plays the role of an effective Planck constant in (4.1), we define

$$\varphi = \varphi_{\rm cl} + \frac{1}{\sqrt{J}}\delta\varphi\,, \qquad \psi = \psi_{\rm cl} + \frac{1}{\sqrt{J}}\delta\psi\,, \tag{4.9}$$

together with the field redefinitions

$$f_1 = -\sin(2\psi_{\rm cl})\delta\varphi\,, \qquad f_2 = \delta\psi\,. \tag{4.10}$$

The result is a fluctuation Lagrangian [15, 19, 23] that can be usefully written as [20]

$$\mathcal{L}_{LL} = 2\,f_2\,\dot{f}_1 - \frac{\tilde{\lambda}}{2}\Big[\,f_1'^2 + f_2'^2 - V_1(\sigma)\,f_1^2 - V_2(\sigma)\,f_2^2\,\Big]. \tag{4.11}$$

In the formula above the potentials are given by

$$V_1(\sigma) = 4\mathrm{w}\left[1 + 4k^2 - 6k^2\,\mathrm{sn}^2(C\sigma, k^2)\right], \qquad V_2(\sigma) = 4\mathrm{w}\left[1 - 2k^2\,\mathrm{sn}^2(C\sigma, k^2)\right]. \tag{4.12}$$

The time-independence of the potentials allows to Fourier-transform $\partial_\tau \to i\,\omega$, after which the fluctuation equations following from (4.11) form a non-trivial matrix eigenvalue problem for the characteristic frequencies ω[3]

$$-\,f_2''(\sigma) - V_2(\sigma)f_2 = i\omega f_1\,, \tag{4.13}$$
$$f_1''(\sigma) + V_1(\sigma)f_1 = i\omega f_2. \tag{4.14}$$

[3] As in [20], the time has been rescaled by $\tilde{\lambda}$, which we will restore in the final expressions.

This system can be solved perturbatively in the elliptic modulus k^2 (or equivalently in the momenta ratio J_2/J), as it was done in [20]. The main result of our analysis is the *analytically exact diagonalization* of this non-trivial spectral problem. To this end, we start by decoupling (4.13)–(4.14) into two fourth-order equations

$$f_2'''' + [V_1(\sigma) + V_2(\sigma)]\, f_2'' + 2V_2'(\sigma) f_2' + \left[V_2''(\sigma) + V_1(\sigma)V_2(\sigma)\right] f_2 = 4\omega^2 f_2, \quad (4.15)$$

$$f_1'''' + [V_1(\sigma) + V_2(\sigma)]\, f_1'' + 2V_1'(\sigma) f_1' + \left[V_1''(\sigma) + V_1(\sigma)V_2(\sigma)\right] f_1 = 4\omega^2 f_1. \quad (4.16)$$

We can rewrite the first equation of the system (4.13)–(4.14) as the equation for $f \equiv f_2$[4]

$$\mathcal{O}^{(4)}\, f(x) = 0\,, \qquad \mathcal{O}^{(4)} = \partial_x^4 + 2\left(1 + 2k^2 - 4k^2\,\mathrm{sn}^2(x)\right)\partial_x^2 - 8k^2\,\mathrm{sn}(x)\,\mathrm{cn}(x)\,\mathrm{dn}(x)\,\partial_x + 1 - \Omega^2\,, \quad (4.17)$$

with

$$x \equiv C\sigma = 2\sqrt{\mathrm{w}}\sigma, \qquad \Omega = \frac{\omega}{2\mathrm{w}} \equiv \frac{\omega\,\pi^2}{2\,\mathbb{K}^2}\,. \quad (4.18)$$

Although for the one-spin solution $\nu = 0$ the fourth-order order equation factorizes into two second-order order operators, for the general case this seems not possible. The system of equation is written now as a *fourth-order differential equation* (4.17) *with doubly periodic elliptic coefficient functions*[5] with period $2L = 4\mathbb{K}$ (following from the 2π-periodicity of the closed string) and only one regular singular pole, in Fuchsian classification. A first (incomplete) attempt to study this kind of equations was done by Mittag-Leffler in [24], and to our knowledge not much else is known in literature. In Appendix C we will present a systematic study of the eigenvalue problem associated to this equation, showing that the corresponding determinant can be computed analytically.

Before doing that, we show that this class of operators is of more general interest, as it appears governing the (at least, bosonic) spectrum of fluctuations above the folded string with two angular momenta (S, J) [2], and thus it can likely be of help for the study of a large variety of problems involving a coupled system of fluctuations over elliptic string solutions.[6]

4.1.2 Bosonic Action for the (S, J)-String

In this section we review the semiclassical analysis about the classical closed string solution studied in [2]. The string ansatz in $AdS_5 \times S^5$, equipped with metric

---

[4] We omit the dependence of the Jacobi functions on the modulus k^2 in the following.

[5] See Appendix A for the relevant terminology.

[6] This observation is based on the already noticed similarity between the fluctuation spectra over the minimal surfaces corresponding to space-like Wilson loops of [3] and the one of [2, 22].

$$ds^2 = -\cosh^2 \rho dt^2 + d\rho^2 + \sinh^2 \rho d\Omega_3\,, + d\psi_1^2 + \cos^2\psi_1(d\psi_2^2 + \cos^2\psi_2 d\Omega_3')\,, \tag{4.19}$$

$$d\Omega_3 = d\beta_1^2 + \cos^2\beta_1(d\beta_2^2 + \cos^2\beta_2 d\beta_3^2)\,, \qquad \beta_3 \equiv \phi\,, \tag{4.20}$$

$$d\Omega_3' = d\psi_3^2 + \cos^2\psi_3(d\psi_4^2 + \cos^2\psi_4 d\psi_5^2)\,, \qquad \psi_5 \equiv \varphi\,, \tag{4.21}$$

reads for a state rotating with spin S in AdS_5 and angular momentum J in S^5[7]

$$t = \kappa\tau\,, \qquad \phi = \bar{w}\tau\,, \qquad \varphi = \nu\tau\,, \qquad \kappa, \bar{w}, \nu = \text{constant}\,, \tag{4.22}$$

$$\rho = \rho(\sigma) = \rho(\sigma + 2\pi), \;\; \beta_u = 0, \; (u = 1, 2), \;\; \psi_s = 0, \; (s = 1, 2, 3, 4). \tag{4.23}$$

The equations of motion and the Virasoro constraints impose the one-dimensional sinh-Gordon equation for the radial function $\rho(\sigma)$, which is solved in terms of the Jacobi sine,

$$\rho'^2 = \kappa^2\cosh^2\rho - \bar{w}^2\sinh^2\rho - \nu^2\,, \qquad k^2 = \frac{\kappa^2 - \nu^2}{\bar{w}^2 - \nu^2}\,, \tag{4.24}$$

$$\rho'^2(\sigma) = (\kappa^2 - \nu^2)\,\mathrm{sn}^2(\sqrt{\bar{w}^2 - \nu^2}\,\sigma + \mathbb{K}(k^2)\,|\,k^2)\,. \tag{4.25}$$

For notational convenience we will leave again the elliptic modulus k^2 implicit in what follows. Notice that when $\nu = 0$, the string rotates only in the AdS subspace and corresponds to the case studied in [1]. If instead we set $\bar{w} = 0$ and $\nu = \kappa$, it shrinks to a point rotating in S^5 [21]. The conserved charges are the energy E and the two momenta S, J associated to the invariance of the background metric under t, ϕ, φ-translations respectively

$$E \equiv \sqrt{\lambda}\kappa \int_0^{2\pi} \frac{d\sigma}{2\pi}\cosh^2\rho(\sigma)\,, \qquad S \equiv \sqrt{\lambda}\bar{w}\int_0^{2\pi}\frac{d\sigma}{2\pi}\sinh^2\rho(\sigma)\,, \tag{4.26}$$

$$J \equiv \sqrt{\lambda}\nu\int_0^{2\pi}\frac{d\sigma}{2\pi}\,, \qquad E = \kappa + \frac{\kappa}{\bar{w}}S\,.$$

The quadratic fluctuations over the folded string solution (4.22)–(4.25) are described in conformal gauge by the following Lagrangian [2]

$$\begin{aligned}\mathcal{L}_B^{\text{folded}} &= -\partial_a\tilde{t}\partial^a\tilde{t} - \mu_t^2\tilde{t}^2 + \partial_a\tilde{\phi}\partial^a\tilde{\phi} + \mu_\phi^2\tilde{\phi}^2 + \partial_a\tilde{\rho}\partial^a\tilde{\rho} + \mu_\rho^2\tilde{\rho}^2 \\ &\quad + 4\,\tilde{\rho}(\kappa\sinh\rho\,\partial_0\tilde{t} - \bar{w}\cosh\rho\,\partial_0\tilde{\phi}) \\ &\quad + \partial_a\tilde{\beta}_u\partial^a\tilde{\beta}_u + \mu_\beta^2\tilde{\beta}_u^2 + \partial_a\tilde{\varphi}\pi^a\tilde{\varphi} + \partial_a\tilde{\psi}_s\pi^a\tilde{\psi}_s + \nu^2\tilde{\psi}_s^2\,.\end{aligned} \tag{4.27}$$

[7] Compared to [2, 22], we use different conventions to label parameters in the classical solution.

Here tilded fields are fluctuations over the background (4.22)–(4.23). The fluctuations are non-trivially coupled both in their bosonic sector [2] and in the fermionic one [4], regardless of the gauge choice. In addition to this, their masses are non-constant

$$\mu_t^2 = 2\rho'^2 - \kappa^2 + \nu^2, \qquad \mu_\phi^2 = 2\rho'^2 - \bar{w}^2 + \nu^2, \tag{4.28}$$
$$\mu_\rho^2 = 2\rho'^2 - \bar{w}^2 - \kappa^2 + 2\nu^2, \qquad \mu_\beta^2 = 2\rho'^2 + \nu^2$$

and depend non-trivially on the classical field ρ satisfying the equation of motion (4.24).

The $\tilde{\beta}_u$ fields, transverse to the motion of the classical solution and decoupled from the other but with non-trivial mass, give a contribution to the one-loop partition function that has been evaluated exactly in [22]. The remaining three AdS_3 fields (t, ρ, ϕ) and the φ field in S^5 are non-trivially coupled through the Virasoro constraints. Their equations of motion read

$$(\partial_\tau^2 - \partial_\sigma^2)\,\tilde{t} + \mu_t^2\,\tilde{t} + 2\,\kappa\,\sinh\rho\,\partial_\tau\tilde{\rho} = 0\,, \tag{4.29}$$
$$(\partial_\tau^2 - \partial_\sigma^2)\,\tilde{\rho} + \mu_\rho^2\,\tilde{\rho} + 2\,(\kappa\,\sinh\rho\,\partial_\tau\tilde{t} - \bar{w}\,\cosh\rho\,\partial_\tau\tilde{\phi}) = 0\,, \tag{4.30}$$
$$(\partial_\tau^2 - \partial_\sigma^2)\,\tilde{\phi} + \mu_\phi^2\,\tilde{\phi} + 2\,\bar{w}\,\cosh\rho\,\partial_\tau\tilde{\rho} = 0\,, \tag{4.31}$$
$$(\partial_\tau^2 - \partial_\sigma^2)\,\tilde{\varphi} = 0\,. \tag{4.32}$$

From the conformal gauge conditions (Virasoro constraints) it follows

$$-\kappa\,\cosh^2\rho\,\partial_\tau\,\tilde{t} + (\bar{w}^2 - \kappa^2)\,\sinh\rho\,\cosh\rho\,\tilde{\rho} + \nu\,\partial_\tau\tilde{\varphi} + \rho'\,\partial_\sigma\,\tilde{\rho} + \bar{w}\,\sinh^2\rho\,\partial_\tau\,\tilde{\phi} = 0\,, \tag{4.33}$$
$$-\kappa\,\cosh^2\rho\,\partial_\sigma\,\tilde{t} + \bar{w}\,\sinh^2\rho\,\partial_\sigma\tilde{\phi} + \nu\,\partial_\sigma\tilde{\varphi} + \rho'\,\partial_\tau\,\tilde{\rho} = 0\,. \tag{4.34}$$

Since the ρ-background does not depend on τ and since the above equations are linear we may consider to pass to the Fourier space, i.e. replacing $\tilde{t} \to e^{i\,\omega\,\tau}\tilde{t}$ and $\phi \to e^{i\,\omega\,\tau}\tilde{\phi}$. Then we can solve the Virasoro constraints for $\tilde{t}$ and $\tilde{\phi}$ and substituting in (4.29)–(4.31) we get only one non-trivial equation[8]

$$(\partial_\sigma^2 + \omega^2)\,\frac{1}{\rho'}\,\mathcal{O}^{(4)}\,\tilde{\rho} = 0 \tag{4.35}$$

with the definitions

$$\mathcal{O}^{(4)} = \frac{1}{\rho'}(\partial_\sigma^2 + \omega^2 - V(\sigma))\,\rho'^2\,(\partial_\sigma^2 + \omega^2)\frac{1}{\rho'} - 4\,\nu^2\,\omega^2\,, \tag{4.36}$$
$$V(\sigma) = 2\,\rho'^2 + 2\frac{(\kappa^2 - \nu^2)(\bar{w}^2 - \nu^2)}{\rho'^2}\,. \tag{4.37}$$

[8]This structure was already understood by the authors of [22].

Changing to Euclidean signature ($\omega^2 \to -\omega^2$) and introducing the new variables

$$x = \sqrt{\bar{w}^2 - \nu^2}\sigma, \qquad \bar{\Omega}^2 = \frac{\omega^2}{\bar{w}^2 - \nu^2}, \tag{4.38}$$

the operator takes the canonical form

$$\mathcal{O}^{(4)} = \partial_x^4 + 2[-\bar{\Omega}^2 + k^2 + 1 - 4k^2\mathrm{sn}^2(x)]\partial_x^2 - 8k^2\mathrm{sn}(x)\mathrm{cn}(x)\mathrm{dn}(x)\partial_x \tag{4.39}$$
$$+ [(\bar{\Omega}^2 + 1 + k^2)^2 - 4k^2] + \frac{4\nu^2\bar{\Omega}^2}{\bar{w}^2 - \nu^2}.$$

The operator is very similar to the one (4.17) emerging in the LL quantum model, displaying however a significative difference as it cannot be seen as a "traditional" eigenvalue problem, as $\bar{\Omega}$ does not only appear in the constant term but also in the coefficient of the second-order derivative.

On a parallel note, we remark that a similar analysis can be done for the fermionic fluctuation Lagrangian. However, the eigenvalue equation for (the square of) the Dirac-like operator is a fourth-order differential equation whose coefficients are *not* meromorphic elliptic functions, and we are not currently able to solve it with the tools in Appendix C.

4.2 Bosonic One-Loop Partition Functions

We are now ready to use the analysis performed in Appendix C for the computation of determinants of the fluctuation operators in Sect. 4.1.

4.2.1 One-Loop Energy for the (J_1, J_2)-String

The fourth-order differential operator in (4.17), governing the fluctuations of the LL quantum model (4.11) is easily seen to be of the type (C.17) once the following identifications are performed

$$\alpha_0 = 2(1+2k^2), \qquad \alpha_1 = -8, \qquad \beta_0 = \beta_1 = 0,$$
$$\beta_2 = -4, \qquad \gamma_0 = 1 - \frac{\omega^2}{4\mathrm{w}^2}, \qquad \gamma_1 = \gamma_2 = \gamma_3 = 0. \tag{4.40}$$

Using the $n = 1$ consistency equations (C.34), one finds (here $\alpha \equiv \bar{\alpha}$)

$$\lambda = \pm k\sqrt{\mathrm{sn}^2(\alpha) - 1}, \tag{4.41}$$

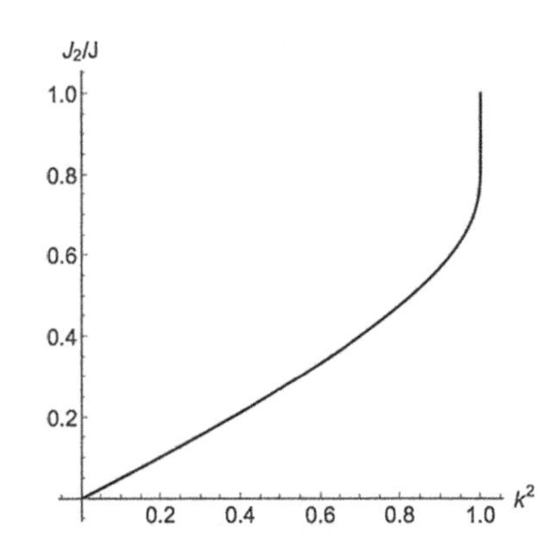

Fig. 4.1 Relation (4.7) between the auxiliary variable k^2 and the ratio of string angular momenta $J_2/J = J_2/(J_1 + J_2)$. The limiting cases $k = 0$ and $k = 1$ correspond to a string rotating only in one plane ($J_2 = 0$ and $J_1 = 0$ respectively)

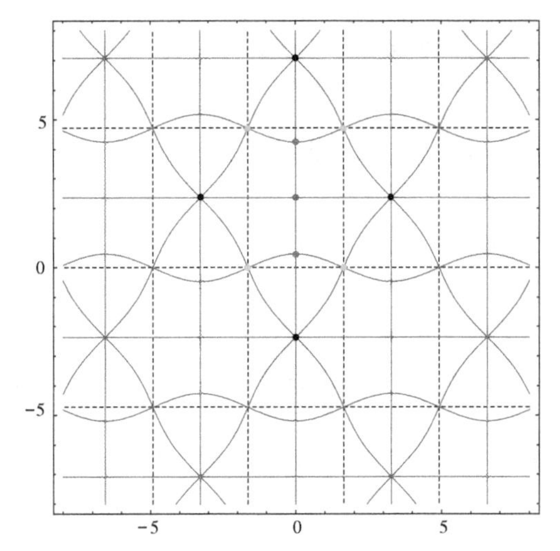

Fig. 4.2 In the complex α-plane one can plot the lines where $\Omega^2(\alpha)$ is real. The "physical" four linear independent solution live on these lines. The *green dots* represent places where $\Omega^2 = 0$, *red* for $\Omega^2 = 4k^2k'^2$, *blue* for $\Omega^2 = 1$ and *black* for poles. We chose $k = 0.4$

and the relation between Ω and α is

$$\Omega^2_{\mp}(\alpha) = 4k^2 \mathrm{cn}^2(\alpha)\,[ik\,\mathrm{sn}(\alpha) \mp \mathrm{dn}(\alpha)]^2 \ . \tag{4.42}$$

It seems advantageous to consider $\alpha \in \mathbb{C}$ as the independent parameter and therefore Ω as a doubly periodic function of α as in (4.42) (Fig. 4.1). There exist four values of α, which correspond to one value of Ω. As far as the *physical spectrum* is concerned, we look for all complex values of α that correspond to a *real* Ω^2.

The analysis in Appendix B.1 of [25] was devoted to this study and it is here summarized in Appendix C.5. In Fig. 4.2 the lines where $\Omega^2(\alpha)$ is real are plotted in the complex α-plane. The "physical" four linear independent solutions of the fourth-order differential operator (4.15)–(4.15) live on these lines, and they correspond to the different colours in the fundamental domain shown in Fig. 4.3.

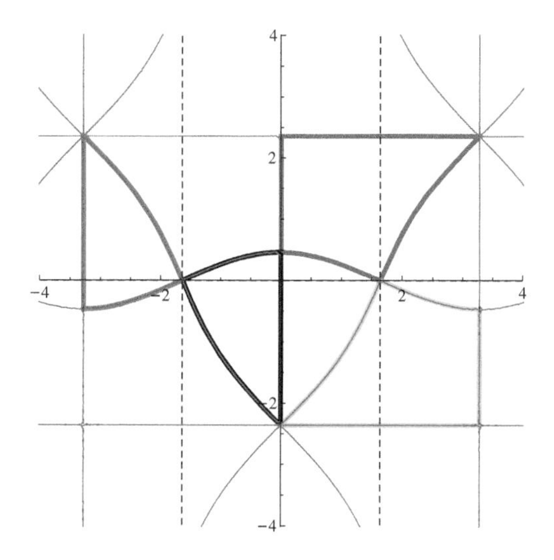

Fig. 4.3 In the fundamental domain the four independent solutions of the fourth-order differential operator (4.17) are marked with different colors. Walking on such a closed path, Ω^2 runs from $-\infty$ to $+\infty$. We chose $k = 0.4$

To exemplify the procedure, let us focus on a choice of string momenta (J_1, J_2) such that $0 < k < 1/\sqrt{2}$ through the last formula in (4.7). The other case $1/\sqrt{2} < k < 1$ can be discussed analogously.

For a given $\Omega^2 \in \mathbb{R}$ the four independent solutions read

$$f_i(x, \Omega, k) = \frac{H(x + \alpha_i)}{\Theta(x)} e^{-x[Z(\alpha_i) - ik\mathrm{cn}(\alpha_i)]}, \qquad i = 1, \ldots 4, \tag{4.43}$$

where the α_i have to be chosen according to the range of Ω according to the list in Appendix C.5.

The quasi-momenta p_i are obtained from the periodicity of $f_i(x + 2\mathbb{K}) = e^{2\mathbb{K}ip_i} f_i(x)$

$$p_i(\Omega, k) = iZ(\alpha_i, k) + k\mathrm{cn}(\alpha_i, k) - \frac{\pi}{2\mathbb{K}}. \tag{4.44}$$

By construction,[9] only two of the quasi-momenta are independent, which we will call p_1 and p_2. The exact discriminant of the Landau–Lifshitz model defined by (4.11)–(4.15), using (C.16) with $2L = 4\mathbb{K}$, reads then

$$\Delta_{LL}(\Omega, k) = 16 \sin^2\left(2\mathbb{K}\, p_1(\Omega, k)\right) \sin^2\left(2\mathbb{K}\, p_2(\Omega, k)\right). \tag{4.45}$$

We can also immediately recover the characteristic frequencies of the problem, found in [20] using operator methods up to second-order perturbation theory, by simply looking at the zeroes of the discriminant (4.45), where the quasi momenta are built with the α_i in the branch $\Omega^2 > 0$ and are Taylor-expanded around $k = 0$. It

[9] See discussion around (C.11)–(C.12).

is enough to look to the factor in (4.45) involving p_1, whose expansion re-expressed in terms of the ω is

$$p_1 = \sqrt{2\omega+1} - \frac{\omega}{2\sqrt{2\omega+1}}k^2 + \frac{(-10\omega^4 - 11\omega^3 + 6\omega + 2)}{32\omega^2(2\omega+1)^{3/2}}k^4 + \frac{(-22\omega^7 - 39\omega^6 - 19\omega^5 + 2\omega^4 + 10\omega^3 + 16\omega^2 + 10\omega + 2)}{64\omega^4(2\omega+1)^{5/2}}k^6 + O(k^8)\,. \quad (4.46)$$

Inserting it into (4.45) and requiring the vanishing of the expression order by order in small k^2, one finds the (squared) frequencies to be

$$\omega^2 = \frac{1}{4}(n^2-1)^2 + \frac{1}{4}(1-n^2)k^2 + \frac{(3n^4 - 2n^2 + 15)}{64(1-n^2)}k^4 - \frac{(n^8 + n^6 + 7n^4 + 27n^2 + 28)}{128\,(n^2-1)^3}k^6 + O\left(k^8\right), \quad (4.47)$$

where we do not report higher orders, although it would be straightforward to calculate them. The first three orders of the expansion above coincide with the ones of [20].

The one-loop correction to the $SU(2)$ LL string energy can be of course obtained perturbatively via a regularized sum over the frequencies given above [20] or equivalently in our approach from the Euclidean LL partition function Z_{LL}. In the latter case, the one-loop shift to the energy is the one-loop worldsheet effective action $\Gamma^{(1)}$ divided by the total time interval $\mathcal{T}$

$$E^{(1)} = \frac{\Gamma^{(1)}}{\mathcal{T}} = -\frac{\log Z_{LL}}{\mathcal{T}} = -\frac{1}{\mathcal{T}}\log\left(\text{Det}^{-1/2}\mathcal{O}_{LL}\right), \qquad \mathcal{T} = \int_{-\infty}^{\infty} d\tau \quad (4.48)$$

which using (4.45) can be explicitly written as[10]

$$\Gamma^{(1)} = -\log Z_{LL} = \frac{\mathcal{T}}{2}\int_{-\infty}^{\infty}\frac{d\Omega}{2\pi}\log\left[16\sin^2\left(2\mathbb{K}\,p_1(\Omega,k)\right)\sin^2\left(2\mathbb{K}\,p_2(\Omega,k)\right)\right]. \quad (4.49)$$

Above, the Euclidean setting requires the quasi-momenta p_i to be built out of the α_i in the branch $\Omega^2 < 0$ (C.35).

The integral (4.49) is divergent, which is not surprising because of absence in the LL action of fermionic and some bosonic modes which are crucial for UV finiteness, and one has to consider a suitable regularization. A meaningful choice is to refer the functional determinant to the $k = 0$ case, corresponding to the configuration $(J_1 = 0, J_2)$. Indeed this limit, as discussed in [20], represents a nearly point-like string and

[10] It is convenient to use the rescaled frequency Ω (4.18) as integration variable.

the correction to the ground-state energy should vanish. Hence, we consider instead the well-behaved integral expression[11]

$$\Gamma^{(1)}_{\text{reg}} = \frac{\mathcal{T}}{2}\int_{-\infty}^{\infty}\frac{d\Omega}{2\pi}\log\left[\frac{\sin^2\left(2\mathbb{K}\,p_1(\Omega,k)\right)\,\sin^2\left(2\mathbb{K}\,p_2(\Omega,k)\right)}{\sin^2\left(\pi\,p_1(\Omega,0)\right)\,\sin^2\left(\pi\,p_2(\Omega,0)\right)}\right] \tag{4.50}$$

where in the denominator the quasi-momenta $p_i(\Omega,0)$ are computed at $k=0$. The regularized expression can be computed numerically for a given k, i.e. for arbitrary (J_1, J_2). In order to analytically perform the above integral over Ω, we can resort to the *short-string* expansion $k\to 0$. The starting point is the expression of the quasi-momentum function (4.44), evaluated on the four functions α_i given in (C.35) (which differ from the ones considered in (4.46) because now for the Euclidean partition function we need the physical branch $\Omega^2<0$)

$$p_1 = -i\sqrt{-1-i\sqrt{\Omega^2}} + \frac{\left(2-i\sqrt{\Omega^2}\right)\sqrt{1+i\sqrt{\Omega^2}}}{8\Omega^2}k^4 + O\left(k^6\right), \tag{4.51}$$

$$p_2 = +i\sqrt{-1+i\sqrt{\Omega^2}} + \frac{\left(2+i\sqrt{\Omega^2}\right)\sqrt{1-i\sqrt{\Omega^2}}}{8\Omega^2}k^4 + O\left(k^6\right) \tag{4.52}$$

and then we integrate over Ω the corresponding expressions computed order by order in k^2 to get

$$\Gamma^{(1)}_{\text{reg}} = \sum_{i=0}^{\infty}\Gamma^{(1)}_{i,\text{reg}}k^{2i}\,. \tag{4.53}$$

Each term in the series must be further regularized, which is of course expected as only certain bosonic degrees of freedom and no fermionic ones participate in the effective LL action. One can use two different ways of regularizing (one inspired by zeta-function regularization and one with standard cutoff in Appendix B.4 of [25])

$$E^{(1)} = \frac{\Gamma^{(1)}_{\text{reg}}}{\mathcal{T}} = \frac{1}{4}k^2 + \frac{1}{16}\left(1-\frac{\pi^2}{3}\right)k^4 + O(k^6)\,, \tag{4.54}$$

which reproduces the expression for the k^2-expansion of the one-loop energy in [20]. It is interesting to notice that this result follows smoothly by our standard regularization of the LL string effective action, while in [20] it is implied by a zeta-function regularization supplemented by a general prescription for the vacuum energy in terms of characteristic frequencies of a mixed system of oscillators [26].

From the last relation in (4.7), the short string limit $k^2\to 0$ reads in terms of the physical parameter J_2/J as

---

[11] We cannot discard the existence of another regularization that still enforces the vanishing of the energy of the BPS configuration ($J_1=0$, J_2). We thank Radu Roiban for this remark.

$$\frac{J_2}{J} = \frac{k^2}{2} + \frac{k^4}{16} + O(k^5), \qquad k^2 = \frac{2\,J_2}{J} - \frac{1}{2}\left(\frac{J_2}{J}\right)^2, \tag{4.55}$$

and the expression for the energy becomes

$$E^{(1)} = \frac{\tilde{\lambda}}{2}\left(\frac{J_2}{J} + \left(\frac{1}{4} - \frac{\pi^2}{6}\right)\left(\frac{J_2}{J}\right)^2\right) + O\left(\left(\frac{J_2}{J}\right)^3\right), \tag{4.56}$$

where we restored the $\tilde{\lambda}$ dependence. The first three terms in the formula above are in agreement with [20].

4.2.2 One-Loop Energy for the (S, J)-String

The fourth-order differential operator in (4.39)[12] is again of the type (C.17) with the identification

$$\begin{aligned} \alpha_0 &= 2(\bar{\Omega}^2 + k^2 + 1), \qquad \alpha_1 = -8, \qquad \beta_0 = \beta_1 = 0, \qquad \beta_2 = -4 \\ \gamma_0 &= [(-\bar{\Omega}^2 + 1 + k^2)^2 - 4k^2] - \frac{4\nu^2\bar{\Omega}^2}{\bar{w}^2 - \nu^2}, \qquad \gamma_1 = \gamma_2 = \gamma_3 = 0, \end{aligned} \tag{4.57}$$

where $\bar{\Omega}$ is defined in (4.38). Using the consistency equations (C.34) one finds (here $\bar{\alpha} \equiv \alpha$)

$$\lambda = \pm\sqrt{k^2 \mathrm{sn}^2(\alpha) - \bar{\Omega}^2}\,, \tag{4.58}$$

where the relation between $\bar{\Omega}$ and α is

$$8k^4 \mathrm{sn}^4(\alpha) - 4(1 + k^2 + \bar{\Omega}^2)k^2 \mathrm{sn}^2(\alpha) \pm 8k^2 \mathrm{sn}(\alpha)\mathrm{cn}(\alpha)\mathrm{dn}(\alpha)\sqrt{k^2\mathrm{sn}^2(\alpha) - \bar{\Omega}^2} - \frac{4\nu^2\bar{\Omega}^2}{\bar{w}^2 - \nu^2} = 0\,. \tag{4.59}$$

Again, we are interested in all values of the complex parameter α that correspond to real values of $\bar{\Omega}^2$, see Appendix C.6 here and Appendix C.2 in [25]. The "physical" four linear independent solutions of (4.39) live on the straight and ellipse-like lines in Fig. 4.4.

For a given real value of $\bar{\Omega}^2$ the linear independent solutions of the fourth-order differential equation (4.39) are

$$f_{1,2}(x, \bar{\Omega}) = \frac{H(x \pm \alpha_1)}{\Theta(x)} e^{\mp x[Z(\alpha_1) + \lambda(\alpha_1)]}, \tag{4.60}$$

$$f_{3,4}(x, \bar{\Omega}) = \frac{H(x \pm \alpha_2)}{\Theta(x)} e^{\mp x[Z(\alpha_2) + \lambda(\alpha_2)]}, \tag{4.61}$$

[12]In this section we are working in Minkowski signature, so that (4.57) is obtained from (C.17) by analytically continuing the frequencies.

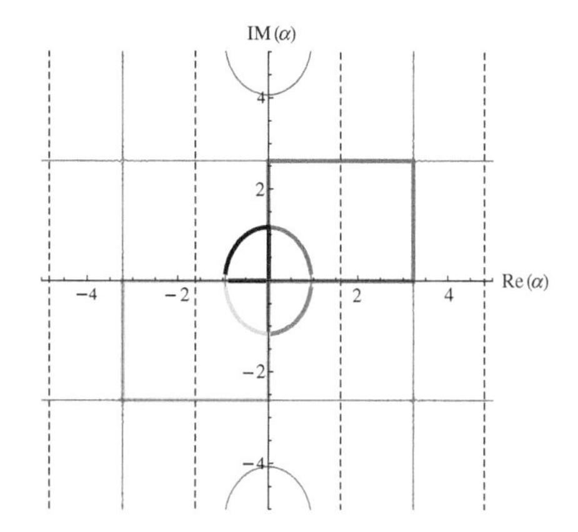

Fig. 4.4 The places in the fundamental domain of the complex α where the four linear independent solutions (4.60)–(4.60) (marked with different colours) for the fourth-order differential operator (4.39) live. Here we set $k = 3$, $w = 6$, $\nu = 2.5$

where the α_i as functions of $\bar{\Omega}$ have to be chosen according to the list in Appendix C.6. Using their explicit expressions of the functions above in $f_i(x + 2\mathbb{K}) = e^{2\mathbb{K}ip_i} f_i(x)$, the quasi-momenta are then obtained as

$$p_n(\bar{\Omega}) = \pm i \left[Z(\alpha_n) + \frac{k\,\text{sn}(\alpha_n)}{(\kappa^2 - \nu^2)\text{sn}^2(\alpha_n) + \nu^2} \left(\kappa\bar{w} \pm \sqrt{\kappa^2 - \nu^2}\sqrt{\bar{w}^2 - \nu^2}\,\text{cn}(\alpha_n)\text{dn}(\alpha_n) \right) \right] + \frac{\pi}{2\mathbb{K}} . \tag{4.62}$$

The discriminant is again given by

$$\Delta_\nu(\Omega, k) = 16 \sin^2(L\, p_1) \sin^2(L\, p_2) , \tag{4.63}$$

with $2L = 4\mathbb{K}$.

As a first check of the correctness of the procedure, one can take the *long-string* limit $k \to 1$, $\bar{w}^2 \to \kappa^2$ and look at the zeroes of (4.63) choosing the positive-frequency range (C.50) and (C.52) for the α_i in (4.62). In Appendix C.1 of [25] we obtain the characteristic frequencies

$$\omega_n = \sqrt{n^2 + 2\kappa^2 \pm 2\sqrt{\kappa^4 + n^2\nu^2}}, \tag{4.64}$$

which is the same result as found in [27].

Since we are missing the fermionic counterpart of (4.63), we cannot proceed with the exact evaluation of the full one-loop partition function on the folded two-spin solution. However, we observe that a nice consequence of our procedure is the possibility of making a non-trivial, analytical statement on the equivalence of partition functions in conformal and static gauge in the single-spin case $\nu = 0$. While here the fermionic determinant can be given exactly for all values of the spin [22], it is only the bosonic partition function in static gauge—where fluctuations are naturally decoupled—which has been written down in an analytically exact closed form, and reads [22]

$$\log Z^{\rm bos}_{\rm static\ gauge} = -\frac{\mathcal{T}}{2}\int_{-\infty}^{\infty}\frac{d\omega}{2\pi}\log\left(\mathrm{Det}\mathcal{O}_\phi\,\mathrm{Det}^2\mathcal{O}_\beta\mathrm{Det}^5\mathcal{O}_0\right) \tag{4.65}$$

where

$$\mathrm{Det}\mathcal{O}_\phi = 4\sinh^2[2\tilde{\mathbb{K}}Z(\alpha_\phi|\tilde{k}^2)]\,,\ \mathrm{Det}\mathcal{O}_\beta = 4\sinh^2\left[2\mathbb{K}\,Z(\alpha_\beta|k^2)\right]\,,\ \mathrm{Det}\mathcal{O}_0 = 4\sinh^2(\pi\omega) \tag{4.66}$$

and

$$\mathrm{sn}(\alpha_\phi\,|\tilde{k}^2) = \frac{1}{\tilde{k}}\sqrt{1+\left(\frac{\pi\omega}{2\tilde{\mathbb{K}}}\right)^2}\,,\quad \mathrm{sn}(\alpha_\beta|k^2) = \frac{1}{k}\sqrt{1+k^2+\left(\frac{\pi\omega}{2\mathbb{K}}\right)^2}\,, \tag{4.67}$$

with $\tilde{k}^2 = 4k/(1+k)^2$ and $\tilde{\mathbb{K}} = \mathbb{K}(\tilde{k}^2)$.

The analysis in Sect. 4.1.2 shows that, in conformal gauge, the spectral problem associated to the mixed-mode, 3×3 matrix differential operator corresponding to (4.29)–(4.31) can be evaluated, see (4.35), via the product of a free determinant times the determinant of the fourth order differential operator (4.39), and thus

$$\log Z^{\rm bos}_{\rm conformal\ gauge} = -\frac{\mathcal{T}}{2}\int_{-\infty}^{\infty}\frac{d\omega}{2\pi}\log\left(\mathrm{Det}\mathcal{O}_{\nu=0}\,\mathrm{Det}^2\mathcal{O}_\beta\,\mathrm{Det}^4\mathcal{O}_0\right)\,, \tag{4.68}$$

where in the counting of massless operators we already have taken into account the two conformal gauge massless ghosts [28], and

$$\mathrm{Det}\mathcal{O}_{\nu=0} = 16\sinh^2\left[2\mathbb{K}\left(Z(\alpha|k^2)+\frac{1+\mathrm{cn}(\alpha|k^2)\mathrm{dn}(\alpha|k^2)}{\mathrm{sn}(\alpha|k^2)}\right)\right]\sinh^2(2\mathbb{K}\bar{\Omega})\,, \tag{4.69}$$

with (switching to Euclidean signature)

$$\mathrm{sn}^2(\alpha|k^2) = \frac{-4\bar{\Omega}^2}{(1+k^2-\bar{\Omega}^2)^2-4k^2}\,. \tag{4.70}$$

One can see that the second factor in (4.69) corresponds to the same massless boson mode of (4.66) (recalling (4.38) and that for $\nu=0$ it is $\bar{w}=\frac{2\mathbb{K}}{\pi}$), while for the first factor one should use the transformation (C.35) for the Jacobi zeta function which, writing $\tilde{\alpha} = \alpha/(1+\tilde{k}') + i\mathbb{K}'/(1+\tilde{k}')$, leads to the identity

$$2\mathbb{K}\left[Z(\alpha|k^2)+\frac{1+\mathrm{cn}(\alpha|k^2)\mathrm{dn}(\alpha|k^2)}{\mathrm{sn}(\alpha|k^2)}\right] = 2\tilde{\mathbb{K}}Z(\tilde{\alpha}|\tilde{k}^2)+i\pi\,. \tag{4.71}$$

This establishes analytically the equivalence of static and conformal gauge bosonic determinants (4.65)–(4.68).

References

1. S.S. Gubser, I.R. Klebanov, A.M. Polyakov, A semi-classical limit of the gauge/string correspondence. Nucl. Phys. B **636**, 99 (2002), arXiv:hep-th/0204051
2. S. Frolov, A.A. Tseytlin, Semiclassical quantization of rotating superstring in $AdS_5 \times S^5$. JHEP **0206**, 007 (2002), arXiv:hep-th/0204226
3. N. Drukker, V. Forini, Generalized quark-antiquark potential at weak and strong coupling. JHEP **1106**, 131 (2011), arXiv:1105.5144
4. V. Forini, V.G.M. Puletti, O. Ohlsson, Sax, The generalized cusp in $AdS_4 \times \mathbb{CP}^3$ and more one-loop results from semiclassical strings. J. Phys. A **46**, 115402 (2013), arXiv:1204.3302
5. M. Beccaria, G. Dunne, G. Macorini, A. Tirziu, A. Tseytlin, Exact computation of one-loop correction to energy of pulsating strings in $AdS_5 \times S^5$. J. Phys. A **A44**, 015404 (2011), arXiv:1009.2318
6. V. Forini, Quark-antiquark potential in AdS at one loop. JHEP **1011**, 079 (2010), arXiv:1009.3939
7. M. Kruczenski, Spin chains and string theory. Phys. Rev. Lett. **93**, 161602 (2004), arXiv:hep-th/0311203
8. M. Kruczenski, A. Ryzhov, A.A. Tseytlin, Large spin limit of $AdS_5 \times S^5$ string theory and low-energy expansion of ferromagnetic spin chains. Nucl. Phys. B **692**, 3 (2004), arXiv:hep-th/0403120
9. B.J. Stefanski A.A. Tseytlin, Large spin limits of AdS/CFT and generalized Landau-Lifshitz equations. JHEP **0405**, 042 (2004), arXiv:hep-th/0404133
10. J.A. Minahan, K. Zarembo, The Bethe ansatz for $\mathcal{N} = 4$ superYang-Mills. JHEP **0303**, 013 (2003), arXiv:hep-th/0212208
11. H. Bethe, Zur Theorie der Metalle. I. Eigenwerte und Eigenfunktionen der linearen Atomkette. Z. Phys. **71**, 205 (1931). (in German)
12. J. Plefka, Spinning strings and integrable spin chains in the AdS/CFT correspondence. Living Rev. Rel. **8**, 9 (2005), arXiv:hep-th/0507136
13. M. Staudacher, Review of AdS/CFT integrability, Chapter III.1: Bethe Ansätze and the R-matrix formalism. Lett. Math. Phys. **99**, 191 (2012), arXiv:1012.3990
14. N. Beisert, J.A. Minahan, M. Staudacher, K. Zarembo, Stringing spins and spinning strings. JHEP **0309**, 010 (2003), arXiv:hep-th/0306139
15. N. Beisert, S. Frolov, M. Staudacher, A.A. Tseytlin, Precision spectroscopy of AdS/CFT. JHEP **0310**, 037 (2003), arXiv:hep-th/0308117
16. S. Frolov, A.A. Tseytlin, Rotating string solutions: AdS/CFT duality in nonsupersymmetric sectors. Phys. Lett. B **570**, 96 (2003), arXiv:hep-th/0306143
17. G. Arutyunov, S. Frolov, J. Russo, A.A. Tseytlin, Spinning strings in $AdS_5 \times S^5$ and integrable systems. Nucl. Phys. B **671**, 3 (2003), arXiv:hep-th/0307191
18. A. Mikhailov, Speeding strings. JHEP **0312**, 058 (2003), arXiv:hep-th/0311019
19. M. Kruczenski, A.A. Tseytlin, Semiclassical relativistic strings in S^5 and long coherent operators in $\mathcal{N} = 4$ SYM theory. JHEP **0409**, 038 (2004), arXiv:hep-th/0406189
20. J. Minahan, A. Tirziu, A.A. Tseytlin, $1/J$ corrections to semiclassical AdS/CFT states from quantum Landau-Lifshitz model. Nucl. Phys. B **735**, 127 (2006), arXiv:hep-th/0509071
21. D.E. Berenstein, J.M. Maldacena, H.S. Nastase, Strings in flat space and pp waves from $\mathcal{N} = 4$ superYang-Mills. JHEP **0204**, 013 (2002), arXiv:hep-th/0202021
22. M. Beccaria, G.V. Dunne, V. Forini, M. Pawellek, A.A. Tseytlin, Exact computation of one-loop correction to energy of spinning folded string in $AdS_5 \times S^5$. J. Phys. A **43**, 165402 (2010), arXiv:1001.4018
23. S. Frolov, A.A. Tseytlin, Multispin string solutions in $AdS_5 \times S^5$. Nucl. Phys. B **668**, 77 (2003), arXiv:hep-th/0304255
24. M.G. Mittag-Leffler, Ueber die Integration der Hermiteschen Differentialgleichungen der dritten und vierten Ordnung, bei denen die Unendlichkeitsstellen der Integrale von der ersten Ordnung sind. Annali di Matematica **11**, 65 (1882). (in German)

25. V. Forini, V.G.M. Puletti, M. Pawellek, E. Vescovi, One-loop spectroscopy of semiclassically quantized strings: bosonic sector. J. Phys. A **48**, 085401 (2015), arXiv:1409.8674
26. M. Blau, M. O'Loughlin, G. Papadopoulos, A.A. Tseytlin, Solvable models of strings in homogeneous plane wave backgrounds. Nucl. Phys. B **673**, 57 (2003), arXiv:hep-th/0304198
27. S. Frolov, A. Tirziu, A.A. Tseytlin, Logarithmic corrections to higher twist scaling at strong coupling from AdS/CFT. Nucl. Phys. B **766**, 232 (2007), arXiv:hep-th/0611269
28. N. Drukker, D.J. Gross, A.A. Tseytlin, Green-Schwarz string in $AdS_5 \times S^5$: Semiclassical partition function. JHEP **0004**, 021 (2000), arXiv:hep-th/0001204

Chapter 5
Towards Precision Holography for Latitude Wilson Loops

In recent years a new wealth of exact results has become available for path-integrals in supersymmetric QFTs on curved manifolds by means of *supersymmetric localization*. It was inaugurated by N. Nekrasov for 4d $\mathcal{N} = 2$ gauge theories in the Omega background [1] and, in particular, by V. Pestun for $\mathcal{N} = 2$ SYM on S^4 [2]. The mathematical principles behind localization in supersymmetric theories [3–6] are a powerful synthesis of previous *equivariant localisation* theorems for ordinary integrals that have a symmetry with fixed points [7]. Localization relies on supersymmetry to show that a path-integral invariant under a fermionic charge receives contributions only from the set of fixed points (*localization locus*) of the action of that symmetry group [8–11]. The power of this method is to produce all-loop expectation values of (local or non-local) operators by reducing (*localizing*) the original infinite-dimensional integral to a less-dimensional one, often a matrix integral that is amenable to an evaluation in terms of elementary functions. In that, it offers a way to extract physical information on a QFT in a manner that is independent from the possible integrable properties of the theory. The reader can consult [12, 13] for a collection of reviews to scratch the surface of the vast literature on the subject.

Operators preserving a fraction of the (global[1]) supersymmetry of the theory—commonly referred to as *Bogomol'nyi-Prasad-Sommerfield (BPS)* operators—are computable within this framework. As in the seminal paper [2], some of the most studied cases are supersymmetric Wilson loops in $\mathcal{N} = 4$ SYM. In Sects. 5.1 and 5.2 we collect some background material for the interested reader. In particular, we want to highlight Sect. 5.2.1, where we recall that localization predicts the vacuum expectation value of the 1/2-BPS circular Wilson loop for all values of the coupling. Here we also review the (unsuccessful) efforts in trying to match the strong-coupling expansion of this field theory result using sigma-model perturbation theory. In Sects. 5.4–5.8 we revisit this delicate issue by turning to the study of the 1/4-BPS Wilson loops of Sect. 5.2.2, which comprise the circular loop as a limiting case. Our

[1] See below (5.2) for Wilson loops.

E. Vescovi, *Perturbative and Non-perturbative Approaches to String Sigma-Models in AdS/CFT*, Springer Theses, DOI 10.1007/978-3-319-63420-3_5

driving motivation is that, as we explain at length in Sect. 5.3, we can strengthen our stringy computational tools by carefully considering such discrepancies in a setting where we are supported by exact results from the gauge theory side of AdS/CFT.

5.1 Review of Supersymmetric Wilson Loops in $\mathcal{N}=4$ SYM

In Sect. 1.3 we have already introduced supersymmetric *Maldacena-Wilson loop operators* in $\mathcal{N}=4$ SYM[2]

$$\mathcal{W}[\mathcal{C}] \equiv \frac{1}{\dim\mathcal{R}} \mathrm{tr}_{\mathcal{R}} \mathcal{P} \exp\left[\int \left(i A_\mu (x(\tau))\, \dot{x}^\mu(\tau) + |\dot{x}(\tau)|\, \theta^I(\tau)\, \phi_I(x(\tau))\right) d\tau\right]. \quad (5.1)$$

They are defined by a generalized holonomy that contains couplings to the scalar fields ϕ_I $(I = 1, \ldots 6)$, in addition to the gauge connection A_μ $(\mu = 1, \ldots 4)$, while preserving gauge invariance of the operator (5.1) and reparametrization invariance of the closed path $\{x^\mu(\tau)\} \equiv \mathcal{C} \subset \mathbb{R}^4$. All fields are in a representation $\mathcal{R}$ of the gauge group $SU(N)$. The scalar couplings $\{\theta^I(\tau)\} \subset \mathbb{R}^6$ draw a path, not necessarily closed or related to the $\mathcal{C}$, on a five-sphere $\sum_{I=1}^6 \left(\theta^I(\tau)\right)^2 = 1$. The inclusion of the scalars improves the ultraviolet properties of the operators when compared to the purely bosonic ones: for any smooth and non-self-intersecting $\mathcal{C}$ their expectation values are divergence-free and invariant under non-singular conformal transformations. Finally, the dimension of the gauge-group representation $\dim\mathcal{R}$ ensures that the operator has unit-normalized expectation value in the free theory ($g_{\mathrm{YM}} = 0$, or $\lambda = 0$ in the planar limit).

The constraint on θ^I guarantees the existence of 16 out of 32 supercharges in the 4d $\mathcal{N}=4$ superconformal algebra that locally preserve the Wilson loop (here $\{\Gamma_\mu, \Gamma_{I+4}\}$ are $SO(10)$ Dirac matrices)

$$\delta_{\epsilon(\tau)}\mathcal{W} = 0 \quad \Longleftrightarrow \quad \left(i\Gamma_\mu \dot{x}^\mu(\tau) + |\dot{x}(\tau)|\, \theta^I(\tau)\, \Gamma_{I+4}\right) \epsilon(\tau) = 0\,. \quad (5.2)$$

They are called *locally 1/2-BPS* Wilson loops because in (5.2) we can find 16 distinct fermionic transformations $\delta_{\epsilon(\tau)}$ generated by Majorana-Weyl spinors $\epsilon(\tau)$ depending on the loop point. However, only rigid supersymmetry is a symmetry of the action. One typically seeks some restrictions on $x^\mu(\tau)$ and $\theta^I(\tau)$ to make the solution $\epsilon(\tau) = \epsilon$ unique along the loop. This is the case of *globally BPS* Wilson loops. Such operators are classified according to the fraction of global supercharges leaving

[2] Here we consider the spacetime in Euclidean signature, for which the path-ordered exponential is not a pure complex phase and there is no unitarity bound $\langle\mathcal{W}\rangle \leq 1$, see comments in [14]. We recommend reading the review in [15], which also deals with loop operators in ABJM theory.

them invariant: they are (globally) 1/2-BPS when they preserve 16 supercharges, and similarly for other fractions.[3]

A systematic classification of Wilson loops locally preserving at least one fermionic charge was made in [16, 17]. Some of these operators have been known earlier and fall under two classes: *Zarembo loops* and *Drukker-Giombi-Ricci-Trancanelli (DGRT) loops*, according to the relationship between $x^\mu(\tau)$ and $\theta^I(\tau)$. They are both globally 1/16-BPS at least: the former type for any spacetime contour in $\mathbb{R}^4$, whereas the latter one only when $\mathcal{C}$ lies entirely in $S^3 \subset \mathbb{R}^4$.

- The *Zarembo loops* [18] eliminate the dependence of $\epsilon(\tau)$ in (5.2) on the loop parameter τ by means of a constant 4×6 "projection" matrix $M^I{}_\mu$ that assigns every tangent vector in $\mathbb{R}^4$ to a point in the coupling space $S^5 \subset \mathbb{R}^6$

$$\theta^I(\tau) = M^I{}_\mu \frac{\dot{x}^\mu(\tau)}{|\dot{x}(\tau)|}, \qquad \sum_{I=1}^{6} M^I{}_\mu M^I{}_\nu = \delta_{\mu\nu}. \tag{5.3}$$

This choice reduces (5.2) to a τ-independent constraint $\left(i\,\Gamma_\mu + M^I{}_\mu \Gamma_{I+4}\right)\epsilon = 0$ and makes the operator invariant only under super-Poincaré generators. One can also prove a correlation between the amount of unbroken supersymmetries and the dimensionality of the linear subspace of $\mathbb{R}^4$ where the loop is embedded into.

Hyperplane of lowest dimension embedding $\mathcal{C}$	Supercharges unbroken by $\mathcal{W}$	Amount of SUSY of $\mathcal{W}$
$\mathbb{R}^4$	2	1/16 BPS
$\mathbb{R}^3$	4	1/8 BPS
$\mathbb{R}^2$	8	1/4 BPS
$\mathbb{R}^2$	16	1/2 BPS

(5.4)

The only 1/2-BPS case is the *straight Wilson line* $x^\mu(\tau) = (\tau, 0, 0, 0)$ passing through a point at infinity and coupling to the same scalar. As it preserves all the 16 Poincaré supercharges, supersymmetry-based arguments guarantee that its expectation value is coupling-independent and it equals to one.[4] Non-renormalization theorems were also proven for the more general 1/8-BPS case using superspace formalism [22, 23] and topological arguments [22–25].

[3]By this counting over the full set of 16 super-Poincaré $\mathcal{Q}$'s and 16 superconformal $\mathcal{S}$'s generators, an operator that breaks all the $\mathcal{Q}$'s while preserving at least one $\mathcal{S}$ is still called *supersymmetric*.

[4]In gauge theory this is equivalent to the statement that gauge and scalar propagators coincide in Feynman gauge and cancel order by order in perturbation theory. At strong coupling, it is easy to check that the dual classical worldsheet has zero regularized area [19, 20] (see 3.100 in this thesis), but the vanishing of the subleading corrections is less transparent. The one-loop order is is rather subtle because one needs an *ad hoc* prescription to subtract divergences [20, 21].

- A richer variety of results is found for the *Drukker-Giombi-Ricci-Trancanelli* Wilson loops [26, 27]. The construction of the scalar couplings ($i = 1, 2, 3$ and $I, J = 1, \dots 6$)

$$|\dot{x}(\tau)|\,\theta_i(\tau) = \frac{1}{2}\sigma_i^R(\tau)\,M^i{}_I\,, \qquad \theta_{i+3}(\tau) = 0\,, \qquad \sum_{I=1}^{6} M^i{}_I M^j{}_I = \delta_{ij}\,. \quad (5.5)$$

is valid for any loop on a three-sphere $\sum_{i=1}^4 x_\mu^2 = 1$ and based on the interpretation of the "effective" gauge connection in the exponent of (5.1) as a non-trivial topological twist of $\mathcal{N} = 4$ SYM. We can represent the $SU(2)$ right-invariant one-forms σ_i^R on such S^3 as

$$\sigma_i^R(\tau) = 2\epsilon_{ijk}\,x^j(\tau)\,\dot{x}^k(\tau) - x^i(\tau)\dot{x}^4(\tau) + x^4(\tau)\dot{x}^i(\tau) \quad (5.6)$$

where ϵ_{ijk} is the totally antisymmetric symbol with $\epsilon_{123} = 1$ and the sphere radius is set equal to one thanks to conformal symmetry.

The choice (5.5)–(5.6) endows the operators with two global supercharges, a combination of both Poincaré and conformal supersymmetries. When the loop lies on a great sphere $S^2 \subset S^3$, extra relations between loop variables $x^\mu(\tau)$ and their derivatives enhance supersymmetry to eight unbroken supercharges (1/8 BPS). We will see below that DGRT loops on a two-sphere show peculiar localization properties that allow for the exact evaluation of their expectation values.

5.2 Localization of DGRT Wilson Loops on S^2

We shall focus on the subset of DGRT Wilson loops in the fundamental representation of $SU(N)$ along non-intersecting closed curves $\mathcal{C}$ on $S^2 = \{x^\mu : \sum_{i=1}^3 x_i^2 = 1\,,\ x_4 = 0\}$. The holonomy of the generalized connection in the Wilson loop operator (5.5) with (5.5)–(5.6)

$$\mathcal{W}[\mathcal{C}] = \frac{1}{N}\,\mathrm{tr}\mathcal{P}\exp\left[\oint \left(i A_i\left(x(\tau)\right)\dot{x}^i(\tau) + m^i(\tau)\,\phi_i(x(\tau))\right)d\tau\right] \quad (5.7)$$

has a triplet of couplings $m_i(\tau) = \epsilon_{ijk}x^j(\tau)\,\dot{x}^k(\tau)$ that draws a curve on a "dual" two-sphere, as explained in caption of Fig. 5.1. They are engineered to preserve the same four supercharges independently of the contour.

Perturbative computations [26, 27, 29] suggested that the expectation values and their quantum correlators are reproduced exactly by analogous observables in 2d bosonic Yang-Mills (YM_2) on S^2, in the sector where instantonic contributions of the

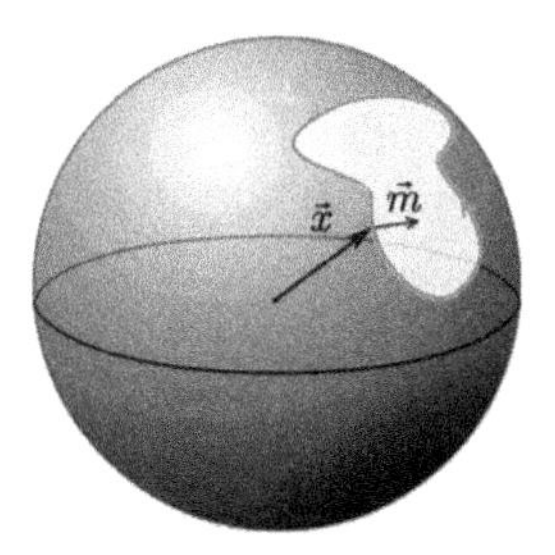

Fig. 5.1 Spacetime path of a DGRT Wilson loop on a two-sphere. The position vector $\vec{x} = (x_1, x_2, x_3)$ stems from the origin of $\mathbb{R}^3 = \{x^\mu : x_4 = 0\}$ and sweeps the loop $\mathcal{C}$ (*red line*). The local couplings to the three scalars can be suggestively rewritten in a geometric fashion as the vector $\vec{m} = \vec{x} \times \dot{\vec{x}}$ tangent to the sphere and orthogonal to $\mathcal{C}$. Picture taken from [28]

2d gauge field are excluded [30]. The proposal was extended in [31] to all correlators of 1/8-BPS DGRT loops and *chiral primary operators* (CPOs) of the form

$$\mathcal{O}_J(x) \equiv \mathrm{tr}\left(x_i\, \phi^i(x) + i\phi^4(x)\right)^J , \qquad x^i \in S^2 , \qquad i = 1, 2, 3 . \tag{5.8}$$

They are 1/2-BPS operators carrying J charge units under a $U(1)$ subgroup of the $SU(4)$ R-symmetry group. By construction they are position-dependent combinations of scalar fields that render the n-point functions $\langle \mathcal{O}_{J_1}(x_1) \ldots \mathcal{O}_{J_n}(x_n) \rangle$ independent of the insertion points, tree-level exact and globally supersymmetric with four supercharges on S^2 [32].

When also Wilson loops are present, mixed correlators are no longer protected from radiative corrections, but retain two supercharges [31, 33] which allow to set up the machinery of supersymmetric localization. The authors of [31] supported the conjecture that mixed correlation functions

$$\langle \mathcal{W}_{\mathcal{R}_1}(\mathcal{C}_1)\, \mathcal{W}_{\mathcal{R}_2}(\mathcal{C}_2) \ldots \mathcal{O}_{J_1}(x_1)\, \mathcal{O}_{J_2}(x_2) \ldots \rangle , \qquad \mathcal{C}_1, \mathcal{C}_2, \ldots \subset S^2 , \;\; x_1, x_2, \ldots \in S^2 \tag{5.9}$$

exactly equal the expectation value

$$\left\langle \widetilde{\mathcal{W}}_{\mathcal{R}_1}(\mathcal{C}_1)\, \widetilde{\mathcal{W}}_{\mathcal{R}_2}(\mathcal{C}_2) \ldots \mathrm{tr}\left(i *_{2d}\; \tilde{F}(x_1)\right)^{J_1} \mathrm{tr}\left(i *_{2d}\; \tilde{F}(x_2)\right)^{J_2} \ldots \right\rangle_{\text{0-instanton}} \tag{5.10}$$

in ordinary bosonic Yang-Mills on S^2 with coupling constant $\tilde{g}^2_{\mathrm{YM}} = -g^2_{\mathrm{YM}}/(2\pi)$, where only perturbative contributions around the trivial vacuum $\tilde{A}_{\tilde{\mu}} = 0$ with $\tilde{\mu} = 1, 2$ (up to gauge transformations) are included. Under this map, DGRT loops are replaced by their bosonic counterparts in YM_2, while the CPOs by integer powers of the 2d Hodge star on S^2 of the YM_2 field strength. The statement holds at any value of the coupling and does *not* assume the planar limit in (5.9).

Bosonic YM on arbitrary 2d Riemann surfaces can be exactly solved using localization methods [34]. The computation of (5.10) in the zero-instanton sector[5] is eventually captured by a Gaussian Hermitian multi-matrix model [31]

$$\frac{1}{\tilde{Z}}\int [dX_1]\,[dX_2]\ldots[dY_1]\,[dY_2]\ldots \mathrm{tr}_{\mathcal{R}_1} e^{X_1}\mathrm{tr}_{\mathcal{R}_2} e^{X_2}\ldots \mathrm{tr}\,Y_1^{J_1}\,\mathrm{tr}\,Y_2^{J_2}\ldots e^{-S_{\mathrm{mm}}[X,Y]}$$
$$\text{with}\quad \tilde{Z}\equiv\int [dX_1]\,[dX_2]\ldots[dY_1]\,[dY_2]\ldots e^{-S_{\mathrm{mm}}[X,Y]}\,. \tag{5.11}$$

The action S_{mm} is a quadratic form (hence the name *Gaussian*) in the self-adjoint matrices X's and Y's with coefficients depending only on the topology (the relative positions of the local operators with respect to the loops) and geometry (the areas singled out by the loops on the spherical surface) of the system. In particular, the matrix model is insensitive to the absolute positions of the loops, so the expectation values (5.9) and (5.10) do not depend on the insertion points $x_1, x_2, \ldots$.

What is surprising about the localization result (5.9) is that the quantum average of an arbitrary number of local and non-local operators in a non-trivial field theory in four dimensions is captured by a zero-dimensional integral (5.11). This is more impressive from the viewpoint of diagrammatical methods, in which even low-order results for (5.9) arise from non-trivial rearrangements of finite parts and divergence cross-cancellations—e.g. up to three loops in [36]—especially when the geometric configuration of the loops and operators on S^2 lacks some symmetry—e.g. a Wilson loop placed on a great circle of S^2 and a CPO displaced from the symmetry axis [15, 37].

The conjecture above has been thoroughly scrutinized at weak/strong coupling in a number of special cases [31, 38–44],[6] extended to't Hooft loop operators [45] and instrumental in deriving an all-loop expression for a *non-BPS* quantity called *Bremsstrahlung function* [46, 47] beautifully tested against its integrability prediction [48, 49], see also [15, 50]. The rich interplay between localization and AdS/CFT holography, covering some of these results, is contextualized in the review [51].

Of particular interest for the rest of the chapter is the restriction to a subsector of Wilson loop operators (5.7) which is amenable to a direct evaluation of their Gaussian matrix model for any g_{YM} and N.

5.2.1 1/2-BPS Circular Wilson Loop

The most symmetric configuration in (5.7) is a geodesic line along a great circle of S^2 [14, 26, 52]

[5]It can be computed with the matrix integral technology in [35] and references therein.

[6]However there exists a claim of disagreement in the subleading order of correlators of latitude loops at strong coupling [43].

$$\mathcal{W}_{\text{circle}} = \frac{1}{N}\text{tr}\mathcal{P}\exp\left[\int\left(iA_\mu x^\mu(\tau)+\phi^3\right)d\tau\right], \qquad x^\mu(\tau)=(\cos\tau,\ \sin\tau,\ 0,\ 0)\ . \tag{5.12}$$

The operator preserves the bosonic group $SL(2,\mathbb{R})\times SU(2)\times SO(5)$ of the $PSU(2,2|4)$ symmetry of $\mathcal{N}=4$ SYM. Together with the 16 supercharges of the loop, it forms the orthosymplectic group $OSp(4^*|4)$. This is the same symmetry of the Wilson line below (5.4), but the explicit embedding inside $PSU(2,2|4)$ is different (see [26]) and the operator has non-trivial expectation value

$$\langle\mathcal{W}_{\text{circle}}\rangle = \frac{1}{\tilde{Z}}\int[dM]\ \text{tr}\left(e^M\right)e^{-\frac{2}{g_{\text{YM}}^2}\text{tr}M^2} \quad \text{with} \quad \tilde{Z}=\int[dM]\ e^{-\frac{2}{g_{\text{YM}}^2}\text{tr}M^2} \tag{5.13}$$

$$= \frac{1}{N}L^1_{N-1}\left(-\frac{g_{\text{YM}}^2}{4}\right)e^{\frac{g_{\text{YM}}^2}{8}} \tag{5.14}$$

$$= \frac{2}{\sqrt{\lambda}}I_1\left(\sqrt{\lambda}\right)+\frac{\lambda}{48\,N^2}I_2\left(\sqrt{\lambda}\right)+\frac{\lambda^2}{1280\,N^4}I_4\left(\sqrt{\lambda}\right)+O\left(N^{-6}\right). \tag{5.15}$$

The matrix integral from (5.11) evaluates to a generalized Laguerre polynomial L_n^m which in the't Hooft limit reduces to a power series in N^{-2} with coefficients being modified Bessel functions I_α of the first kind.[7]

The matrix model (5.13) was first argued from a two-loop analysis at $\lambda \ll 1$ in the planar limit [53], which showed that interacting Feynman diagrams do not contribute up to this order in Feynman gauge. Postulating that a similar mechanism carries over to any loop order, the matrix integral arises because the combined gauge/scalar propagator between two points on the circle is constant and interacting diagrams sum to zero. Summing over all constant *ladder* diagrams[8] is a combinatorics problem that leads to the zero-dimensional integral (5.13).

Postulating that all interacting diagrams vanish, an elegant argument [54] supported (5.13) at finite N using the observation that an "anomalous" conformal map transforms the trivial straight Wilson line (below (5.4)) into a circle. A subtle change in their global properties when one loop point is taken to infinity is responsible for the non-trivial expectation value of the circular loop.

A rigorous derivation of (5.13) was shown in [2] by reformulating $\mathcal{N}=4$ SYM on a four-sphere in such a way that the path-integral localizes on a simple Gaussian matrix model. The circular Wilson loop can be conformally mapped to S^4 and computed as an observable in the matrix model, leading to the expected result (5.13).

The matrix model of the circular loop is consistent with the AdS/CFT prediction as reproduced by string theory, namely the leading behaviour at large λ and N [53] and the behaviour at leading λ and all N^{-1}-corrections from D3-branes [55, 56] and D5-branes [57, 58].

The subleading correction in λ at large N was studied via a semiclassical string calculation: the one-loop contribution, encoding fluctuations above the classical solu-

[7] A derivation and other practical expansions are in [53, 54].

[8] These are diagrams that "stretch" across the circular loop without carrying interaction vertices.

tion (schematically, in the form of 3.98), was formally written down in [19], explicitly evaluated in [20] (see also [59]) using the Gel'fand-Yaglom method (appendix B), reconsidered in [60] with a different choice of boundary conditions and reproduced again in [21] with the heat kernel method (see also Sect. 1.4.1). The semiclassical result obtained from the string sigma-model ("sm") [20, 21]

$$\log\langle\mathcal{W}_{\text{circle}}\rangle_{\text{sm}} = \sqrt{\lambda} - \frac{3}{4}\log\lambda + \log c + \frac{1}{2}\log\frac{1}{2\pi} + O(\lambda^{-1/2}) \tag{5.16}$$

was determined up to an unknown contribution of ghost zero modes (the constant c, see comments below) originating from the one-loop effective action contribution *and* an unknown, overall numerical factor in the measure of the partition function. However, no agreement was found with the analogue subleading correction in the strong coupling expansion for $\lambda \gg 1$ and $N = \infty$ from localization ("loc") (5.15)

$$\log\langle\mathcal{W}_{\text{circle}}\rangle_{\text{loc}} = \log\tfrac{2}{\sqrt{\lambda}}I_1(\sqrt{\lambda}) = \sqrt{\lambda} - \frac{3}{4}\log\lambda + \frac{1}{2}\log\frac{2}{\pi} + O(\lambda^{-1/2})\,, \tag{5.17}$$

The term proportional to $\log\lambda$ in (5.17) is argued to originate from the $SL(2,\mathbb{R})$ ghost zero modes on the disc [54]. The discrepancy between (5.16) and (5.17) occurs in the λ-independent part.

The situation becomes even worse when considering a loop (5.12) winding k-times around itself [20, 61], where also the functional dependence on k is failed by the one-loop string computation. Different group representations of the circular Wilson loops were also considered: for the k-symmetric and k-antisymmetric representations, whose gravitational description is given in terms of D3- and D5-branes respectively, the first stringy correction [62] does not match the localization result. Interestingly, the Bremsstrahlung function of $\mathcal{N} = 4$ SYM derived from localization arguments (see end of Sect. 5.2 is instead correctly reproduced [63] through a one-loop computation around the classical cusp solution [14, 26].

5.2.2 1/4-*BPS* Latitude Wilson Loops

The family of circles of constant radius on S^2 [26] (also [64]) can be parametrized by

$$\mathcal{W}_{\text{latitude}} = \frac{1}{N}\text{tr}\mathcal{P}\exp\left[\int\left(i A_\mu \dot{x}^\mu(\tau) + m^i(\tau)\,\phi_i\right)d\tau\right] \tag{5.18}$$

with, as in Fig. 5.2,

$$\begin{aligned} x^\mu(\tau) &= (\sin\phi_0\cos\tau,\ \sin\phi_0\sin\tau,\ \cos\phi_0,\ 0)\,, \\ m^i(\tau) &= \sin\phi_0\,(-\cos\phi_0\cos\tau,\ -\cos\phi_0\sin\tau,\ \sin\phi_0)\,. \end{aligned} \tag{5.19}$$

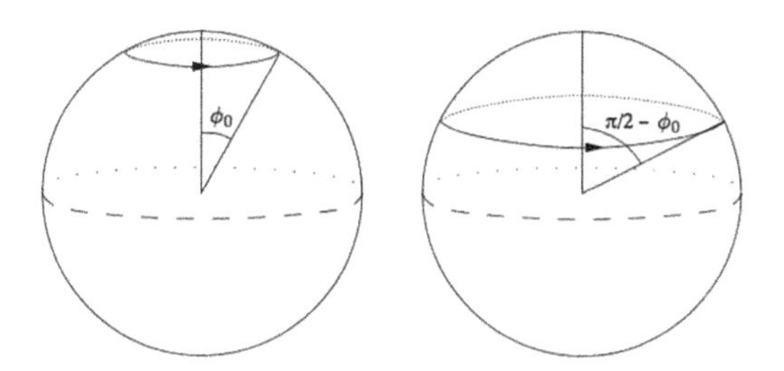

Fig. 5.2 DGRT latitude Wilson loop in the original parametrization (5.19). In the *left panel*, the spacetime path runs along a parallel of S^2 (hence the name *latitude*), located at angular distance ϕ_0 from the north pole. The loop lies on a spherical surface, here embedded in flat space with Cartesian coordinates x_1, x_2, x_3. In the right panel, the path in the scalar-coupling space m_1, m_2, m_3 is induced by the spatial loop $x^i(\tau)$ through the supersymmetric couplings (5.5)–(5.6). Courtesy of [26]

When the loop $x^\mu(\tau)$ coincides with the great circle ($\phi_0 = \frac{\pi}{2}$), the operator couples to a constant direction in the R-symmetry space and one recovers the half-supersymmetric circle in the last section. A generic loop preserves a $SU(2) \times U(1) \times SU(2)$ bosonic subgroup of $PSU(2,2|4)$ and 8 supercharges, which can be organized to form a $SU(2|2)$ supergroup [26].

Circular loops with coupling along latitudes of an S^2 in R-symmetry space [65]

$$\begin{aligned} x^\mu(\tau) &= (\cos\tau,\ \sin\tau,\ 0,\ 0)\ , \\ m^i(\tau) &= (\sin\theta_0\cos\tau,\ \sin\theta_0\sin\tau,\ \cos\theta_0) \end{aligned} \tag{5.20}$$

are still 1/4 BPS and conformally equivalent—upon a dilatation and a translation along the x_3-axis—to the operator defined by (5.19) with $\theta_0 \equiv \frac{\pi}{2} - \phi_0$. We will refer (somewhat improperly) to the operators (5.18) with (5.20) as latitude Wilson loops from now on. The circular limit is then recovered for $\theta_0 = 0$.

Strong evidences [65], later substantially[9] proved using supersymmetric localization [33], relate these operators to the matrix model of the circular loop (5.13)–(5.15) by a rescaling of the YM coupling $g_{\rm YM} \to g_{\rm YM}\cos\theta_0$ (and $\lambda \to \lambda\cos^2\theta_0$ in planar limit).

5.3 Semiclassical Strings for Latitude Wilson Loops

In Sect. 5.2.1 we have seen that the matching between localization (5.17) and sigma-model perturbation theory (5.16) is a thorny issue beyond the supergravity approximation. In order to gain further intuition, we re-examine this issue addressing the problem of how to possibly eliminate the ambiguity related to the partition function

[9]The proof is not completed at the same level of rigour of [2] for the circle: one still needs to compute the one-loop determinants around the localization locus and effectively prove that the Wilson loops localizes in YM_2 on S^2.

measure (the constant c in (5.16)). A similar direction was pursued in parallel in [66], which we will comment on in Sect. 5.7.

To this end, we consider the string dual to a latitude Wilson loop (5.18), parameterized by the angle θ_0 in (5.20). According to (1.11), the expectation value of a latitude loop should equate the partition function for the $AdS_5 \times S^5$ superstring. Therefore, at large λ and infinite N we evaluate the corresponding string one-loop path-integral, analogously to what was done in [20, 61] only for the circular loop at $\theta_0 = 0$. We finally calculate the *ratio* between the partition functions corresponding to latitude loop and a circular one. Our underlying assumption, suggested in [20], is that the measure is actually independent on the geometry of the worldsheet associated to the Wilson loop,[10] and therefore in such ratio measure-related ambiguities should simply cancel. It appears non-trivial to actually prove a background independence of the measure, whose diffeomorphism-invariant definition includes in fact explicitly the worldsheet fields (e.g. the discussion in [69]). This working hypothesis seems however a reasonable one, especially in light of the absence of zero mode in the classical solutions here considered[11] and of similar ratios between pairs of string partition functions—"antiparallel-lines" Wilson loop with straight line [70] and cusped Wilson loop with straight line [63]—where a perfect agreement exists between sigma-model perturbation theory and localization/integrability results (see end of Sect. 5.2).

The ratio of interest follows from the expectation value of the circular loop $\langle \mathcal{W}(\lambda, \theta_0 = 0)\rangle \equiv \langle \mathcal{W}_{\text{circle}}\rangle$ (5.17) and the one of the latitude loop $\langle \mathcal{W}(\lambda, \theta_0)\rangle \equiv \langle \mathcal{W}_{\text{latitude}}\rangle = \langle \mathcal{W}_{\text{circle}}\rangle|_{\lambda\to\lambda\cos^2\theta_0}$ (see text below 5.20)

$$\frac{\langle \mathcal{W}(\lambda, \theta_0)\rangle}{\langle \mathcal{W}(\lambda, 0)\rangle}\bigg|_{\text{loc}} = e^{\sqrt{\lambda}(\cos\theta_0 - 1)}\left[\cos^{-3/2}\theta_0 + O(\lambda^{-1/2})\right] + O\left(e^{-\sqrt{\lambda}}\right), \tag{5.21}$$

where in the large-λ expansion only the dominant exponential contribution is kept. In terms of string effective action $\Gamma(\lambda, \theta_0) \equiv -\log Z(\lambda, \theta_0) = -\log\langle \mathcal{W}(\lambda, \theta_0)\rangle$, this leads to the prediction

$$\log\frac{\langle \mathcal{W}(\lambda, \theta_0)\rangle}{\langle \mathcal{W}(\lambda, 0)\rangle}\bigg|_{\text{loc}} = [\Gamma(\theta_0 = 0) - \Gamma(\theta_0)]_{\text{loc}} = \sqrt{\lambda}(\cos\theta_0 - 1) - \frac{3}{2}\log\cos\theta_0 + O(\lambda^{-1/2}), \tag{5.22}$$

where the leading term comes from the regularized minimal-area surface of the strings dual to these Wilson loops (Sect. 5.4), while the semiclassical fluctuations in the string sigma-model account for the subleading correction (Sects. 5.5 and 5.6). For the latter

[10] About the *topological* contribution of the measure, its relevance in canceling the divergences occurring in evaluating quantum corrections to the string partition function has been first discussed in [19] after the observations of [67, 68]. We use this general argument below, see discussion around (5.79).

[11] In presence of zero modes of the classical solution, a possible dependence of the path-integral measure on the classical solution comes from the integration over collective coordinates associated to them, see arguments in [18].

point, we follow the strategy method followed by [20]—namely splitting the 2d determinants into an infinite product of 1d determinants solved with the Gel'fand-Yaglom method (see Sect. 1.4.1)—but our case is substantially more complicated due to the non-diagonal matrix structure of the fermionic-fluctuation operator for arbitrary θ_0, preventing us from factorizing the value of the fermionic determinants into a product of two contributions. The summation of the 1d Gel'fand-Yaglom determinants is quite difficult, due to the appearance of some Lerch-type functions, and we were not able to obtain a direct analytic result. We resort therefore to a numerical approach.

We anticipate that our analysis shows that the disagreement between sigma-model and localization results (5.22) is *not* washed out yet. Within a certain numerically accuracy, we claim that the discovered θ_0-dependent discrepancy is very well quantified as

$$\log \frac{\langle \mathcal{W}(\lambda, \theta_0)\rangle}{\langle \mathcal{W}(\lambda, 0)\rangle}\bigg|_{\text{sm}} = \sqrt{\lambda}\,(\cos\theta_0 - 1) - \frac{3}{2}\log\cos\theta_0 + \log\cos\frac{\theta_0}{2} + O(\lambda^{-1/2})\,, \tag{5.23}$$

suggesting that the discrepancy from (5.22) should be $\log\cos\frac{\theta_0}{2}$. We will comment on this result at length in Sects. 5.7 and 5.8.

5.4 Classical Solution

The classical string surface dual to latitude loops was first presented in [64] and discussed in details in [26, 65, 66]. Here it is rederived using a strategy described at the end of Sect. 3.2. In $AdS_5 \times S^5$ equipped with coordinates (cf. 3.2-3.3)

$$\begin{aligned} ds^2_{AdS_5\times S^5} = &-\cosh^2\rho\, dt^2 + d\rho^2 + \sinh^2\rho\left(d\chi^2 + \cos^2\chi\, d\psi^2 + \sin^2\chi\, d\varphi_1^2\right) \\ &+ d\theta^2 + \sin^2\theta\, d\phi^2 + \cos^2\theta\left(d\vartheta_1^2 + \sin^2\vartheta_1\left(d\vartheta_2^2 + \sin^2\vartheta_2\, d\varphi_2^2\right)\right) \end{aligned} \tag{5.24}$$

one makes the following ansatz for the classical configuration

$$\begin{aligned} &t = 0, \quad \rho = \rho(\sigma), \quad \chi = 0, \quad \psi = \tau, \quad \varphi_1 = 0, \\ &\theta = \theta(\sigma), \quad \phi = \tau, \quad \vartheta_1 = 0, \quad \vartheta_2 = 0, \quad \varphi_2 = 0\,, \\ &\tau \in [0, 2\pi)\,, \quad \sigma \in [0, \infty)\,. \end{aligned} \tag{5.25}$$

The ansatz (5.25) is for a surface in $H^3 \times S^2 \subset AdS_5 \times S^5$ that does not propagate in time—neither the 3d hyperbolic space H^3 nor the two-sphere S^2 have a timelike direction—but sweeps out a Euclidean surface embedded in a Lorentzian target space. If we demand that $\rho(0) = \infty$ and $\theta(0) = \theta_0$ at the worldsheet boundary ($\sigma = 0$), then the ansatz parametrizes a open string[12] that ends on a unit circle at the AdS_5 boundary and on a latitude located at polar angle θ_0 on $S^2 \subset S^5$. When we impose the string equations of motion, the functions $\rho(\sigma)$ and $\theta(\sigma)$ read

[12] There exist other solutions with more wrapping in S^5, but they are not supersymmetric [65].

$$\sinh\rho(\sigma) = \frac{1}{\sinh\sigma}, \qquad \cosh\rho(\sigma) = \frac{1}{\tanh\sigma},$$
$$\sin\theta(\sigma) = \frac{1}{\cosh(\sigma_0 \pm \sigma)}, \qquad \cos\theta(\sigma) = \tanh(\sigma_0 \pm \sigma)\,. \tag{5.26}$$

We defined the convenient angular parameter $\sigma_0 \in [0, \infty)$

$$\tanh\sigma_0 \equiv \cos\theta_0 \tag{5.27}$$

to sets the position of the latitude at angular position $\theta_0 \in [0, \frac{\pi}{2}]$. Here the angular coordinate θ of S^5 spans the interval $[-\frac{\pi}{2}, \frac{\pi}{2}]$. The double sign in (5.26) accounts for the existence of two solutions, effectively doubling the range of θ_0: the stable (unstable) configuration—with the upper (lower) sign in (5.26)—is the one that minimizes (maximizes) the action functional and wraps the north pole $\theta = 0$ (south pole $\theta = \pi$) of S^5.

The dual gauge-theory operator is a one-parameter class of 1/4-BPS Wilson loops that interpolates between two notable cases. The 1/2-BPS circular case (Sect. 5.2.1) falls under this family when the latitude in S^2 shrinks to a point for $\theta_0 = 0$, which implies $\theta(\sigma) = 0$ and $\sigma_0 = +\infty$ from (5.26)–(5.27). In this case the string propagates only in the H^3 subspace of the AdS space. The other case is the circular 1/4-BPS Zarembo Wilson loop (in Table (5.4)) when the worldsheet extends over a maximal circle of S^2 for $\theta_0 = \frac{\pi}{2}$ and $\sigma_0 = 0$ [18].[13]

For the semiclassical analysis we prefer to work with the stereographic coordinates[14] v^m $(m = 1, 2, 3)$ of $S^3 \subset AdS_5$ and w^n $(n = 1, \ldots 5)$ of S^5

$$ds^2_{AdS_5\times S^5} = -\cosh^2\rho dt^2 + d\rho^2 + \sinh^2\rho\,\frac{dv_m dv_m}{(1+\frac{v^2}{4})^2} + \frac{dw_n dw_n}{(1+\frac{w^2}{4})^2}\,, \tag{5.28}$$
$$v^2 = \sum_{m=1}^{3} v_m v_m\,, \qquad w^2 = \sum_{n=1}^{5} w_n w_n\,,$$

where the classical solution reads

$$\begin{aligned} &t = 0, \quad \rho = \rho(\sigma), \quad v_1 = 2\sin\tau, \quad v_2 = 2\cos\tau, \quad v_3 = 0\,,\\ &w_1 = w_2 = 0, \quad w_3 = 2\cos\theta(\sigma), \quad w_4 = 2\sin\theta(\sigma)\sin\tau, \quad w_5 = 2\sin\theta(\sigma)\cos\tau\,. \end{aligned} \tag{5.29}$$

[13] See also [71] for an analysis of the contribution to the string partition function due to (broken) zero modes of the solution in [18].

[14] The reason is that the initial angular coordinates (5.24) in the background (5.25) would not yield a bosonic quadratic Lagrangian in the standard form for the kinetic terms of the eight physical fields.

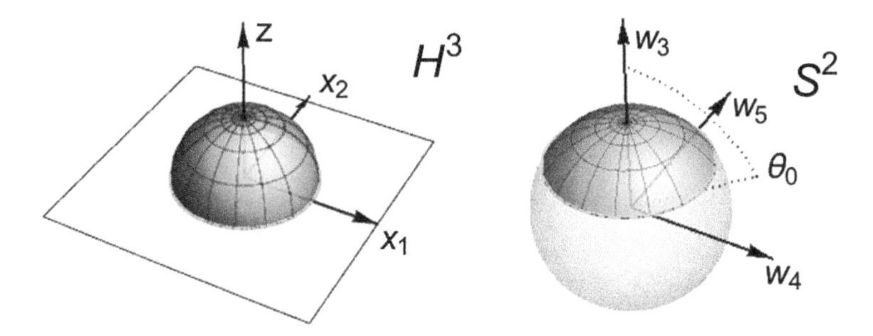

Fig. 5.3 Stable classical solution dual to the latitude loops (5.31) with upper sign in (5.26). Lines of constant τ and σ are meridians and latitudes on the geodesic dome inside H^3 and on S^2. The worldsheet (*orange*) is diffeomorphic to a disk, ends on a unit circle and a latitude with $\theta_0 \in [0, \frac{\pi}{2}]$ (both *green* in the two subspaces) and reaches the north pole $(w_3, w_4, w_5) = (1, 0, 0)$ of the two-sphere. Selecting the lower sign in (5.26) produces an unstable solution covering the shaded region of the sphere between the latitude and the south pole $(w_3, w_4, w_5) = (-1, 0, 0)$. The axial symmetry around z and w_3 is a manifestation of the $SO(2)$ symmetry discussed in the text

Alternatively, we could translate the solution into the AdS Poincaré patch (3.4)

$$ds^2_{AdS_5 \times S^5} = \frac{-dx_0^2 + dx_1^2 + dx_2^2 + dx_3^2 + dz^2}{z^2} + \frac{dw_n dw_n}{(1 + \frac{w^2}{4})^2}, \tag{5.30}$$

with

$$\begin{aligned} &x_1 = \tanh \rho(\sigma) \cos \tau, \qquad x_2 = \tanh \rho(\sigma) \sin \tau, \qquad x_0 = x_3 = 0, \qquad z = \frac{1}{\cosh \rho(\sigma)}, \\ &w_1 = w_2 = 0, \quad w_3 = 2 \cos \theta(\sigma), \quad w_4 = 2 \sin \theta(\sigma) \sin \tau, \quad w_5 = 2 \sin \theta(\sigma) \cos \tau. \end{aligned} \tag{5.31}$$

In Sect. 5.2.2 we recalled that the gauge-theory operator is invariant under a $SU(2|2)$ subgroup of the superconformal group $PSU(2, 2|4)$ of $\mathcal{N} = 4$ SYM. The bosonic symmetries $SU(2) \times U(1) \times SU(2) \subset SU(2|2)$ have a geometric interpretation as the symmetry group $SO(3) \times SO(2) \times SO(3)$ of the classical surface. It is easy to see the action of the three factors by expressing the solution into the embedding coordinates (after (3.4) and some index reshuffling)

$$\begin{aligned} &X_0 = \cosh \rho(\sigma), \quad X_1 = \sinh \rho(\sigma) \cos \tau, \quad X_2 = \sinh \rho(\sigma) \sin \tau, \quad X_3 = X_4 = X_5 = 0, \\ &Y_1 = \sinh \theta(\sigma) \cos \tau, \quad Y_2 = \sinh \theta(\sigma) \sin \tau, \quad Y_3 = \cosh \theta(\sigma), \quad Y_4 = Y_5 = Y_6 = 0. \end{aligned} \tag{5.32}$$

The action of $SO(3) \times SO(3)$ rotates the linear subspaces (X_3, X_4, X_5) and (Y_4, Y_5, Y_6) where the string motion does not occur, while the $SO(2)$ is a translation in τ that rotates the planes (X_1, X_2) and (Y_1, Y_2) simultaneously.

The induced metric on the worldsheet Σ and its Ricci curvature depend on the latitude angle θ_0 through the conformal factor $\Omega(\sigma)^2 \equiv \sinh^2 \rho(\sigma) + \sin^2 \theta(\sigma)$ $(\sigma^i = (\tau, \sigma))$

$$ds^2_{\Sigma} = h_{ij}d\sigma^i d\sigma^j = \Omega^2(\sigma)\left(d\tau^2 + d\sigma^2\right), \qquad {}^{(2)}R = -\frac{2\,\partial_\sigma^2 \log\Omega(\sigma)}{\Omega^2(\sigma)}. \tag{5.33}$$

From the bosonic action (3.7)

$$S_B = T\int d\tau d\sigma\sqrt{h} \equiv \int d\tau d\sigma \mathcal{L}_B\,, \tag{5.34}$$

the leading contribution to the string partition function comes from the regularized on-shell action

$$S_B^{(0)}(\theta_0) = \frac{\sqrt{\lambda}}{2\pi}\int_0^{2\pi} d\tau \int_{\epsilon_0}^{\infty} d\sigma\ \Omega^2(\sigma) = \sqrt{\lambda}\left(\mp\cos\theta_0 + \frac{1}{\epsilon} + O(\epsilon)\right). \tag{5.35}$$

Following [20] we have chosen to distinguish the cutoff ϵ_0 in the worldsheet coordinate from the cutoff $\epsilon = \tanh\epsilon_0$ in the Poincaré radial coordinate z (5.30). The pole in the IR cutoff ϵ in (5.36) keeps track of the boundary singularity of the AdS metric and it is proportional to the circumference of the boundary circle. The standard regularization scheme (Sect. 3.5.1) consists in subtracting a term

$$S_B^{(0)}(\theta_0) - \sqrt{\lambda}\chi_b(\theta_0) \quad\rightarrow\quad S_B^{(0)}(\theta_0) - \sqrt{\lambda}\chi_b(\theta_0) = \mp\sqrt{\lambda}\cos\theta_0\,, \tag{5.36}$$

which is proportional to the boundary part of the Euler number

$$\chi_b(\theta_0) = \frac{1}{2\pi}\int_0^{2\pi} d\tau\,\sqrt{h}|_{\sigma=\epsilon_0}\ \kappa_g = \frac{1}{\epsilon} + O(\epsilon)\,, \tag{5.37}$$

namely the line integral of the geodesic curvature κ_g of the boundary at $\sigma = \epsilon_0$. The upper-sign solution dominates the string path-integral and is responsible for the leading exponential behaviour in (5.21) and so, in the following, we will restrict to the upper sign in (5.26).

5.5 One-Loop Fluctuation Determinants

We focus on the semiclassical expansion of the string partition function around the stable classical solution (5.29) (taking upper signs in 5.26) and the determinants of the differential operators describing the semiclassical fluctuations around it as in (3.98).

In Sect. 1.4.1 we illustrated the standard methods to quantify the fluctuation determinants. They can be easily computed on the classical surface at $\theta_0 = 0$ with the heat kernel method [19, 21] since the worldsheet reduces to the H^2 geometry which is maximally symmetric (the three isometries form the bosonic subgroup $SL(2,\mathbb{R})$ of the full symmetry group of $OSp(4^*|4)$ mentioned in Sect. 5.2.1). The lack of readily available heat kernel expressions in the general case $\theta_0 \neq 0$ motivates to look for a different way.

The heat kernel method seems to be feasible in a perturbative approach $\theta_0 \sim 0$ when the worldsheet geometry is nearly H^2. This way of proceeding is currently scrutinized in [72] and outlined in Chap. 8. Here we exploit instead the only $SO(2)$ isometry generated by time translations ∂_τ, separate the 2d determinants into infinitely-many 1d spectral problems (Sect. 1.4.1) and proceed with the Gel'fand-Yaglom method (Appendix B).

We will see that our worldsheet operators are singular in $\sigma \in [0, \infty)$ as their principal symbols diverge at $\sigma = 0$. Also, the interval is non-compact, making the spectra continuous and more difficult to deal with. We consequently introduce an IR cutoff at $\sigma = \epsilon_0$ (related to the $\epsilon = \tanh \epsilon_0$ cutoff in z) and one at large values of $\sigma = R$ [20]. While the former is necessary in order to tame the near-boundary singularity, the latter has to be regarded as a mere regularization artifact descending from a small fictitious boundary on the top of the surfaces in H^3 and S^2 (Fig. 5.3). Indeed it will disappear in the one-loop effective action.

5.5.1 Bosonic Sector

The bosonic fluctuation Lagrangian was assembled in Sect. 5.2 of [73] using the equations for embedded manifolds reported in this thesis in Sect. 3.3:

$$\mathcal{L}_B^{(2)} \equiv \Omega^2(\sigma)\, y^T\, \mathcal{O}_B\,(\theta_0)\; y\,, \tag{5.38}$$

where the differential operator $\mathcal{O}_B\,(\theta_0)$ is an 8×8 matrix acting upon the vector of fluctuation fields orthogonal to the worldsheet $y \equiv (y_1\,, \ldots y_8)$. In components it reads

$$\mathcal{O}_B\,(\theta_0) = \mathrm{diag}(\mathcal{O}_1,\, \mathcal{O}_1,\, \mathcal{O}_1,\, \mathcal{O}_2,\, \mathcal{O}_2,\, \mathcal{O}_2,\, \mathcal{O}_{3+},\, \mathcal{O}_{3-})\,(\theta_0)\;, \tag{5.39}$$

where, going to Fourier space ($\partial_\tau \to i\omega$),

$$\mathcal{O}_1 \equiv \frac{1}{\Omega^2(\sigma)}\left[-\partial_\sigma^2 + \omega^2 + \frac{2}{\sinh^2\sigma}\right] \tag{5.40}$$

$$\mathcal{O}_2\,(\theta_0) \equiv \frac{1}{\Omega^2(\sigma)}\left[-\partial_\sigma^2 + \omega^2 - \frac{2}{\cosh^2(\sigma+\sigma_0)}\right] \tag{5.41}$$

$$\mathcal{O}_{3\pm}\,(\theta_0) = \frac{1}{\Omega^2(\sigma)}\left[-\partial_\sigma^2 + \omega^2 - 2 + 3\tanh^2(2\sigma+\sigma_0) \mp 2\omega\tanh(2\sigma+\sigma_0)\right] \tag{5.42}$$

In what follows we assume that we rescale away the conformal factor $\sqrt{h} = \Omega^2(\sigma)$ (as in the analogous computations of [20, 63, 70]) which will not affect the final determinant ratio (5.72) (see discussions in Appendix A of [19] and in [20, 63, 70]) and is actually instrumental for the analysis in appendices B.3 and B.4.

The operator $\mathcal{O}_1$ does not depend on θ_0, and indeed it also appears among the circular Wilson loop fluctuation operators [20]. While its contribution formally cancels in the ratio (5.22), we report it below along with the others for completeness. Both $\mathcal{O}_2(\theta_0)$ and $\mathcal{O}_{3\pm}(\theta_0)$ become massless operators in the circular Wilson loop limit, which is clear for the latter upon an integer shift in ω,[15] irrelevant for the determinant at given frequency, as long as we do not take products over frequencies into consideration. Thus, in this limit one recovers the mass spectrum of the circle (three bosons with mass squared 2 and five massless bosons) and the bosonic partition function of [20].

The eight physical bosonic fields contribute to the one-loop partition function as

$$\mathrm{Det}\mathcal{O}_B(\theta_0) = \mathrm{Det}^3\mathcal{O}_1\,\mathrm{Det}^3\mathcal{O}_2(\theta_0)\,\mathrm{Det}\mathcal{O}_{3+}(\theta_0)\,\mathrm{Det}\mathcal{O}_{3-}(\theta_0)\,. \tag{5.43}$$

To reconstruct the complete bosonic contribution after going to Fourier space, we formally rewrite (5.43) as the infinite product over all the determinants taken at fixed ω

$$\mathrm{Det}_\omega\mathcal{O}_B(\theta_0) = \mathrm{Det}^3_\omega\mathcal{O}_1\,\mathrm{Det}^3_\omega\mathcal{O}_2(\theta_0)\,\mathrm{Det}_\omega\mathcal{O}_{3+}(\theta_0)\,\mathrm{Det}_\omega\mathcal{O}_{3-}(\theta_0)\,,. \tag{5.44}$$

The evaluation of one-dimensional spectral problems is outlined in appendix B.4. The fields satisfy Dirichlet boundary conditions at the endpoints of the compactified interval $\sigma \in [\epsilon_0, R]$. Then we take the limit of the value of the regularized determinants for $R \to \infty$ at fixed ω and ϵ_0. As evident from the expressions below, the limit on the physical IR cutoff (ϵ in z or equivalently ϵ_0 in σ) would drastically change the ω-dependence at this stage and thus would spoil the product over the frequencies. It is a crucial, a posteriori, observation that it is only keeping ϵ_0 *finite* while sending R to infinity that one precisely reproduces the expected large-ω (UV) divergences [19, 73]. This comes at the price of more complicated results for the bosonic (and especially fermionic) determinants. Afterwards we will remove the IR divergence in the one-loop effective action by referring the latitude to the circular solution.

[15] In the language of [73], this shift corresponds to a different choice of orthonormal vectors that are orthogonal to the string surface.

The solutions of the differential equations governing the different determinants are singular for small subset of frequencies. We shall treat apart these special values when reporting the solutions. The solutions of the initial value problems associated to the operators $\mathcal{O}_1$ in (5.40), $\mathcal{O}_2(\theta_0)$ in (B.23)–(B.24) and $\mathcal{O}_{3+}(\theta_0)$ in (5.42) yield the determinants

$$\mathrm{Det}_\omega \mathcal{O}_1 = \begin{cases} e^{|\omega|(R-\epsilon_0)} \dfrac{(|\omega| + \coth \epsilon_0)}{2|\omega|(|\omega|+1)} & \omega \neq 0 \\ R \coth \epsilon_0 & \omega = 0 \end{cases} \tag{5.45}$$

$$\mathrm{Det}_\omega \mathcal{O}_2(\theta_0) = \begin{cases} e^{|\omega|(R-\epsilon_0)} \dfrac{(|\omega| + \tanh(\sigma_0 + \epsilon_0))}{2|\omega|(|\omega|+1)} & \omega \neq 0 \\ R \tanh(\sigma_0 + \epsilon_0) & \omega = 0 \end{cases} \tag{5.46}$$

$$\mathrm{Det}_\omega \mathcal{O}_{3+}(\theta_0) = \begin{cases} \dfrac{e^{R(\omega-1)-\sigma_0-(\omega+1)\epsilon_0}\left(\omega + (\omega+1)e^{2\sigma_0+4\epsilon_0} - 1\right)}{2\left(\omega^2-1\right)\sqrt{1+e^{2\sigma_0+4\epsilon_0}}} & \omega \geq 2 \\ \dfrac{R e^{\sigma_0+2\epsilon_0}}{\sqrt{1+e^{2\sigma_0+4\epsilon_0}}} & \omega = 1 \\ \dfrac{e^{-R(\omega-1)+\sigma_0+(\omega+1)\epsilon_0}}{2(1-\omega)\sqrt{1+e^{2\sigma_0+4\epsilon_0}}} & \omega \leq 0 \end{cases} \tag{5.47}$$

and in view of the relation $\mathcal{O}_{3-}(\theta_0) = \mathcal{O}_{3+}(\theta_0)\,|_{\omega \to -\omega}$, which follows from (5.42), we can easily deduce

$$\mathrm{Det}_\omega \mathcal{O}_{3-}(\theta_0) = \begin{cases} \dfrac{e^{R(\omega+1)+\sigma_0+(-\omega+1)\epsilon_0}}{2(1+\omega)\sqrt{1+e^{2\sigma_0+4\epsilon_0}}} & \omega \geq 0 \\ \dfrac{R e^{\sigma_0+2\epsilon_0}}{\sqrt{1+e^{2\sigma_0+4\epsilon_0}}} & \omega = -1 \\ \dfrac{e^{-R(\omega+1)-\sigma_0-(-\omega+1)\epsilon_0}\left(-\omega + (-\omega+1)e^{2\sigma_0+4\epsilon_0} - 1\right)}{2\left(\omega^2-1\right)\sqrt{1+e^{2\sigma_0+4\epsilon_0}}} & \omega \leq -2\,. \end{cases} \tag{5.48}$$

Notice that a shift of $\omega \to \omega - 1$ in $\mathrm{Det}_\omega \mathcal{O}_{3+}(\theta_0)$ and $\omega \to \omega + 1$ in $\mathrm{Det}_\omega \mathcal{O}_{3-}(\theta_0)$ gives back the symmetry around $\omega = 0$ in the distribution of power-like and exponential large-R divergences which characterizes the other determinants (5.45) and (5.46). Such a shift—also useful for the circular Wilson loop limit as discussed below (5.44)—does not affect the determinant, and we will perform it in Sect. 5.6.

It is easy to take the limit $\sigma_0 \to \infty$ of the bosonic and fermionic determinants to directly obtain the solutions of the spectral problem for the circular loop case $\theta_0 = 0$. The result for $\mathrm{Det}_\omega \mathcal{O}_1$ in (5.45) stays obviously the same, while the limits of (5.46), (5.47), (5.48) and (5.65)–(5.71) become

$$\mathrm{Det}_\omega \mathcal{O}_2(\theta_0=0)=\begin{cases}\frac{e^{|\omega|(R-\epsilon_0)}}{2|\omega|} & \omega\neq 0\\ R & \omega=0\end{cases}\tag{5.49}$$

$$\mathrm{Det}_\omega \mathcal{O}_{3+}(\theta_0=0)=\begin{cases}\frac{e^{(R-\epsilon_0)(\omega-1)}}{2(\omega-1)} & \omega\geq 2\\ R & \omega=1\\ \frac{e^{-(R-\epsilon_0)(\omega-1)}}{2(1-\omega)} & \omega\leq 0\end{cases}\tag{5.50}$$

$$\mathrm{Det}_\omega \mathcal{O}_{3-}(\theta_0=0)=\begin{cases}\frac{e^{(R-\epsilon_0)(\omega+1)}}{2(1+\omega)} & \omega\geq 0\\ R & \omega=-1\\ -\frac{e^{-(R-\epsilon_0)(\omega+1)}}{2(\omega+1)} & \omega\leq -2\,.\end{cases}\tag{5.51}$$

5.5.2 *Fermionic Sector*

The fluctuation analysis in the fermionic sector can be easily carried out here with Sect. 3.4 and the $SO(10)$ Dirac algebra in appendix D.1. We refer to Sect. 5.2 of [73] for details. After gauge-fixing κ-symmetry (3.78), the Lagrangian[16] becomes

$$\mathcal{L}_F^{(2)}=2i\,\Omega^2(\sigma)\,\bar{\Psi}\,\mathcal{O}_F\,(\theta_0)\,\Psi\tag{5.52}$$

with

$$\begin{aligned}\mathcal{O}_F\,(\theta_0)=&\,\frac{i}{\Omega(\sigma)}\left(\Gamma_4\partial_\tau+\Gamma_3\partial_\sigma-a_{34}(\sigma)\Gamma_3+a_{56}(\sigma)\Gamma_{456}\right)\\&+\frac{1}{\Omega(\sigma)^2}\left(\sinh^2\rho(\sigma)\Gamma_{012}+\sin^2\theta(\sigma)\Gamma_{0123456}\right),\end{aligned}\tag{5.53}$$

$$a_{34}(\sigma)=-\frac{1}{2}\frac{d}{d\sigma}\log\Omega(\sigma)\,,\qquad a_{56}(\sigma)=\frac{1}{4}\frac{d}{d\sigma}\log\frac{\cosh\rho(\sigma)+\cos\theta(\sigma)}{\cosh\rho(\sigma)-\cos\theta(\sigma)}\,.$$

In the $\theta_0\to 0$ limit (hence $\theta(\sigma)\to 0$), one gets

$$\mathcal{O}_F\,(\theta_0=0)=i\sinh\sigma\Gamma_4\partial_\tau+i\sinh\sigma\Gamma_3\partial_\sigma-\frac{i}{2}\cosh\sigma\Gamma_3+\frac{i}{2}\sinh\sigma\Gamma_{456}+\Gamma_{012}\,,\tag{5.54}$$

which coincides with the operator found in the circular Wilson loop analysis of [20] (see (5.17) therein), once we go back to Minkowski signature and reabsorb the

[16] In the free Lagrangian $\mathcal{L}_F^{(2)}$ the spinor field Ψ couples only to the classical background (5.25), which lies in the timeslice $t=0$ and has Euclidean signature (5.33). This may cause some issues with the fact that the Green-Schwarz action and the Majorana condition are only defined for a worldsheet of Lorentzian signature. Notice that the analytic continuation of the AdS time t does not affect the signature of the classical solution. Here we simply think of doing the expansion for imaginary worldsheet time τ and only at the end Wick-rotate back to Euclidean signature (5.25).

connection-related Γ_{456}-term via the τ-dependent rotation $\Psi \to \exp\left(-\frac{\tau}{2}\Gamma_{56}\right)\Psi$. In Fourier space this results in a shift of the integer fermionic frequencies ω by one half, turning periodic fermions into anti-periodic ones. In the general case (5.53) we cannot eliminate all the connection-related terms $-a_{34}(\sigma)\Gamma_3+a_{56}(\sigma)\Gamma_{456}$, since the associated normal bundle (3.30) is not *flat* in the sense explained below (3.52). Performing anyway the above τ-rotation at the level of (5.53) has the merit of simplifying the circular limit making a direct connection with known results. This is how we will proceed: for now, we continue with the analysis of the fermionic operator in the form (5.53) without performing any rotation. Then, in Sect. 5.6, we shall take into account the effect of this rotation by relabelling the fermionic Fourier modes in terms of a suitable choice of half-integers.

The analysis of the fermionic operator (5.53) drastically simplifies by means of the projectors

$$\mathcal{P}^{\pm}_{12} \equiv \frac{\mathbb{I}_{32} \pm i\Gamma_{12}}{2}, \quad \mathcal{P}^{\pm}_{56} \equiv \frac{\mathbb{I}_{32} \pm i\Gamma_{56}}{2} \quad \text{and} \quad \mathcal{P}^{\pm}_{89} \equiv \frac{\mathbb{I}_{32} \pm i\Gamma_{89}}{2}, \tag{5.55}$$

which commute with the operator itself and leaves invariant the spinor constraint D.7 and the gauge fixing (3.78). The fermionic operator and the 10d spinor are projected onto 8 orthogonal subspaces labeled by the triplet $\{p_{12}, p_{56}, p_{89} = -1, 1\}$

$$\mathcal{O}_F(\theta_0) = \bigoplus_{p_{12},p_{56},p_{89}=-1,1} \mathcal{O}_F^{p_{12},p_{56},p_{89}}(\theta_0)\,, \tag{5.56}$$

$$\Psi = \bigotimes_{p_{12},p_{56},p_{89}=-1,1} \Psi^{p_{12},p_{56},p_{89}}\,, \tag{5.57}$$

where each 2×2 operator

$$\begin{aligned}\mathcal{O}_F^{p_{12},p_{56},p_{89}}(\theta_0) \equiv\ & \frac{i}{\Omega(\sigma)}\left(\Gamma_4\partial_\tau + \Gamma_3\partial_\sigma - a_{34}(\sigma)\Gamma_3 - i p_{56} a_{56}(\sigma)\Gamma_4\right) \\ & + \frac{1}{\Omega^2(\sigma)}\left(-i p_{12}\sinh^2\rho(\sigma)\Gamma_0 - p_{12}p_{56}\sin^2\theta(\sigma)\Gamma_{034}\right)\end{aligned} \tag{5.58}$$

acts on the eigenstates $\Psi^{p_{12},p_{56},p_{89}}$ of $\{\mathcal{P}^{\pm}_{12}, \mathcal{P}^{\pm}_{56}, \mathcal{P}^{\pm}_{89}\}$ with eigenvalues $\{\frac{1\pm p_{12}}{2}, \frac{1\pm p_{56}}{2}, \frac{1\pm p_{89}}{2}\}$. Notice that the operator defined in (5.58) does not depend on the label p_{89}. The spectral problem reduces to the computation of eight 2d functional determinants[17] (cf. 1.20)

[17] A non-trivial matrix structure is also encountered in the fermionic sector of the circular Wilson loop [20], but the absence of a background geometry in S^5 leads to a simpler gamma structure. It comprises only three gamma combinations (Γ_0, Γ_4, Γ_{04}), whose algebra allows their identification with the three Pauli matrices without the need of labelling the subspaces.

$$
\begin{aligned}
\mathrm{Det}\mathcal{O}_F\,(\theta_0) &= \prod_{p_{12},p_{56},p_{89}=\pm 1} \mathrm{Det}\mathcal{O}_F^{p_{12},p_{56},p_{89}}\,(\theta_0) \\
&= \prod_{\omega\in\mathbb{Z}} \mathrm{Det}_\omega[(\mathcal{O}_F^{1,1,1}(\omega))^2]^2 \mathrm{Det}_\omega[(\mathcal{O}_F^{1,1,1}(-\omega))^2]^2\ .
\end{aligned}
\tag{5.59}
$$

The second line follows from a deeper look at the properties of $\mathcal{O}_F^{p_{12},p_{56},p_{89}}$ (Appendix C.1 [74]). In Fourier space we have

$$
\begin{aligned}
\mathcal{O}_F^{1,1,1}(\theta_0) \equiv \Big[&\frac{i}{\Omega(\sigma)}\big(-i\omega\sigma_2+\sigma_1\partial_\sigma - a_{34}(\sigma)\sigma_1 + i a_{56}(\sigma)\sigma_2\big) \\
&+\frac{1}{\Omega^2(\sigma)}\big(\sinh^2\rho(\sigma)\sigma_3 - \sin^2\theta(\sigma)\mathbb{I}_2\big)\Big]\otimes M \equiv \tilde{\mathcal{O}}_F^{1,1,1}\otimes M\ ,
\end{aligned}
\tag{5.60}
$$

where $M=\sigma_2\otimes\mathbb{I}_4\otimes\sigma_1$. For simplicity of notation, from now on we will denote with $O_F^{1,1,1}(\theta_0)$ the first factor in the definition above. In a similar spirit to the analysis of the bosonic sector, we start to find the solutions of the homogeneous problem

$$
\mathcal{O}_F^{1,1,1}\,(\theta_0)\,\bar{f}(\sigma)=0\,,\qquad\qquad \bar{f}(\sigma)\equiv(f_1(\sigma),\,f_2(\sigma))^T\ . \tag{5.61}
$$

Decoupling the system of equations

$$
\Big(-\sin^2\theta(\sigma)+\sinh^2\rho(\sigma)\Big)\,f_1(\sigma)+i\Omega(\sigma)\,(\partial_\sigma-\omega-a_{34}(\sigma)+a_{56}(\sigma))\,f_2(\sigma)=0\,, \tag{5.62}
$$

$$
\Big(-\sin^2\theta(\sigma)-\sinh^2\rho(\sigma)\Big)\,f_2(\sigma)+i\Omega(\sigma)\,(\partial_\sigma+\omega-a_{34}(\sigma)-a_{56}(\sigma))\,f_1(\sigma)=0\,, \tag{5.63}
$$

we arrive at a Schrödinger-type equation

$$
f_1''(\sigma)-\Big(\frac{1}{2\sinh^2\sigma}-\frac{1}{2\cosh^2(\sigma+\sigma_0)}+\Big(\frac{1}{2\tanh\sigma}+\frac{\tanh(\sigma+\sigma_0)}{2}-\omega\Big)^2\Big)\,f_1(\sigma)=0 \tag{5.64}
$$

for a fictitious particle on a semi-infinite line and subject to a supersymmetric potential[18] $V(\sigma)=-W'(\sigma)+W^2(\sigma)$ derived from the prepotential $W(\sigma)=\frac{1}{2\tanh\sigma}+\frac{\tanh(\sigma+\sigma_0)}{2}-\omega$. Traces of supersymmetry are not surprising, as they represent a vestige of the supercharges unbroken by the classical background.

As for the bosonic case, we do not report the solutions of the equations above and we proceed to the evaluation of the determinants using the results of appendix B.3, namely using Dirichlet boundary conditions for the square of the first-order differential operator. Having in mind the solutions above and how they enter in (B.9) and (B.17), it is clear that already the *integrand* in (B.22) is significantly complicated. A simplification occurs by recalling that our final goal is taking the $R\to\infty$ limit of all determinants and combine them in the ratio of bosonic and fermionic contributions. As stated above in the bosonic analysis and shown explicitly

[18]The same property is showed by (5.26) in [20].

below, for the correct large ω divergences to be reproduced, it is crucial to send $R \to \infty$ while keeping ϵ_0 finite.

The determinant of the operator $O_F^{1,1,1}$ for modes $\omega \neq \{-1, 0, 1\}$ reads for large R

$$\begin{aligned}
\text{Det}_{\omega\geq 2}[(\mathcal{O}_F^{1,1,1})^2] &= \frac{a_0\, e^{2\omega(R-\epsilon_0)}}{\omega^2\,(1+\omega)^2(\omega-1)}\left[a_1\,\Phi\left(e^{-2\epsilon_0},1,\omega\right)+a_2\,\Phi(-e^{-2(\sigma_0+\epsilon_0)},1,\omega)+a_3\right]\\
\text{Det}_{\omega\leq -2}[(\mathcal{O}_F^{1,1,1})^2] &= \frac{b_0\, e^{-2\omega(R-\epsilon_0)}}{\omega\,(1-\omega)^2}\left[b_1\,\Phi\left(e^{-2\epsilon_0},1,-\omega\right)+b_2\,\Phi(-e^{-2(\sigma_0+\epsilon_0)},1,-\omega)+b_3\right]
\end{aligned} \tag{5.65}$$

where $\Phi(z,s,a)$ is the Lerch transcendent (5.76) defined below. For integers values of ω, it can be written in terms of elementary functions, but its expression becomes more and more unhandy as the value of ω increases. The coefficients a_i are

$$\begin{aligned}
a_0 &= e^{-R-\frac{3\sigma_0}{2}}\,\frac{\sinh\epsilon_0\,(\tanh\sigma_0+1)\cosh\,(\sigma_0+\epsilon_0)}{8\sqrt{2}\cosh\,(\sigma_0+2\epsilon_0)}\\
a_1 &= 4\,\text{sech}\sigma_0(\tanh\sigma_0+\omega)^2\\
a_2 &= 4[2\left(1-\omega^2\right)\omega^2\cosh\sigma_0-2\left(1-\omega^2\right)\omega\sinh\sigma_0+\text{sech}\sigma_0\left(\text{sech}^2\sigma_0+\omega^2-1\right)]\\
a_3 &= \tanh^2\sigma_0\,(\coth\epsilon_0+1)\,\text{csch}\epsilon_0\,\text{sech}\,(\sigma_0+\epsilon_0)\,\big[e^{\sigma_0}\,(\cosh\sigma_0-2\sinh\sigma_0-\sinh(2\epsilon_0-\sigma_0))\\
&\quad+\cosh(2\sigma_0+2\epsilon_0))\big]+2\omega\Big[-\omega^2\cosh^2\sigma_0\,\text{csch}\epsilon_0\,\text{sech}(\sigma_0+\epsilon_0)\\
&\quad+\cosh\sigma_0\big(2\omega^2+\omega+3\omega^2\,\text{csch}\epsilon_0\,\cosh(\sigma_0+2\epsilon_0)\,\text{sech}(\sigma_0+\epsilon_0)+\omega\coth^2\epsilon_0+2\coth\epsilon_0-2\big)\\
&\quad+2\left(3\omega\,\cosh\epsilon_0\,\text{sech}(\sigma_0+\epsilon_0)-\sinh\sigma_0\,\left(\omega-2\omega\coth\epsilon_0-\text{csch}^2\epsilon_0\right)-\text{sech}\sigma_0(\coth\epsilon_0+1)\right)\Big]
\end{aligned} \tag{5.66}$$

while the b_i read

$$\begin{aligned}
b_0 &= e^{R-\frac{\sigma_0}{2}}\,\text{sech}^2\sigma_0\,\frac{\sinh\epsilon_0\,(\tanh\sigma_0+1)\cosh\,(\sigma_0+\epsilon_0)}{8\sqrt{2\cosh(\sigma_0+2\epsilon_0)}}\\
b_1 &= -2\\
b_2 &= -2\left[\omega\big(\omega\cosh(2\sigma_0)+\sinh(2\sigma_0)\big)+\omega^2-1\right]\\
b_3 &= -\cosh^2\sigma_0\left[4\omega\tanh(\sigma_0+\epsilon_0)-2\omega\,\coth\epsilon_0+\text{csch}^2\epsilon_0\right]-\omega\\
&\quad-\cosh(2\sigma_0)(\omega+1)-\sinh(2\sigma_0)+\cosh(\epsilon_0-\sigma_0)\text{sech}\,(\sigma_0+\epsilon_0)\,.
\end{aligned} \tag{5.67}$$

The determinants of the lower modes have to be computed separately:

$$\begin{aligned}
\text{Det}_{\omega=1}[(\mathcal{O}_F^{1,1,1})^2] = R\,e^R\,&\frac{e^{-\frac{\sigma_0}{2}}\,(\tanh\sigma_0+1)\sinh\epsilon_0\cosh(\sigma_0+\epsilon_0)}{\left(e^{2\sigma_0}+1\right)^3\sqrt{2\cosh(\sigma_0+2\epsilon_0)}}\Bigg[-2e^{4\sigma_0}\left(\log\frac{e^{2\epsilon_0}-1}{e^{2(\sigma_0+\epsilon_0)}+1}+\right.\\
&\left.+2\sigma_0\right)+\frac{(e^{2\sigma_0}+1)\big(e^{6\sigma_0+4\epsilon_0}+(e^{2\epsilon_0}+1)e^{4\sigma_0+2\epsilon_0}+e^{2\sigma_0}(-5e^{2\epsilon_0}+3e^{4\epsilon_0}+3)+(e^{2\epsilon_0}-1)^2\big)}{(e^{2\epsilon_0}-1)^2(e^{2(\sigma_0+\epsilon_0)}+1)}\Bigg]
\end{aligned} \tag{5.68}$$

$$\mathrm{Det}_{\omega=0}[(\mathcal{O}_F^{1,1,1})^2] = R\, e^R \frac{e^{-\frac{\sigma_0}{2}}(\tanh\sigma_0+1)\sinh\epsilon_0\cosh(\sigma_0+\epsilon_0)}{\left(e^{2\sigma_0}+1\right)^2\sqrt{2\cosh(\sigma_0+2\epsilon_0)}}\Big[-2e^{2\sigma_0}\Big(\log\frac{e^{2\epsilon_0}-1}{e^{2(\sigma_0+\epsilon_0)}+1}+ \\ +2\sigma_0\Big)+\frac{\left(e^{2\sigma_0}+1\right)\left(-e^{2\sigma_0}+3e^{2(\sigma_0+\epsilon_0)}+e^{4(\sigma_0+\epsilon_0)}-e^{2\epsilon_0}+e^{4\epsilon_0}+1\right)}{\left(e^{2\epsilon_0}-1\right)^2\left(e^{2(\sigma_0+\epsilon_0)}+1\right)}\Big] \tag{5.69}$$

$$\mathrm{Det}_{\omega=-1}[(\mathcal{O}_F^{1,1,1})^2] = e^{3R} \frac{e^{-\frac{\sigma_0}{2}}(\tanh\sigma_0+1)\sinh\epsilon_0\cosh(\sigma_0+\epsilon_0)}{8\left(e^{2\sigma_0}+1\right)^2\sqrt{2\cosh(\sigma_0+2\epsilon_0)}}\Big[-2e^{2\sigma_0}\Big(\log\frac{e^{2\epsilon_0}-1}{e^{2(\sigma_0+\epsilon_0)}+1}+ \\ +2\sigma_0\Big)+\frac{\left(e^{2\sigma_0}+1\right)\left(e^{4\sigma_0}\left(2e^{2\epsilon_0}-1\right)+e^{2\sigma_0}\left(7e^{2\epsilon_0}-2e^{4\epsilon_0}-3\right)+e^{2\epsilon_0}\right)}{\left(e^{2\epsilon_0}-1\right)^2\left(e^{2(\sigma_0+\epsilon_0)}+1\right)}\Big]. \tag{5.70}$$

The expressions above considerably simplify in the circular loop limit $\sigma_0 \to 0$:

$$\mathrm{Det}_\omega\Big[\Big(\mathcal{O}_F^{1,1,1}(\theta_0=0)\Big)^2\Big] = \begin{cases} \frac{e^{(R-\epsilon_0)(2\omega-1)}\left(\omega(e^{2\epsilon_0}-1)+1\right)}{4(\omega-1)\omega^2\,(e^{2\epsilon_0}-1)} & \omega\geq 2 \\ \frac{R\,e^{R+\epsilon_0}}{2(e^{2\epsilon_0}-1)} & \omega=0,1 \\ \frac{e^{3(R-\epsilon_0)}\,(2e^{2\epsilon_0}-1)}{16(e^{2\epsilon_0}-1)} & \omega=-1 \\ \frac{e^{-(R-\epsilon_0)(2\omega-1)}\left((\omega-1)e^{2\epsilon_0}-\omega\right)}{4(\omega-1)^2\omega\,(e^{2\epsilon_0}-1)} & \omega\leq -2\,. \end{cases} \tag{5.71}$$

5.6 One-Loop Partition Functions

We now combine together the determinants evaluated in the previous sections and present the one-loop partition functions for the open strings dual to the latitude ($\theta_0 \neq 0$) and the circular loop ($\theta_0 = 0$). We begin with the explanation of the summation procedure over the Fourier frequencies and eventually calculate the ratio of partition functions.

In the bosonic sector—as discussed around (5.43) and (5.48)—we pose $\omega = \ell+1$ in $\mathrm{Det}_\omega\mathcal{O}_{3+}$ together with $\omega = \ell - 1$ in $\mathrm{Det}_\omega\mathcal{O}_{3-}$. This relabeling of the frequences provides in (5.47) and (5.48) a distribution of the R-divergences that is centered around $\ell = 0$ (i.e. with a divergence $\sim R$ for $\ell = 0$ and $\sim e^{|\ell|R}$ for $\ell \neq 0$) in the same way (in ω) as for the other bosonic determinants (5.45) and (5.46). This will turn out to be useful while discussing the cancellation of R-dependence.

In the case of fermionic determinants, as motivated by the discussion below (5.54), we will consider (5.65)–(5.70) relabelled using half-integer Fourier modes. In fact, once projected onto the subspace labelled by (p_{12}, p_{56}, p_{89}), the spinor Ψ is an eigenstate of Γ_{56} with eigenvalue $-ip_{56}$ and is a periodic function along τ. The rotation $\Psi \to \exp\left(-\frac{\tau}{2}\Gamma_{56}\right)\Psi$ turns the boundary conditions into anti-periodic and causes a shift of the Fourier modes by $\omega \to \omega + \frac{p_{56}}{2}$. This means that we will consider (5.65)–(5.70) evaluated for $\omega = s + \frac{1}{2}$ and now labeled by the half-integer frequency s.

Recalling also the value of the action (5.36), we write the formal expression of the one-loop string action as in (3.103)

$$Z(\theta_0) = e^{\sqrt{\lambda}\cos\theta_0} \frac{\prod_{s\in\mathbb{Z}+1/2}\left[\mathrm{Det}_s(\mathcal{O}_F^{1,1,1})^2\,\mathrm{Det}_{-s}(\mathcal{O}_F^{1,1,1})^2\right]^{4/2}}{\prod_{\ell\in\mathbb{Z}}\left[\mathrm{Det}_\ell\mathcal{O}_1(\theta_0)\right]^{3/2}\left[\mathrm{Det}_\ell\mathcal{O}_2(\theta_0)\right]^{3/2}\left[\mathrm{Det}_\ell\mathcal{O}_{3+}(\theta_0)\right]^{1/2}\left[\mathrm{Det}_\ell\mathcal{O}_{3-}(\theta_0)\right]^{1/2}}\,. \tag{5.72}$$

To proceed, we rewrite (5.72) as the (still unregularized) sum in the one-loop part $\Gamma^{(1)}(\theta_0)$ of the *effective action* $\Gamma(\theta_0)$

$$\Gamma(\theta_0) \equiv -\log Z(\theta_0) \equiv -\sqrt{\lambda}\cos\theta_0 + \Gamma^{(1)}(\theta_0)\,, \tag{5.73}$$
$$\Gamma^{(1)}(\theta_0) \equiv \sum_{\ell\in\mathbb{Z}}\Omega^B_\ell(\theta_0) - \sum_{s\in\mathbb{Z}+1/2}\Omega^F_s(\theta_0)\,,$$

where the bosonic and fermionic contributions are weighted by the multiplicities of the fluctuation scalars and spinors

$$\Omega^B_\ell(\theta_0) = \frac{3}{2}\log\left[\mathrm{Det}_\ell\mathcal{O}_1(\theta_0)\right] + \frac{3}{2}\log\left[\mathrm{Det}_\ell\mathcal{O}_2(\theta_0)\right] + \frac{1}{2}\log\left[\mathrm{Det}_\ell\mathcal{O}_{3+}\right] + \frac{1}{2}\log[\mathrm{Det}_\ell\mathcal{O}_{3-}]\,,$$
$$\Omega^F_s(\theta_0) = \frac{4}{2}\log\left[\mathrm{Det}_s(\mathcal{O}_F^{1,1,1})^2\right] + \frac{4}{2}\log\left[\mathrm{Det}_{-s}(\mathcal{O}_F^{1,1,1})^2\right]. \tag{5.74}$$

Equation (5.73) has the same form with effectively antiperiodic fermions encountered in [20, 75].

Introducing the small exponential regulator μ, we proceed with the "supersymmetric regularization" of the one-loop effective action illustrated in [75, 76]

$$\Gamma^{(1)}(\theta_0) = \sum_{\ell\in\mathbb{Z}} e^{-\mu|\ell|}\left[\Omega^B_\ell(\theta_0) - \frac{\Omega^F_{\ell+\frac{1}{2}}(\theta_0) + \Omega^F_{\ell-\frac{1}{2}}(\theta_0)}{2}\right]$$
$$+\frac{\mu}{2}\Omega^F_{\frac{1}{2}}(\theta_0) + \frac{\mu}{2}\sum_{\ell\geq 1} e^{-\mu\ell}\left(\Omega^F_{\ell+\frac{1}{2}}(\theta_0) - \Omega^F_{\ell-\frac{1}{2}}(\theta_0)\right). \tag{5.75}$$

In the first sum (where the divergence in ℓ is the same as the one in ω in the original sum) one can remove μ by sending $\mu \to 0$, and use a cutoff regularization in the summation index $|\ell| \leq \Lambda$.

Importantly, the non-physical regulator R disappears in (5.75). While in [20][19] the R-dependence drops out in each summand, here it occurs as a subtle effect of the regularization scheme, and comes in the form of a cross-cancellation between the first and the second line once the sums have been carried out. The difference in the R-divergence cancellation mechanism is a consequence of the different arrangement of fermionic frequencies in our regularization scheme (5.75). In the circular case ($\theta_0 = 0$) this cancellation can be seen analytically, as in (5.81)–(5.82) below. The

[19] In this reference a regularization slightly different from [75, 76] was adopted.

same can be then inferred for the general latitude case, since in the normalized one-loop effective action $\Gamma^{(1)}(\theta_0) - \Gamma^{(1)}(\theta_0 = 0)$ one observes (see below) that the R-dependence drops out in each summand.

A non-trivial consistency check of (5.75) is to confirm that in the large-ℓ limit the expected UV divergences [19, 73]) are reproduced. Importantly, for this to happen one cannot take the limit $\epsilon_0 \to 0$ in the determinants above *before* considering $\ell \gg 1$, which is the reason why we kept dealing with the complicated expressions for fermionic determinants above. Using for the Lerch transcendent in (5.65)

$$\Phi(z,s,a) \equiv \sum_{n=0}^{\infty} \frac{z^n}{(n+a)^s} \tag{5.76}$$

the asymptotic behavior for $|a| \gg 1$ [77] (i.e. $|\ell| \gg 1$ in (5.65))

$$\Phi(z,s,a) \sim \text{sgn}(a) \left(\frac{s(s+1)z\,(z+1)\,a^{-s-2}}{2(1-z)^3} - \frac{s\,z\,a^{-s-1}}{(1-z)^2} + \frac{a^{-s}}{1-z} \right), \tag{5.77}$$

one finds that the leading Λ-divergence is logarithmic and proportional to the volume part of the Euler number[20]

$$\Gamma^{(1)}(\theta_0) = -\chi_v(\theta_0) \sum_{1 \ll |\ell| \leq \Lambda} \frac{1}{2|\ell|} + O(\Lambda^0) = -\chi_v(\theta_0) \log\Lambda + O(\Lambda^0)\,, \qquad \Lambda \to \infty \tag{5.78}$$

where

$$\chi_v(\theta_0) = \frac{1}{4\pi} \int_0^{2\pi} d\tau \int_{\epsilon_0}^{\infty} d\sigma \sqrt{h}\;^{(2)}R = 1 - \frac{1}{\epsilon} + \mathcal{O}(\epsilon)\,, \tag{5.79}$$

and we notice that this limit is independent from σ_0 (θ_0). This divergence should be cancelled via completion of the Euler number with its boundary contribution (5.37) and inclusion of the (opposite sign) measure contribution, as discussed in Sect. 3.5.2 and [19, 20]. Having this argument in mind, we will proceed to remove (5.78) by hand in $\Gamma^{(1)}(\theta_0)$ and in $\Gamma^{(1)}(\theta_0 = 0)$.

[20] This is expected from an analysis of the Seeley-DeWitt coefficients in (3.104), with logarithmic divergencies given by (3.111) instead of (3.113) because in (5.78) we are not including the effect of the non-trivial Jacobian explained below (3.111).

5.6.1 The Circular Loop

The UV-regulated partition function in the circular Wilson loop limit reads

$$\Gamma^{(1)}_{\text{UV-reg}}(\theta_0 = 0) = \sum_{|\ell|\leq\Lambda}\left[\Omega^B_\ell(0) - \frac{\Omega^F_{\ell+\frac{1}{2}}(0) + \Omega^F_{\ell-\frac{1}{2}}(0)}{2}\right] + \chi_v(0)\log\Lambda$$
$$+\frac{\mu}{2}\Omega^F_{\frac{1}{2}}(0) + \frac{\mu}{2}\sum_{\ell\geq 1} e^{-\mu\ell}\left(\Omega^F_{\ell+\frac{1}{2}}(0) - \Omega^F_{\ell-\frac{1}{2}}(0)\right). \quad (5.80)$$

The first line is now convergent and its total contribution evaluates for $\Lambda \to \infty$ to

$$* \sum_{|\ell|\leq\Lambda}\left[\Omega^B_\ell(0) - \frac{\Omega^F_{\ell+\frac{1}{2}}(0) + \Omega^F_{\ell-\frac{1}{2}}(0)}{2}\right] + \chi_v(0)\log\Lambda = -2R + \log\frac{16\,\Gamma\left(\frac{3}{2}+\frac{1}{2\epsilon}\right)^4}{(1-\epsilon)\sqrt{\epsilon}\,\Gamma\left(2+\frac{1}{\epsilon}\right)^3}, \quad (5.81)$$

* where Γ is the Euler gamma function. The R-dependence in (5.81) cancels against the $O(\mu^0)$ contribution stemming from the regularization-induced sum in the second line of (5.80)

$$\frac{\mu}{2}\sum_{\ell\geq 1} e^{-\mu\ell}\left(\Omega^F_{\ell+\frac{1}{2}}(0) - \Omega^F_{\ell-\frac{1}{2}}(0)\right) = 2R - 2\,\text{arctanh}\,\epsilon\,.$$

Summing the two lines above in the limit $\epsilon \to 0$, the result is precisely as in [20]

$$\Gamma^{(1)}_{\text{UV-reg}}(\theta_0 = 0) = \frac{1}{\epsilon}\left(\log\frac{\epsilon}{4} + 1\right) + \frac{1}{2}\log(2\pi)\,, \quad (5.82)$$

despite the different frequency arrangement we commented on. We have checked that the same result is obtained employing zeta-function regularization in the sum over ℓ. The same finite part was found in [21] via heat kernel methods. The $\log\epsilon/\epsilon$-divergence appearing in (5.82) will be cancelled in the ratio (5.22). In [20] this subtraction was done by considering the ratio between the circular and the straight line Wilson loop.

5.6.2 Ratio Between Latitude and Circular Loop

We now illustrate the evaluation of the ratio (5.22)

$$\log\frac{Z(\lambda,\theta_0)}{Z(\lambda,0)} = \sqrt{\lambda}(\cos\theta_0 - 1) + \Gamma^{(1)}_{\text{UV-reg}}(\theta_0 = 0) - \Gamma^{(1)}_{\text{UV-reg}}(\theta_0) \quad (5.83)$$

where $\Gamma^{(1)}_{\text{UV-reg}}(\theta_0 = 0)$ is the effective action (5.80) and $\Gamma^{(1)}_{\text{UV-reg}}(\theta_0)$ is regularized analogously. The complicated fermionic determinants (5.65)–(5.67) make an analytical treatment highly non-trivial, and we proceed numerically in `Mathematica`.

First, we spell out (5.83) as

$$\begin{aligned}
\Gamma^{(1)}_{\text{UV-reg}}(0) - \Gamma^{(1)}_{\text{UV-reg}}(\theta_0) = & \sum_{\ell=-2}^{2} \left[\Omega^B_\ell(0) - \Omega^B_\ell(\theta_0) - \frac{\Omega^F_{\ell+\frac{1}{2}}(0)+\Omega^F_{\ell-\frac{1}{2}}(0)}{2} + \frac{\Omega^F_{\ell+\frac{1}{2}}(\theta_0)+\Omega^F_{\ell-\frac{1}{2}}(\theta_0)}{2} \right] \\
& + \sum_{\ell=3}^{\Lambda} 2 \left[\Omega^B_\ell(0) - \Omega^B_\ell(\theta_0) - \frac{\Omega^F_{\ell+\frac{1}{2}}(0)+\Omega^F_{\ell-\frac{1}{2}}(0)}{2} + \frac{\Omega^F_{\ell+\frac{1}{2}}(\theta_0)+\Omega^F_{\ell-\frac{1}{2}}(\theta_0)}{2} \right] \\
& - (\chi_v(\theta_0) - \chi_v(0)) \log\Lambda + \frac{\mu}{2} \left[\Omega^F_{\frac{1}{2}}(0) - \Omega^F_{\frac{1}{2}}(\theta_0) \right] \\
& + \frac{\mu}{2} \sum_{\ell \geq 1} e^{-\mu\ell} \left[\Omega^F_{\ell+\frac{1}{2}}(0) - \Omega^F_{\ell-\frac{1}{2}}(0) - \Omega^F_{\ell+\frac{1}{2}}(\theta_0) + \Omega^F_{\ell-\frac{1}{2}}(\theta_0) \right]
\end{aligned} \tag{5.84}$$

where we separated the lower modes $|\ell| \leq 2$ from the sum in the second line,[21] and in the latter we have used parity $\ell \to -\ell$. The sum multiplied by the small cutoff μ is zero in the limit $\mu \to 0$.[22] The sum with large cutoff Λ can be then numerically evaluated using the Euler-Maclaurin formula

$$\begin{aligned}
\sum_{\ell=m+1}^{n} f(\ell) = & \int_m^n f(\ell)\, d\ell + \frac{f(n) - f(m)}{2} + \sum_{k=1}^{p} \frac{B_{2k}}{(2k)!} \left[f^{(2k-1)}(n) - f^{(2k-1)}(m) \right] \\
& - \int_m^n f^{(2p)}(\ell)\, \frac{B_{2p}(\{\ell\})}{(2p)!} d\ell\,, \qquad p \geq 1\,,
\end{aligned} \tag{5.85}$$

in which $B_n(x)$ is the n-th Bernoulli polynomial, $B_n \equiv B_n(0)$ is the n-th Bernoulli number, $\{\ell\}$ is the integer part of ℓ, $f(\ell)$ is the summand in the second line of (5.84), so $m = 2$, $n = \Lambda$. After some manipulations to improve the rate of convergence of the integrals, we safely remove $\Lambda \to \infty$ in order to arrive to *normalized effective action*

$$\begin{aligned}
\Delta\Gamma(\theta_0)_{\text{sm}} \equiv & \left[\Gamma^{(1)}_{\text{UV-reg}}(0) - \Gamma^{(1)}_{\text{UV-reg}}(\theta_0) \right]_{\text{sm}} \\
= & \sum_{\ell=-2}^{2} \left[\Omega^B_\ell(0) - \Omega^B_\ell(\theta_0) - \frac{\Omega^F_{\ell+\frac{1}{2}}(0) + \Omega^F_{\ell-\frac{1}{2}}(0)}{2} + \frac{\Omega^F_{\ell+\frac{1}{2}}(\theta_0) + \Omega^F_{\ell-\frac{1}{2}}(\theta_0)}{2} \right] \\
& + \int_2^\infty \left[f(\ell) - \frac{\chi_v(\theta_0) - \chi_v(0)}{\ell} \right] d\ell - (\chi_v(\theta_0) - \chi_v(0)) \log 2 \\
& - \frac{f(2)}{2} - \sum_{k=1}^{3} \frac{B_{2k}}{(2k)!} f^{(2k-1)}(2) - \frac{1}{6!} \int_2^\infty f^{(6)}(\ell)\, B_6(\{\ell\})\, d\ell\,.
\end{aligned} \tag{5.86}$$

---

[21] This is convenient because of the different form for the special modes (5.68)–(5.70) together with the relabeling discussed above.

[22] This can be proved analytically since the summand behaves as $\mu\, e^{-\mu\ell} \ell^{-2}$ for large ℓ. Removing the cutoff makes the sum vanish.

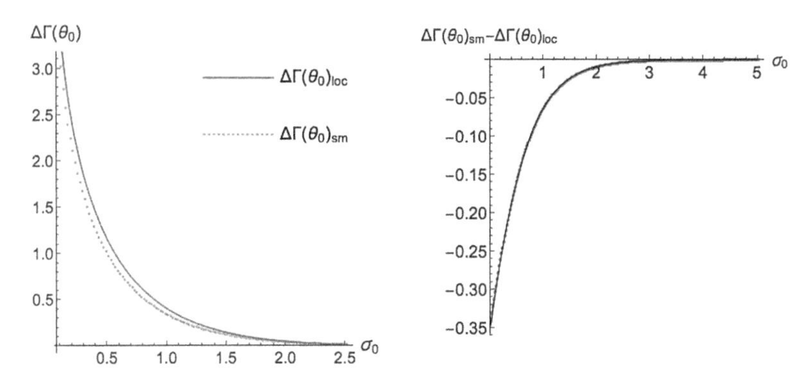

Fig. 5.4 Comparison between string sigma-model perturbation theory and the prediction of supersymmetric localization for the ratio between latitude and circular Wilson loops. In the *left panel*, we *plot* the comparison between $\Delta\Gamma(\theta_0)_{\rm sm}$ in (5.86) (*orange dots*) and $\Delta\Gamma(\theta_0)_{\rm loc}$ in (5.89) (*blue line*). We set $\epsilon_0 = 10^{-7}$, $\Delta\ell = 10^{-9}$. In the *right panel*, we fit the difference (*red dots*) between the two *curves* in the *left panel* and compare it with the test function $-\frac{1}{2}\log(1+e^{-2\sigma_0})$ (*black line*). The interval of σ_0 covers approximately $0.8^\circ \leq \theta_0 \leq 89.4^\circ$

In order to gain numerical stability for large ℓ, above we have set $p = 3$, we have cast the Lerch transcendents inside $\Omega_s^F(\theta_0)$ (see (5.65)) into hypergeometric functions

$$\Phi(z, 1, a) = \frac{{}_2F_1(1, a; a+1; z)}{a}\,, \qquad |z| < 1 \wedge z \neq 0\,, \tag{5.87}$$

and we have approximated the derivatives $f^{(k)}(\ell)$ by finite-difference operators

$$f^{(k)}(\ell) \;\to\; \Delta\ell^{-k}\sum_{i=0}^{k}(-1)^i\binom{k}{i} f\left(\ell + (\tfrac{k}{2} - i)\Delta\ell\right)\,, \qquad \Delta\ell \ll 1\,. \tag{5.88}$$

At this stage, the expression (5.86) is only a function of the latitude parameter σ_0 (or the polar angle θ_0 in (5.27)) and of two parameters—the IR cutoff ϵ_0 and the derivative discretization $\Delta\ell$, both small compared to a given σ_0. We have tuned them in order to confidently extract four decimal digits.

In Fig. 5.4 (left panel) we compare the regularized one-loop effective action obtained from the perturbation theory of the string sigma-model (5.86) to the gauge theory prediction from (5.21)

$$\Delta\Gamma(\theta_0)_{\rm loc} \equiv \left[\Gamma^{(1)}_{\text{UV-reg}}(0) - \Gamma^{(1)}_{\text{UV-reg}}(\theta_0)\right]_{\rm loc} = -\frac{3}{2}\log\tanh\sigma_0 = -\frac{3}{2}\log\cos\theta_0 \tag{5.89}$$

for different values of σ_0. Data points cover almost entirely the finite-angle region between the Zarembo Wilson loop ($\sigma_0 = 0,\ \theta_0 = \frac{\pi}{2}$) and the circular Wilson loop ($\sigma_0 = \infty,\ \theta_0 = 0$). The vanishing of the normalized effective action in the large-σ_0

region is a trivial check of the normalization. As soon as the opposite limit $\sigma_0 = 0$ is approached, the difference (5.86) bends up "following" the localization curve (5.89) but also significantly deviates from it, and the measured discrepancy is incompatible with our error estimation.

Numerics is however accurate enough to quantify the gap between the two plots on a wide range. Figure 5.4 (right panel) shows that, surprisingly, such gap perfectly overlaps a very simple function of σ_0 within the sought accuracy

$$\Delta\Gamma(\theta_0)_{\text{sm}} - \Delta\Gamma(\theta_0)_{\text{loc}} \approx -\frac{1}{2}\log(1+e^{-2\sigma_0}) = \log\cos\tfrac{\theta_0}{2}\,. \tag{5.90}$$

We notice at this point that the *same* simple result above can be obtained taking in (5.84) the limit of $\epsilon \to 0$ *before* performing the sums. As one can check, in this limit UV and IR divergences cancel in the ratio,[23] the special functions in the fermionic determinants disappear and, because in general summands drastically simplify, one can proceed analytically getting the same result calculated in terms of numerics. We remark however that such inversion of the order of sum and limit on the IR cutoff cannot be a priori justified, as it would improperly relate the Λ cutoff with a $1/\epsilon$ cutoff (e.g. forcing ℓ to be smaller than $1/\epsilon$).

As emphasized above, in this limit the effective actions for the latitude and circular case separately do not reproduce the expected UV divergences. Therefore, the fact that in this limit the summands in the difference (5.84) show a special property of convergence and lead to the exact result is a priori highly not obvious, rendering the numerical analysis carried out in this section a rather necessary step. On a related note, the simplicity of the result (5.89) and the possibility of getting an analytical result for the maximal circle $\theta_0 = 0$ suggest that the summation (5.75) could have been performed analytically also in the latitude case $\theta_0 \neq 0$, but we have not investigated this direction.

5.7 Comparison with Recent Developments

The final result (5.23) is inconsistent with the gauge-theory analysis (5.21), insinuating that something subtle is missing in our procedure. Before commenting upon the possible origin of our discrepancy, we want to mention that the same finite remnant was found in [66] soon after our paper [74]. Although conceptually similar to ours, the work of [66] has some technical differences that are interesting to highlight.

- *Role of group theory in the fluctuation spectrum.* It is possible to show that a different choice of the tangent and normal vectors used to define the bosonic fluctuations (3.27) and the local fermionic rotation S (3.80)–(3.81) can induce a shift in the Fourier frequencies ω. In the analysis of Sect. 5.6 this was crucial to arrange

[23] This is also due to the volume part of the Euler number $\chi_v(\theta_0)$ being independent of σ_0 up to ϵ corrections, see (5.79).

the angular modes $\omega \to (\ell, s)$ so that the first sum in (5.75) could be regularized with a symmetric cutoff $|\ell| \leq \Lambda$ and the unphysical R-divergences eventually canceled.
The same procedure can also find a theoretical explanation (Sect. 4.3 of [66]) that takes into account the symmetry group $SU(2|2)$ of the dual latitude Wilson loop. Indeed, the string fluctuation fields fit into multiplets of this supergroup[24] and supersymmetry arguments suggest to sum over multiplets labeled by the $U(1)$-charge q.[25] The origin of the shifts is tracked back to the change of index $q \to E$ and the observation that the frequency—called E in [66]—does not coincide with the quantum number q for some fields. Note that the argument specifically addresses how to *organize* the sum over modes, but not how to *compute* the summands, namely the 1d determinants at fixed frequency.

- *Evaluation of fermionic determinants*. In [66] the 1d fluctuation determinants are still evaluated with the Gel'fand-Yaglom method. The spectral problem is solved for the first-order fermionic operator using the theorem (B.8), opposed to the second-order one that results from squaring it using the corollary in Appendix B.3. This alternative route, also proposed in the conclusions of [74], is tied to a new choice of boundary conditions[26] for the Dirac-like *first-order* operator. As a by product, it simplifies the expressions for the 1d fermionic determinants from the Lerch transcendent for the squared operator to hyperbolic functions for the not-squared one.
- *Ratio of determinants*. In Sect. 5.6.2 we effectively evaluated the ratio between contributions of one bosonic ϕ_B and one fermionic field ϕ_F

$$\frac{\mathrm{Det}_\omega^{1/4}(\mathcal{O}^2_{\phi_F}(\theta_0))}{\mathrm{Det}_\omega(\mathcal{O}_{\phi_B}(\theta_0))} \tag{5.91}$$

on the same classical solution at given θ_0, eventually leading us to perform the sum over ω only numerically. The strength of our approach is twofold. Firstly, we checked that our regularization scheme cancels the unphysical R-divergences and reproduces the expected UV logarithmic divergence in Λ. Secondly, we can extract the expectation value of one *single* latitude loop by multiplying the value of the circular loop (5.82) and the ratio latitude/circle (5.23).
This situation is in contrast with the choice operated by [66], which considers instead the ratio between modes for the same (bosonic ϕ_B or fermionic ϕ_F) field

$$\frac{\mathrm{Det}_\omega(\mathcal{O}_{\phi_B}(\theta_0))}{\mathrm{Det}_\omega(\mathcal{O}_{\phi_B}(0))} \quad \text{and} \quad \frac{\mathrm{Det}_\omega^{1/2}(\mathcal{O}_{\phi_F}(\theta_0))}{\mathrm{Det}_\omega^{1/2}(\mathcal{O}_{\phi_F}(0))} \tag{5.92}$$

---

[24] The $SU(2|2)$ quantum numbers of the fields are summarized in (4.26) of [66]. For vanishing θ_0 it was known that they fill representations of the symmetry group $OSp(4^*|4)$ of the half-supersymmetric circular Wilson loop [78].

[25] We are referring to formulas (4.27) and (4.28) in [66].

[26] See Eq. (5.50) in [66].

on different classical solutions. Together with the choice of working with the first-order fermionic operator, this strategy allows for an analytical evaluation of the one-loop correction. This led the authors of [66] (Sect. 6.3 and 7 therein) to the intriguing observation that the expected result (5.89) seems to be captured by a *finite* number of Fourier modes (after removing the IR cutoff $\epsilon_0 \to 0$)

$$\left[\Gamma^{(1)}(0) - \Gamma^{(1)}(\theta_0)\right]_{\text{conjecture}} = -\frac{3}{2}\log\left(\frac{\text{Det}_{\omega=0}\mathcal{O}_2(\theta_0)}{\text{Det}_{\omega=0}\mathcal{O}_2(\theta_0=0)}\right) = -\frac{3}{2}\log\cos\theta_0\,, \quad (5.93)$$

namely those bosonic modes (5.46) and (5.49) with vanishing angular frequency ω and charged under the second $SU(2)$ factor in the symmetry group $SU(2) \times U(1) \times SU(2) \subset SU(2|2)$ preserved by the latitude Wilson loop operator.

It is possible that this conjecture will be either confirmed or adjusted once the one-loop semiclassical analysis done for the latitude will be extended [79] to other DGRT Wilson loops or some mixed correlators of the form (5.9).

5.8 Unresolved Subtleties in Sigma-Model Perturbation Theory

Both the setups of [66, 74] for the evaluation of the one-loop determinants (summing over 1d determinants evaluated with the Gel'fand-Yaglom approach and a fictitious boundary) do *not* reach the agreement with the localization result. One reasonable expectation is that the disagreement should be cured by a change of the world-sheet computational setup, tailored so to naturally lend itself to a regularization scheme equivalent to the one (implicitly) assumed by the localization calculation.[27] In principle, a number of reasons could explain the observed discrepancy, which we summarize as follows.

- *Path-integral measure ambiguities.* The expectation that considering the ratio of string partition functions dual to Wilson loops with the same worldsheet topology should cancel measure-related ambiguities is founded on the assumption that the partition function measure is actually *not* depending on the particular classical solution considered. Although motivated in light of literature examples similar in spirit (see Sect. 5.3), this remains an assumption, and it is not possible to exclude a priori a geometric interpretation for the observed discrepancy.

[27] This resembles the quest for an "integrability-preserving" regularization scheme, different from the most natural one suggested by worldsheet field theory considerations, in the worldsheet calculations of light-like cusps in $\mathcal{N} = 4$ SYM [80] and ABJM theory [81].

- *Fermionic boundary conditions*. One possibility is a choice of boundary conditions for the fermionic spectral problem different from the standard ones here adopted for the squared fermionic operator. The quest for an alternative choice is likely to be tied to a search of supersymmetry-preserving boundary conditions on the lines of [82].
- *Two-dimensional spectral methods*. Ideally, one should evaluate determinants in a diffeomorphism-preserving regularization scheme that treats the worldsheet coordinates τ and σ on equal footing, unlike the standard procedure employed here and explained in Sect. 1.4.1. An important step in this direction can be achieved either indirectly on the lines of [83, 84]—where one accounts for the splitting τ and σ while summing over the angular frequencies—or with heat kernel techniques—a fully 2d approach currently under investigation in [72] and explained in Chap. 8.

References

1. N.A. Nekrasov, Seiberg-Witten prepotential from instanton counting. Adv. Theor. Math. Phys. **7**, 831 (2003), arXiv:hep-th/0206161
2. V. Pestun, Localization of gauge theory on a four-sphere and supersymmetric Wilson loops. Commun. Math. Phys. **313**, 71 (2012). arXiv:0712.2824
3. E. Witten, Supersymmetry and Morse theory. J. Diff. Geom. **17**, 661 (1982)
4. N. Berline, M. Vergne, Classes caracteristiques equivariantes. Formule de localisation en cohomologie equivariante. CR Acad. Sci. Paris **295**, 539 (1982)
5. N. Berline, M. Vergne, Zeros d'un champ de vecteurs et classes characteristiques equivariantes. Duke Math. J. **50**, 539 (1983)
6. M.F. Atiyah, R. Bott, The moment map and equivariant cohomology. Topology **23**, 1 (1984)
7. J.J. Duistermaat, G.J. Heckman, On the Variation in the cohomology of the symplectic form of the reduced phase space. Invent. Math. **69**, 259 (1982)
8. M. Marino, Lectures on localization and matrix models in supersymmetric Chern-Simons-matter theories. J. Phys. A **44**, 463001 (2011), arXiv:1104.0783
9. S. Cremonesi, An introduction to localisation and supersymmetry in curved space, PoS Modave 2013, 002, in *Proceedings, 9th Modave Summer School in Mathematical Physics* (2013), p. 002
10. V. Pestun, M. Zabzine, Introduction to localization in quantum field theory, arXiv:1608.02953, http://inspirehep.net/record/1480380/files/arXiv:1608.02953.pdf
11. V. Pestun, Review of localization in geometry, arXiv:1608.02954, http://inspirehep.net/record/1480381/files/arXiv:1608.02954.pdf
12. J. Teschner, in *New Dualities of Supersymmetric Gauge Theories* (Springer, Cham, Switzerland, 2016)
13. V. Pestun et al., in *Localization Techniques in Quantum Field Theories*, arXiv:1608.02952
14. N. Drukker, D.J. Gross, H. Ooguri, Wilson loops and minimal surfaces. Phys. Rev. D **60**, 125006 (1999). arXiv:hep-th/9904191
15. M. Preti, *Studies on Wilson loops, correlators and localization in supersymmetric quantum field theories*, PhD Thesis, https://inspirehep.net/record/1592204/files/Thesis_main.pdf
16. A. Dymarsky, V. Pestun, Supersymmetric Wilson loops in $\mathcal{N} = 4$ SYM and pure spinors. JHEP **1004**, 115 (2010). arXiv:0911.1841
17. V. Cardinali, L. Griguolo, D. Seminara, Impure aspects of supersymmetric Wilson loops. JHEP **1206**, 167 (2012). arXiv:1202.6393
18. K. Zarembo, Supersymmetric Wilson loops. Nucl. Phys. B **643**, 157 (2002). arXiv:hep-th/0205160

19. N. Drukker, D.J. Gross, A.A. Tseytlin, Green-Schwarz string in $AdS_5 \times S^5$: Semiclassical partition function. JHEP **0004**, 021 (2000). arXiv:hep-th/0001204
20. M. Kruczenski, A. Tirziu, Matching the circular Wilson loop with dual open string solution at 1-loop in strong coupling. JHEP **0805**, 064 (2008). arXiv:0803.0315
21. E.I. Buchbinder, A.A. Tseytlin, $1/N$ correction in the D3-brane description of a circular Wilson loop at strong coupling. Phys. Rev. D **89**, 126008 (2014). arXiv:1404.4952
22. Z. Guralnik, S. Kovacs, B. Kulik, Less is more: Non-renormalization theorems from lower dimensional superspace. Int. J. Mod. Phys. A **20**, 4546 (2005). arXiv:hep-th/0409091 (in *Non-perturbative quantum chromodynamics. Proceedings, 8th Workshop, Paris, France, June 7–11, 2004*, pp. 4546–4553)
23. Z. Guralnik, B. Kulik, Properties of chiral Wilson loops. JHEP **0401**, 065 (2004). arXiv:hep-th/0309118
24. A. Dymarsky, S.S. Gubser, Z. Guralnik, J.M. Maldacena, Calibrated surfaces and supersymmetric Wilson loops. JHEP **0609**, 057 (2006). arXiv:hep-th/0604058
25. A. Kapustin, E. Witten, Electric-magnetic duality and the geometric Langlands program. Commun. Num. Theor. Phys. **1**, 1 (2007). arXiv:hep-th/0604151
26. N. Drukker, S. Giombi, R. Ricci, D. Trancanelli, Supersymmetric Wilson loops on S^3. JHEP **0805**, 017 (2008). arXiv:0711.3226
27. N. Drukker, S. Giombi, R. Ricci, D. Trancanelli, More supersymmetric Wilson loops. Phys. Rev. D **76**, 107703 (2007). arXiv:0704.2237
28. D. Correa, J. Maldacena, A. Sever, The quark anti-quark potential and the cusp anomalous dimension from a TBA equation. JHEP **1208**, 134 (2012). arXiv:1203.1913
29. N. Drukker, S. Giombi, R. Ricci, D. Trancanelli, Wilson loops: from four-dimensional SYM to two-dimensional YM. Phys. Rev. D **77**, 047901 (2008). arXiv:0707.2699
30. A. Bassetto, L. Griguolo, Two-dimensional QCD, instanton contributions and the perturbative Wu-Mandelstam-Leibbrandt prescription. Phys. Lett. B **443**, 325 (1998). arXiv:hep-th/9806037
31. S. Giombi, V. Pestun, Correlators of local operators and 1/8 BPS Wilson loops on S^2 from 2d YM and matrix models. JHEP **1010**, 033 (2010). arXiv:0906.1572
32. N. Drukker, J. Plefka, Superprotected n-point correlation functions of local operators in $\mathcal{N} = 4$ super Yang-Mills. JHEP **0904**, 052 (2009). arXiv:0901.3653
33. V. Pestun, Localization of the four-dimensional $\mathcal{N} = 4$ SYM to a two-sphere and 1/8 BPS Wilson loops. JHEP **1212**, 067 (2012). arXiv:0906.0638
34. E. Witten, On quantum gauge theories in two-dimensions. Commun. Math. Phys. **141**, 153 (1991)
35. M. Marino, *Les Houches lectures on matrix models and topological strings*. arXiv:hep-th/0410165
36. A. Bassetto, L. Griguolo, F. Pucci, D. Seminara, Supersymmetric Wilson loops at two loops. JHEP **0806**, 083 (2008). arXiv:0804.3973
37. M. Bonini, L. Griguolo, M. Preti, Correlators of chiral primaries and 1/8 BPS Wilson loops from perturbation theory. JHEP **1409**, 083 (2014). arXiv:1405.2895
38. G.W. Semenoff, K. Zarembo, More exact predictions of SUSYM for string theory. Nucl. Phys. B **616**, 34 (2001). arXiv:hep-th/0106015
39. K. Zarembo, Open string fluctuations in $AdS_5 \times S^5$ and operators with large R charge. Phys. Rev. D **66**, 105021 (2002). arXiv:hep-th/0209095
40. G.W. Semenoff, D. Young, Exact 1/4 BPS Loop: chiral primary correlator. Phys. Lett. B **643**, 195 (2006). arXiv:hep-th/0609158
41. D. Young, BPS Wilson loops on S^2 at Higher loops. JHEP **0805**, 077 (2008). arXiv:0804.4098
42. A. Bassetto, L. Griguolo, F. Pucci, D. Seminara, S. Thambyahpillai et al., Correlators of supersymmetric Wilson-loops, protected operators and matrix models in $\mathcal{N} = 4$ SYM. JHEP **0908**, 061 (2009). arXiv:0905.1943
43. A. Bassetto, L. Griguolo, F. Pucci, D. Seminara, S. Thambyahpillai, D. Young, Correlators of supersymmetric Wilson loops at weak and strong coupling. JHEP **1003**, 038 (2010). arXiv:0912.5440

44. S. Giombi, V. Pestun, Correlators of Wilson loops and local operators from multi-matrix models and strings in AdS. JHEP **1301**, 101 (2013). arXiv:1207.7083
45. S. Giombi, V. Pestun, The 1/2 BPS 't Hooft loops in $\mathcal{N} = 4$ SYM as instantons in 2d Yang-Mills. J. Phys. A **46**, 095402 (2013). arXiv:0909.4272
46. D. Correa, J. Henn, J. Maldacena, A. Sever, An exact formula for the radiation of a moving quark in $\mathcal{N} = 4$ super Yang Mills. JHEP **1206**, 048 (2012). arXiv:1202.4455
47. B. Fiol, B. Garolera, A. Lewkowycz, Exact results for static and radiative fields of a quark in $\mathcal{N} = 4$ super Yang-Mills. JHEP **1205**, 093 (2012). arXiv:1202.5292
48. N. Gromov, A. Sever, Analytic solution of Bremsstrahlung TBA. JHEP **1211**, 075 (2012). arXiv:1207.5489
49. N. Gromov, F. Levkovich-Maslyuk, G. Sizov, Analytic solution of Bremsstrahlung TBA II: turning on the sphere angle. JHEP **1310**, 036 (2013). arXiv:1305.1944
50. M. Bonini, L. Griguolo, M. Preti, D. Seminara, Bremsstrahlung function, leading Luscher correction at weak coupling and localization. JHEP **1602**, 172 (2016). arXiv:1511.05016
51. K. Zarembo, *Localization and AdS/CFT Correspondence*. arXiv:1608.02963
52. D.E. Berenstein, R. Corrado, W. Fischler, J.M. Maldacena, The Operator product expansion for Wilson loops and surfaces in the large N limit. Phys. Rev. D **59**, 105023 (1999). arXiv:hep-th/9809188
53. J. Erickson, G. Semenoff, K. Zarembo, Wilson loops in $\mathcal{N} = 4$ supersymmetric Yang-Mills theory. Nucl. Phys. B **582**, 155 (2000). arXiv:hep-th/0003055
54. N. Drukker, D.J. Gross, An Exact prediction of $\mathcal{N} = 4$ SUSYM theory for string theory. J. Math. Phys. **42**, 2896 (2001). arXiv:hep-th/0010274
55. J. Gomis, F. Passerini, Holographic Wilson Loops. JHEP **0608**, 074 (2006). arXiv:hep-th/0604007
56. J. Gomis, F. Passerini, Wilson Loops as D3-Branes. JHEP **0701**, 097 (2007). arXiv:hep-th/0612022
57. S. Yamaguchi, Wilson loops of anti-symmetric representation and D5-branes. JHEP **0605**, 037 (2006). arXiv:hep-th/0603208
58. S.A. Hartnoll, S.P. Kumar, Higher rank Wilson loops from a matrix model. JHEP **0608**, 026 (2006). arXiv:hep-th/0605027
59. M. Sakaguchi, K. Yoshida, A Semiclassical string description of Wilson loop with local operators. Nucl. Phys. B **798**, 72 (2008). arXiv:0709.4187
60. C. Kristjansen, Y. Makeenko, More about one-loop effective action of open superstring in $AdS_5 \times S^5$. JHEP **1209**, 053 (2012). arXiv:1206.5660
61. R. Bergamin, A.A. Tseytlin, Heat kernels on cone of AdS_2 and k-wound circular Wilson loop in $AdS_5 \times S^5$ superstring. J. Phys. A **49**, 14LT01 (2016), arXiv:1510.06894
62. A. Faraggi, J.T. Liu, L.A.P. Zayas, G. Zhang, One-loop structure of higher rank Wilson loops in AdS/CFT. Phys. Lett. B **740**, 218 (2015). arXiv:1409.3187
63. N. Drukker, V. Forini, Generalized quark-antiquark potential at weak and strong coupling. JHEP **1106**, 131 (2011). arXiv:1105.5144
64. N. Drukker, B. Fiol, On the integrability of Wilson loops in $AdS_5 \times S^5$: Some periodic ansatze. JHEP **0601**, 056 (2006). arXiv:hep-th/0506058
65. N. Drukker, 1/4 BPS circular loops, unstable world-sheet instantons and the matrix model. JHEP **0609**, 004 (2006). arXiv:hep-th/0605151
66. A. Faraggi, L.A. Pando Zayas, G.A. Silva, D. Trancanelli, Toward precision holography with supersymmetric Wilson loops. JHEP **1604**, 053 (2016). arXiv:1601.04708
67. S. Forste, D. Ghoshal, S. Theisen, Stringy corrections to the Wilson loop in $\mathcal{N} = 4$ superYang-Mills theory. JHEP **9908**, 013 (1999). arXiv:hep-th/9903042
68. S. Forste, D. Ghoshal, S. Theisen, *Wilson loop via AdS/CFT duality*, arXiv:hep-th/0003068 (in: *Proceedings, TMR Meeting on Quantum Aspects of Gauge Theories, Supersymmetry and Unification*, [PoStmr99,018(1999)])
69. R. Roiban, A. Tirziu, A.A. Tseytlin, Two-loop world-sheet corrections in $AdS_5 \times S^5$ superstring. JHEP **0707**, 056 (2007). arXiv:0704.3638

70. V. Forini, Quark-antiquark potential in AdS at one loop. JHEP **1011**, 079 (2010). arXiv:1009.3939
71. A. Miwa, Broken zero modes of a string world sheet and a correlation function between a 1/4 BPS Wilson loop and a 1/2 BPS local operator. Phys. Rev. D **91**, 106003 (2015). arXiv:1502.04299
72. V. Forini, A.A. Tseytlin, E. Vescovi, *Perturbative computation of string one-loop corrections to Wilson loop minimal surfaces in $AdS_5 \times S^5$*, JHEP 1703, 003 (2017), arXiv:1702.02164
73. V. Forini, V.G.M. Puletti, L. Griguolo, D. Seminara, E. Vescovi, Remarks on the geometrical properties of semiclassically quantized strings. J. Phys. A **48**, 475401 (2015). arXiv:1507.01883
74. V. Forini, V. Giangreco M. Puletti, L. Griguolo, D. Seminara, E. Vescovi, Precision calculation of 1/4-BPS Wilson loops in $AdS_5 \times S^5$. JHEP **1602**, 105 (2016). arXiv:1512.00841
75. A. Dekel, T. Klose, Correlation function of circular wilson loops at strong coupling. JHEP **1311**, 117 (2013). arXiv:1309.3203
76. S. Frolov, I. Park, A.A. Tseytlin, On one-loop correction to energy of spinning strings in S^5. Phys. Rev. D **71**, 026006 (2005). arXiv:hep-th/0408187
77. C. Ferreira, J.L. López, Asymptotic expansions of the Hurwitz-Lerch zeta function. J. Math. Anal. Appl. **298**, 210 (2004)
78. A. Faraggi, L.A. Pando Zayas, The spectrum of excitations of holographic Wilson loops, JHEP **1105**, 018 (2011). arXiv:1101.5145
79. J. Aguilera-Damia, A. Faraggi, L.A. Pando Zayas, V. Rathee, G.A. Silva, D. Trancanelli, E. Vescovi, in preparation
80. S. Giombi, R. Ricci, R. Roiban, A. Tseytlin, C. Vergu, Quantum $AdS_5 \times S^5$ superstring in the AdS light-cone gauge. JHEP **1003**, 003 (2010). arXiv:0912.5105
81. T. McLoughlin, R. Roiban, A.A. Tseytlin, Quantum spinning strings in $AdS_4 \times \mathbb{CP}^3$: Testing the Bethe Ansatz proposal. JHEP **0811**, 069 (2008). arXiv:0809.4038
82. N. Sakai, Y. Tanii, Supersymmetry in two-dimensional anti-de sitter space. Nucl. Phys. B **258**, 661 (1985)
83. G.V. Dunne, K. Kirsten, Functional determinants for radial operators. J. Phys. A **39**, 11915 (2006). arXiv:hep-th/0607066
84. K. Kirsten, Functional determinants in higher dimensions using contour integrals. arXiv:1005.2595

Chapter 6
Light-Like Cusp Anomaly and the Interpolating Function in ABJM

A powerful attribute that the planar AdS_4/CFT_3 system (1.5) shares with the higher-dimensional AdS_5/CFT_4 duality (1.1) in the planar limit is its conjectured integrability [1–4]. However, the explicit realization of the integrable structure is non-trivial, due to significative peculiarities of the former case.

The first significative difference is the absence of maximal supersymmetry of the $AdS_4 \times \mathbb{CP}^3$ background, which complicates the construction of the corresponding superstring action. In particular, in Sect. 2.2 we saw that there is an issue that prevents the $\frac{OSp(4|6)}{SO(1,3)\times U(3)}$ supercoset action from consistently describing the motion of the superstring occurring only in the AdS_4 subspace. Moreover, the comparison of string-theory calculations with weak-coupling results is complicated by the correction [5] to the string tension (1.9) which plays a role starting from two loops in sigma-model perturbation theory.

Another difference is that the interpolation between weak and strong coupling is much more intricate in this case. In Sect. 1.2 we recalled that the integrability structures of $\mathcal{N} = 4$ SYM and ABJM are described in terms of spin-chains that represent single-trace operators in these theories. The form of the dispersion relation of the fundamental excitations (*magnons*) of the spin-chain is constrained in either dualities. In $\mathcal{N} = 4$ SYM the energy of a single magnon [6]

$$\epsilon_{\text{YM}}(p) = \sqrt{1 + 4\,h^2(\lambda_{\text{YM}})\sin^2\frac{p}{2}} \tag{6.1}$$

is fixed up to an undetermined factor $h(\lambda_{\text{YM}})$ called ($\mathcal{N} = 4$ SYM) *interpolating function* and depending only on the 't Hooft parameter. Its explicit form turns out to be trivial at all orders, $h(\lambda_{\text{YM}}) = \frac{\sqrt{\lambda_{\text{YM}}}}{4\pi}$, as shown in [7–9][1] by computing the *Bremsstrahlung function* via an extrapolation on results of supersymmetric localiza-

[1]See also discussions in [10–12].

E. Vescovi, *Perturbative and Non-perturbative Approaches to String Sigma-Models in AdS/CFT*, Springer Theses, DOI 10.1007/978-3-319-63420-3_6

tion and via integrability. A similar constraint holds in the ABJM theory

$$\epsilon_{\rm ABJM}(p) = \sqrt{\frac{1}{4} + 4\,h^2(\lambda_{\rm ABJM})\sin^2\frac{p}{2}} \tag{6.2}$$

with the crucial difference now that the (ABJM) *interpolating function* $h(\lambda_{\rm ABJM})$ appears to be *not* protected from quantum corrections:

$$\begin{aligned} h^2(\lambda_{\rm ABJM}) &= \lambda^2_{\rm ABJM} - \frac{2\pi^3}{3}\lambda^4_{\rm ABJM} + O(\lambda^6_{\rm ABJM}) && \lambda_{\rm ABJM} \ll 1\,, \\ h(\lambda_{\rm ABJM}) &= \sqrt{\frac{\lambda_{\rm ABJM}}{2}} - \frac{\log 2}{2\pi} + O(\lambda^{-1/2}_{\rm ABJM}) && \lambda_{\rm ABJM} \gg 1\,. \end{aligned} \tag{6.3}$$

The first few orders of the weak-coupling expansion were computed in [1, 13–15] and in [16–18]. The leading [13, 14] and subleading behaviour [19–21] were also computed at strong coupling.

The ABJM interpolating function plays the role of the effective coupling constant of integrability in the AdS_3/CFT_4 system because many other quantities like the S-matrix, the Bethe ansatz, the universal scaling function are related to those of the AdS_5/CFT_4 system by appropriately replacing $\frac{\sqrt{\lambda_{\rm YM}}}{4\pi} \to h(\lambda_{\rm ABJM})$. For this reason, the predictive power of the (conjectured) integrability of ABJM theory deeply relies on the knowledge of the non-trivial relation between $h(\lambda_{\rm ABJM})$ and the gauge-theory coupling $\lambda_{\rm ABJM}$. A conjecture for the exact expression of the interpolating function was put forward [22] in the form of an implicit equation for it

$$\lambda_{\rm ABJM} = \frac{\sinh\left(2\pi h(\lambda_{\rm ABJM})\right)}{2\pi}\,{}_3F_2\left(\frac{1}{2},\frac{1}{2},\frac{1}{2};1,\frac{3}{2};-\sinh^2\left(2\pi h(\lambda_{\rm ABJM})\right)\right) \tag{6.4}$$

by comparing two all-order calculations in ABJM theory: an integrability-based result expressed in terms of the effective coupling $h(\lambda)$ (the "slope function" [23] derived via integrability as exact solution of a quantum spectral curve [4]) and a supersymmetric localization prediction (a 1/6-BPS Wilson loop in ABJM theory, studied in [24–26], see also [27] for short summary). The expansion at weak/strong coupling of (6.4)

$$\begin{aligned} h(\lambda_{\rm ABJM}) &= \lambda_{\rm ABJM} - \frac{\pi^2}{3}\lambda^3_{\rm ABJM} + \frac{5\pi^4}{12}\lambda^5_{\rm ABJM} - \frac{893\pi^6}{1260}\lambda^7_{\rm ABJM} + O(\lambda^9_{\rm ABJM}) && \lambda_{\rm ABJM} \ll 1\,, \\ h(\lambda_{\rm ABJM}) &= \sqrt{\frac{1}{2}\left(\lambda_{\rm ABJM} - \frac{1}{24}\right)} - \frac{\log 2}{2\pi} + O\left(e^{-2\pi\sqrt{2\lambda_{\rm ABJM}}}\right) && \lambda_{\rm ABJM} \gg 1\,. \end{aligned} \tag{6.5}$$

reproduces the known coefficients (6.3) above. As already noticed in [22], a rigorous justification of (6.4) would require the comparison between the localization results of [25, 26] and the ABJM Bremsstrahlung function [28–31], similarly to the strategy adapted for $h(\lambda_{\rm YM})$ of $\mathcal{N}=4$ SYM explained above.

The main aim of this chapter is to extended the computation of (6.3) to two loops at strong coupling in order to give further support to the all-loop proposal for $h(\lambda)$. The strong-coupling behaviour of the interpolating function can computed in sigma-model perturbation theory by means of the *(universal) scaling function* $f_{\rm ABJM}(\lambda_{\rm ABJM})$ of the ABJM theory.[2] Our strategy will be to compute the ABJM scaling function in sigma-model perturbation theory—as a function of the string tension T (hence $\lambda_{\rm ABJM}$ via (1.9))—and then compare with its integrability prediction from asymptotic Bethe ansatz [2]—naturally depending on the "integrability" coupling $h(\lambda_{\rm ABJM})$—as we explain below.

The evaluation of the first two strong-coupling corrections to the ABJM cusp anomalous dimension will lead to

$$f_{\rm ABJM}(\lambda_{\rm ABJM}) = \sqrt{2\lambda_{\rm ABJM}} - \frac{5\log 2}{2\pi} - \left(\frac{K}{4\pi^2}+\frac{1}{24}\right)\frac{1}{\sqrt{2\lambda_{\rm ABJM}}} + O(\lambda_{\rm ABJM}^{-1})\,. \tag{6.6}$$

This can be seen by plugging the coefficients (6.31) and (6.56) into (6.23) below. An important ingredient in the calculation is the correction to the effective string tension (1.9) [5] which must be considered for the first time at this order in sigma-model perturbation theory.

The integrability prediction for the same observable $f_{\rm ABJM}(\lambda_{\rm ABJM})$ derives from the formal equivalence of the *Beisert-Eden-Staudacher (BES)* equation [32] for the $\mathcal{N}=4$ case and the ABJM case

$$f_{\rm ABJM}(\lambda_{\rm ABJM}) = \frac{1}{2}\, f_{\rm YM}(\lambda_{\rm YM})\bigg|_{\frac{\sqrt{\lambda_{\rm YM}}}{4\pi}\to h(\lambda_{\rm YM})}\,, \tag{6.7}$$

which implies

$$f_{\rm ABJM}(\lambda_{\rm ABJM}) = 2h(\lambda_{\rm ABJM}) - \frac{3\log 2}{2\pi} - \frac{K}{8\pi^2}\frac{1}{h(\lambda_{\rm ABJM})} + \cdots\,, \tag{6.8}$$

where $f_{\rm YM}(\lambda_{\rm YM})$ is the scaling function of $\mathcal{N}=4$ SYM and $K\approx 0.916$ is the Catalan constant.

All this leads to the main result of the chapter: the comparison between (6.6) and (6.8) yields our result for the strong-coupling expansion of the interpolating function up to two-loop order

---

[2]The scaling function is identified with twice the null cusp anomalous dimension, governing the UV divergences of a light-like Wilson cusp. This is explained at length in the particular case of $\mathcal{N}=4$ SYM in Sect. 7.1 in the next chapter.

$$h(\lambda_{\rm ABJM}) = \sqrt{\frac{\lambda_{\rm ABJM}}{2}} - \frac{\log 2}{2\pi} - \frac{1}{48\sqrt{2\lambda_{\rm ABJM}}} + O(\lambda_{\rm ABJM}^{-1}) \qquad \lambda_{\rm ABJM} \gg 1 \tag{6.9}$$

which extends the one-loop result in the second line of (6.3) and that agrees[3] with the prediction of the interpolating function (6.5) above, once expanded for large $\lambda_{\rm YM}$.

We also recall that the leading order $h(\lambda_{\rm ABJM}) \approx \sqrt{\lambda_{\rm ABJM}/2}$ can be already extracted from the leading coefficient of (6.6) given in [33] by using the formula (6.8). We also note that the first subleading correction $-\log 2/(2\pi)$ to $h(\lambda_{\rm ABJM})$, on which some debate existed [34], can be analogously computed from the scaling function at one loop [19–21, 35–44] in sigma-model perturbation theory as the energy of closed spinning strings in the large-spin limit or similar means. An interesting observation that only descends from our two-loop computation (6.9) is the fact that the two-loop correction to $h(\lambda_{\rm ABJM})$ is exclusively due to the "anomalous" correction $-1/24$ in (1.9).

On a different note, our result (6.6) is very similar to the expression of the $\mathcal{N}=4$ SYM scaling function, which will be reported in (7.6) (with $g = \sqrt{\lambda_{\rm YM}}/(4\pi)$ defined in (7.3)):

$$f_{\rm YM}(\lambda_{\rm YM}) = \frac{\sqrt{\lambda_{\rm YM}}}{\pi} - \frac{3\log 2}{\pi} - \frac{K}{\pi\sqrt{\lambda_{\rm YM}}} + O(\lambda_{\rm YM}^{-1})\,. \tag{6.10}$$

To ease the comparison, it looks convenient to define the shifted coupling $\tilde\lambda_{\rm ABJM} \equiv \lambda_{\rm ABJM} - \frac{1}{24}$ and rewrite (6.6) in terms of it as

$$f_{\rm ABJM}\left(\tilde\lambda_{\rm ABJM}\right) = \sqrt{2\tilde\lambda_{\rm ABJM}} - \frac{5\log 2}{2\pi} - \frac{K}{4\pi^2\sqrt{2\tilde\lambda_{\rm ABJM}}} + O(\tilde\lambda_{\rm ABJM}^{-1})\,, \tag{6.11}$$

where the change in the transcendentality pattern is due to the corresponding difference in the effective string tensions.

To arrive to the result (6.6) we will use perturbation theory using the open-string approach followed by [45, 46], namely expanding the string partition function for the Euclidean surface ending on a null cusp at the boundary of AdS_4, as also done in the $AdS_5 \times S^5$ setting in [47].

As the classical string lies solely in AdS_4 and higher-order fermions are needed, we must first face the problem, mentioned in Sect. 2.2, of using the correct superstring action. Notice that no issues would be encountered in the action to use at one loop level. There, only the quadratic part of the fermionic Lagrangian is necessary, with a structure which is well-known in terms of the type IIA covariant deriva-

[3]The relation between the gauge-theory 't Hooft coupling $\lambda_{\rm ABJM}$ and the string tension T is sensitive to the regularization choice in the two theories. The consistency of our result suggests that regularization behind the conjectured form of $h(\lambda)$ and the worldsheet regularization in the two-loop calculation effectively capture the same physics, see also [3, 19] for related discussions. We thank Radu Roiban for this remark.

tive restricted by the background RR fluxes.[4] Here instead we will work with the string action (2.39) derived in [51, 52] from the 11d membrane action based on the supercoset $OSp(4|8)/(SO(1,3) \times SO(7))$, performing double dimensional reduction and choosing a κ-symmetry light-cone gauge for which both light-like directions lie in AdS_4.

In general, the mutual consistency of several ingredients—our direct perturbative string calculation, the corrected dictionary of [5], the prediction (6.7)–(6.8) from the Bethe Ansatz [2] and the conjecture of [22] for the interpolating function $h(\lambda_{\rm ABJM})$—provides highly non-trivial evidence in support of the proposal (6.4) for the interpolating function $h(\lambda_{\rm ABJM})$ of ABJM theory, and furnishes an indirect check of the quantum integrability of the $AdS_4 \times \mathbb{CP}^3$ superstring theory in this κ-symmetry light-cone gauge.

6.1 The Null Cusp Vacuum and Fluctuation Lagrangian

In this section we begin with the κ-symmetry gauge-fixed Lagrangian (2.39) discussed at length in Sect. 2.2 and—as discussed in [53] and used in [47, 54, 55]—fix bosonic local symmetry with a "modified" conformal gauge

$$\gamma^{ij} = \text{diag}\left(-e^{4\varphi}, e^{-4\varphi}\right) \tag{6.12}$$

in combination with the standard light-cone gauge

$$x^+ = p^+ \tau\,, \qquad\qquad p^+ = \text{constant}\,. \tag{6.13}$$

Above, we used the fact that $AdS_4 \times \mathbb{CP}^3$ is equipped by the metric given in (2.36), (2.37) and (2.48).

We shall consider the Wick-rotated action S_E formally obtained through the analytic continuations $\tau \to -i\,\tau$, $p^+ \to i p^+$. Having conveniently set the light-cone momentum to $p^+ = 1$, the equations of motion (and the Virasoro constraints) admit an open string solution (*null cusp background*)

$$\begin{aligned} &x^+ = \tau\,, \qquad x^- = -\frac{1}{2\sigma}\,, \qquad x^1 = 0\,, \\ &w \equiv e^{2\varphi} = \sqrt{-2x^+x^-} = \sqrt{\frac{\tau}{\sigma}}\,, \qquad z^a = \bar{z}_a = 0\,, \qquad a = 1,2,3\,, \qquad \tau,\sigma > 0\,. \end{aligned} \tag{6.14}$$

[4] Alternatively, one could still use the coset action of [48, 49]—which is not suitable when strings move confined in AdS [49, 50]—starting with a classical solution spinning both in AdS_4 with spin S and in $\mathbb{CP}^3$ with spin J, and taking on the resulting expression for the one-loop energy a smooth $J \to 0$ limit [36].

that is bounded at $w = 0$ by two light-like lines meeting at a cusp point. The AdS part of the worldsheet is a minimal-area surface extending in a Euclidean AdS_3 subspace (x^+, x^-, w) of AdS_4, and therefore it coincides[5] with the null cusp background in AdS_5 of [45, 47] and later considered in Chap. 7.

In the AdS/CFT dictionary of [56, 57], our aim is to evaluate the expectation value of the cusped Wilson loop in the 3d boundary of AdS_4 as the string path-integral

$$\langle \mathcal{W}_{\rm cusp} \rangle = Z_{\rm string} \equiv \int \mathcal{D}x\, \mathcal{D}w\, \mathcal{D}z\, \mathcal{D}\theta\, \mathcal{D}\eta\; e^{-S_E}\,. \tag{6.15}$$

The semiclassical computation of the partition function relies on expanding S_E around (6.14). An important property of this classical solution is that, taking inspiration from [47], there exists a parametrization of fluctuations[6]

$$\begin{aligned} &x^1 = 2\,, \sqrt{\tfrac{T}{\sigma}}\tilde{x}^1 \qquad && w = \sqrt{\tfrac{T}{\sigma}}\,, \tilde{w} \equiv \sqrt{\tfrac{T}{\sigma}}\, e^{2\tilde{\varphi}} \\ &z^a = \tilde{z}^a\,, && \bar{z}^a = \bar{\tilde{z}}^a\,, \qquad a = 1, 2, 3\,, \\ &\eta = \tfrac{1}{\sqrt{\sigma}}\, \tilde{\eta}\,, && \theta = \tfrac{1}{\sqrt{\sigma}}\, \tilde{\theta} \end{aligned} \tag{6.16}$$

and a redefinition of the worldsheet coordinates $t = \log\tau$ and $s = \log\sigma$ such that the coefficients of the fluctuation Lagrangian become constant. The resulting Lagrangian

$$S_E = \frac{T}{2}\int dt\, ds\, \mathcal{L}_E, \qquad \mathcal{L}_E \equiv \mathcal{L}_B^{(0)} + \mathcal{L}_B + \mathcal{L}_{F^2} + \mathcal{L}_{F^4}\,, \tag{6.17}$$

splits into its classical value $\mathcal{L}_B^{(0)} = \frac{1}{8}$ on the null cusp vacuum (6.14), the bosonic Lagrangian $\mathcal{L}_B$ and the part quadratic $\mathcal{L}_{F^2}$ and quartic $\mathcal{L}_{F^4}$ in fermions

$$\begin{aligned} \mathcal{L}_B =& \left(\partial_t \tilde{x}^1 + \frac{1}{2}\tilde{x}^1\right)^2 + e^{-8\tilde{\varphi}}\left(\partial_s \tilde{x}^1 - \frac{1}{2}\tilde{x}^1\right)^2 + e^{4\tilde{\varphi}}\,(\partial_t\varphi)^2 + e^{-4\tilde{\varphi}}\,(\partial_s\varphi)^2 + \\ &+ \frac{1}{16}\left(e^{4\tilde{\varphi}} + e^{-4\tilde{\varphi}}\right) + e^{4\tilde{\varphi}}\,\tilde{g}_{MN}\,\partial_t\tilde{z}^M\,\partial_t\tilde{z}^N + e^{-4\tilde{\varphi}}\,\tilde{g}_{MN}\,\partial_s\tilde{z}^M\,\partial_s\tilde{z}^N\,, \end{aligned} \tag{6.18}$$

$$\begin{aligned} \mathcal{L}_{F2} =& \, i\Big[\partial_t\tilde{\theta}_a\bar{\tilde{\theta}}^a - \tilde{\theta}_a\partial_t\bar{\tilde{\theta}}^a + \partial_t\tilde{\theta}_4\bar{\tilde{\theta}}^4 - \tilde{\theta}_4\partial_t\bar{\tilde{\theta}}^4 + \partial_t\tilde{\eta}_a\bar{\tilde{\eta}}^a - \tilde{\eta}_a\partial_t\bar{\tilde{\eta}}^a + \partial_t\tilde{\eta}_4\bar{\tilde{\eta}}^4 - \tilde{\eta}_4\partial_t\bar{\tilde{\eta}}^4\Big] + \\ &+ 2ie^{-4\tilde{\varphi}}\Big[\hat{\eta}_a\left(\hat{\partial}_s\bar{\theta}^a - \frac{1}{2}\hat{\bar{\theta}}^a\right) + \left(\hat{\partial}_s\theta_a - \frac{1}{2}\hat{\theta}_a\right)\hat{\bar{\eta}}^a + \frac{1}{2}\left(\partial_s\theta_4\bar{\eta}^4 - \partial_s\eta_4\bar{\theta}^4 + \eta_4\partial_s\bar{\theta}^4 - \right. \\ &\left. - \theta_4\partial_s\bar{\eta}^4\right)\Big] + \partial_t\tilde{z}^M\,\tilde{h}_M + 4i\,e^{-6\tilde{\varphi}}\tilde{C}\left(\partial_s\tilde{x}^1 - \frac{1}{2}\tilde{x}^1\right) - 2ie^{-4\tilde{\varphi}}\,\partial_s\tilde{z}^M\,\tilde{\ell}_M\,, \end{aligned} \tag{6.19}$$

$$\mathcal{L}_{F4} = e^{-8\tilde{\varphi}}\,\tilde{B}\,. \tag{6.20}$$

[5]To see this, compare the expressions for (x^+, x^-, w) in (6.14) with the ones for (x^+, x^-, z) in (7.12), where the latter is shown in Fig. 7.2.

[6]We introduce the factor 2 in the field x^1 to normalize the kinetic term of $\tilde{x}^1$, cf. (7.14).

In the expressions above, $\tilde{B}$, $\tilde{C}$, $\tilde{h}_M$ and $\tilde{\ell}_M$ are understood as the quantities B, C, h_M and ℓ_M in (2.40)–(2.45) where we understood that each field was replaced by the corresponding tilded fluctuation.

Since the Lagrangian has constant coefficients and is thus translationally invariant, the (infinite) world-sheet volume factor V factorizes. The scaling function is then defined via the string partition function as [47]

$$\Gamma \equiv -\log Z = \frac{1}{2} f_{\text{ABJM}}(\lambda_{\text{ABJM}})\, V = \Gamma^{(0)} + \Gamma^{(1)} + \Gamma^{(2)} + \cdots, \quad V = \frac{1}{4} V_2 \equiv \frac{1}{4}\int_{-\infty}^{\infty} dt \int_{-\infty}^{\infty} ds \tag{6.21}$$

where $\Gamma^{(0)} = \frac{T}{16} V_2$ coincides with the value of the action on the background, while $\Gamma^{(1)}$ and $\Gamma^{(2)}$ are the one- and two-loop correction. For the ratio V/V_2 we use the same convention as in [47].[7] From (6.21) we explicitly define $f_{\text{ABJM}}(\lambda_{\text{ABJM}})$ in terms of the free energy Γ

$$f_{\text{ABJM}}(\lambda_{\text{ABJM}}) = \frac{8}{V_2}\, \Gamma\,. \tag{6.22}$$

We are now ready to compute the free energy perturbatively in inverse powers of the effective string tension $g \equiv \frac{T}{2}$. From this we will extract the corresponding strong-coupling perturbative expansion for the scaling function

$$f_{\text{ABJM}}(g) = g\left(1 + \frac{a_1}{g} + \frac{a_2}{g^2} + \ldots\right), \qquad g = \frac{T}{2}, \tag{6.23}$$

where we have factorized the classical result from $\Gamma^{(0)}$ [33] and the string tension T is related to the ABJM 't Hooft coupling via (1.9).

6.2 Cusp Anomaly at One Loop

We start considering one-loop quantum corrections to the free energy $\Gamma^{(1)}$ which are derived expanding the fluctuation Lagrangian (6.17) to second order in the fields. The spectrum of the free bosonic part of the Lagrangian

$$\mathcal{L}_B^{(2)} = \left(\partial_t \tilde{x}^1\right)^2 + \left(\partial_s \tilde{x}^1\right)^2 + \frac{1}{2}\left(\tilde{x}^1\right)^2 + (\partial_t \tilde{\varphi})^2 + (\partial_s \tilde{\varphi})^2 + \tilde{\varphi}^2 + \left|\partial_t \tilde{z}^a\right|^2 + \left|\partial_s \tilde{z}^a\right|^2 \tag{6.24}$$

consists of six real massless scalars (in the complex fluctuations $\tilde{z}^a$ and $\bar{\tilde{z}}^a$ of the $\mathbb{CP}^3$ directions), one real scalar $\tilde{x}^1$ with mass $m^2 = 1/2$ and one real scalar $\tilde{\varphi}$ with mass $m^2 = 1$. This spectrum can be viewed as a simple truncation (in the sense of one less

[7] This is related to coordinate transformations and field redefinitions occurring between the GKP string [58], whose energy is given in terms of $f_{\text{YM}}(\lambda_{\text{YM}})$, and the null cusp solution in the Poincaré patch used here, see discussion in [55].

transverse degree of freedom in the AdS space) of the bosonic one in the $AdS_5 \times S^5$ case [47].

If we accommodate the physical fermions in the sixteen-component vector

$$\psi \equiv \left(\tilde{\theta}_a, \tilde{\theta}_4, \tilde{\tilde{\theta}}^a, \tilde{\tilde{\theta}}^4, \tilde{\eta}_a, \tilde{\eta}_4, \tilde{\tilde{\eta}}^a, \tilde{\tilde{\eta}}^4\right)^T , \tag{6.25}$$

the kinetic operator, defined as

$$\mathcal{L}_F^{(2)} \equiv i\,\psi^T\, K_F \psi\,, \tag{6.26}$$

is the off-diagonal matrix-valued differential operator

$$K_F = \begin{pmatrix} 0 & 0 & -\mathbb{I}_3\,\partial_t & 0 & 0 & 0 & -\mathbb{I}_3\,(\partial_s + \frac{1}{2}) & 0 \\ 0 & 0 & 0 & -\partial_t & 0 & 0 & 0 & -\partial_s \\ -\mathbb{I}_3\,\partial_t & 0 & 0 & 0 & \mathbb{I}_3\,(\partial_s + \frac{1}{2}) & 0 & 0 & 0 \\ 0 & -\partial_t & 0 & 0 & 0 & \partial_s & 0 & 0 \\ 0 & 0 & \mathbb{I}_3\,(\partial_s - \frac{1}{2}) & 0 & 0 & 0 & -\mathbb{I}_3\,\partial_t & 0 \\ 0 & 0 & 0 & \partial_s & 0 & 0 & 0 & -\partial_t \\ \mathbb{I}_3\,(-\partial_s + \frac{1}{2}) & 0 & 0 & 0 & -\mathbb{I}_3\,\partial_t & 0 & 0 & 0 \\ 0 & -\partial_s & 0 & 0 & 0 & -\partial_t & 0 & 0 \end{pmatrix}. \tag{6.27}$$

The determinant of the operator above can be readily computed in momentum space by replacing $\partial_\mu \to i\,p_\mu$ $(\mu = 0, 1)$

$$\text{Det}\,K_F = \left(p^2\right)^2 \left(p^2 + \frac{1}{4}\right)^6 . \tag{6.28}$$

This shows that the spectrum of fermionic excitations contains six degrees of freedom with mass $m^2 = 1/4$ and two with $m^2 = 0$. Interestingly, the massless eigenstates are a linear combination of the η_4- and θ_4-fermions only, namely those fermionic directions corresponding to the broken supercharges of the $AdS_4 \times \mathbb{CP}^3$ background. This is in contrast with the case of the maximally symmetric $AdS_5 \times S^5$ where all the fermions are equally massive, as already noticed in $AdS_4 \times \mathbb{CP}^3$ when studying fluctuations over classical string solutions in AdS_4 [19–21, 44].

Combining bosonic and fermionic contributions, the one-loop free energy is computed as

$$\Gamma^{(1)} \equiv -\log Z^{(1)} \equiv -\log \frac{\text{Det}^{1/2} K_F}{\text{Det}^{6/2}(-\partial_t^2 - \partial_s^2)\text{Det}^{1/2}(-\partial_t^2 - \partial_s^2 + \frac{1}{2})\text{Det}^{1/2}(-\partial_t^2 - \partial_s^2 + 1)} . \tag{6.29}$$

Although the techniques for functional determinants in Appendix B are still applicable, here the computation is straightforward because the homogeneity of the classical background allows to Fourier-transform both time and space direction. Indeed, for this situation the eigenvalues of the bosonic operators (free-wave operators in flat space) are well-known and, combined with (6.28), they yield

$$\Gamma^{(1)} = \frac{V_2}{2} \int \frac{d^2 p}{(2\pi)^2} \log \left[\frac{(p^2)^6 \left(p^2 + \frac{1}{2}\right) \left(p^2 + 1\right)}{(p^2)^2 \left(p^2 + \frac{1}{4}\right)^6} \right] = -\frac{5 \log 2}{16\pi} V_2 . \quad (6.30)$$

The integral is convergent because of the matching of the number of degrees of freedom and the values of their masses. The one-loop correction to the scaling function reads, according to (6.22),

$$a_1 = -\frac{5 \log 2}{2\pi} \quad (6.31)$$

and agrees with previous results [19–21].

6.3 Cusp Anomaly at Two Loops

We proceed to extend the perturbative computation of the scaling function to two-loop order, following the lines of [47]. This amounts to compute the connected vacuum diagrams in the null cusp background (6.14) built out of the propagators and interaction vertices in Appendix E.1. The various contributions combine in the two-loop part of the free energy

$$\Gamma^{(2)} = \langle S_{\text{int}} \rangle - \frac{1}{2} \langle S_{\text{int}}^2 \rangle_c , \quad (6.32)$$

where $S_{\text{int}} = T \int dt ds\, \mathcal{L}_{\text{int}}$ is the interacting part of the action at cubic and quartic order, namely $\mathcal{L}_{\text{int}}$ is sum of the vertices (E.3)–(E.18). The subscript "c" indicates that only connected diagrams are included. Following the conventions of Appendix E.1, we drop tildes in the fluctuation fields. We also omit the string tension T and the volume V_2 in the intermediate steps and reinstate them at the end of the calculation.

6.3.1 *Bosonic Sector*

We first start with the purely bosonic sector, which contains one real boson of squared mass 1, one real boson of squared mass $\frac{1}{2}$ and three complex massless bosons, as seen in Sect. 6.2. At two-loop level the cubic and quartic interactions among these fields give rise to the topologies in Fig. 6.1.

An important feature of the AdS light-cone gauge is that the Lagrangian has diagonal bosonic propagators (E.1), which introduces considerable simplifications in the perturbative computation. To evaluate the momenta integrals, we have to manipulate the tensor structures, involving components of the loop momenta, in the numerators and rewrite them in terms of scalar integrals. The reduction formulas in Appendix E.2 allow to rewrite every integral as linear combinations of

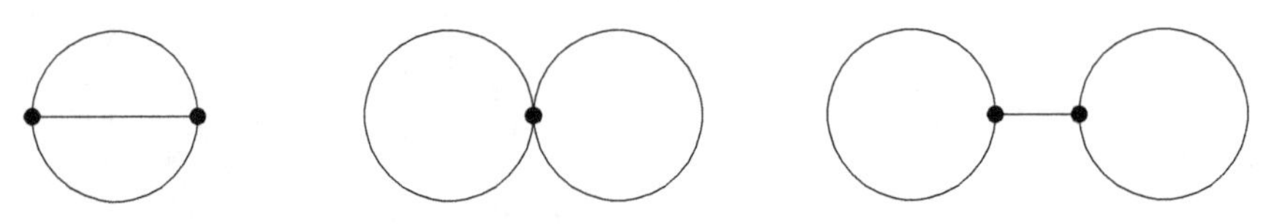

Fig. 6.1 Two-loop connected topologies: *sunset, double bubble* and the non-one-particle-irreducible *double tadpole*

$$I\left(m^2\right) \equiv \int \frac{d^2p}{(2\pi)^2}\,\frac{1}{p^2+m^2}\,, \tag{6.33}$$

which is UV logarithmically divergent and potentially IR singular for vanishing mass, and

$$I\left(m_1^2, m_2^2, m_3^2\right) \equiv \int \frac{d^2p\, d^2q\, d^2r}{(2\pi)^4}\,\frac{\delta^{(2)}(p+q+r)}{(p^2+m_1^2)(q^2+m_2^2)(r^2+m_3^2)}\,, \tag{6.34}$$

which is finite for non-zero masses and suffers from IR infrared singularities otherwise. In our computation we expect all UV divergences to cancel and therefore no divergent integral to appear in the final result. Nonetheless, performing reduction of potentially divergent tensor integrals to scalar ones still implies the choice of a regularization scheme. In our case we use the one adopted in [47, 59, 60]. This prescription consists of performing all manipulations in the numerators in $d = 2$, which has the advantage of simpler tensor integral reductions. In this process we set power UV-divergent massless tadpoles to zero, as in dimensional regularization

$$\int \frac{d^2p}{(2\pi)^2}\,\left(p^2\right)^n = 0\,, \qquad\qquad n \geq 0\,. \tag{6.35}$$

We do not need to choose a regularization prescription for the remaining logarithmically divergent integrals because we will verify that they drop out in the final result.

As an example of a typical contribution, let us have a closer look at the sunset diagram coming from the first vertex in (E.3). Wick contractions and scalar reductions yield the following expression

$$-\frac{1}{2}\langle V^2_{\varphi x^1 x^1}\rangle = -\int \frac{d^2p\, d^2q\, d^2r}{(2\pi)^4}\,\frac{(1+4q_1^2)\,(1+4r_1^2)\,\delta^{(2)}(p+q+r)}{(p^2+1)(q^2+\frac{1}{2})(r^2+\frac{1}{2})} = \frac{1}{2}\,I\left(1, \tfrac{1}{2}, \tfrac{1}{2}\right)\,. \tag{6.36}$$

Such ratio of masses in the integral $I\left(1, \frac{1}{2}, \frac{1}{2}\right)$ was already appeared in [47] and is a particular case of the general class

$$I\left(2\,m^2, m^2, m^2\right) = \frac{K}{8\,\pi^2\,m^2} \tag{6.37}$$

with K being the Catalan constant.

The contribution of the sunset diagram involving the second vertex in (E.3) is proportional to $I(1)^2$, whereas the contribution of the third vertex vanishes

$$-\frac{1}{2}\langle V^2_{\varphi^3}\rangle = 2\,I(1)^2 \qquad -\frac{1}{2}\langle V^2_{\varphi|z|^2}\rangle = 0\,. \tag{6.38}$$

The final contribution of the bosonic sunset diagrams is

$$\Gamma^{(2)}_{\text{bos. sunset}} = \frac{1}{2}\,I\left(1,\tfrac{1}{2},\tfrac{1}{2}\right) + 2\,I(1)^2\,. \tag{6.39}$$

The first two vertices in (E.3) can also be contracted to generate non-1PI graphs, namely double tadpoles. However the resulting diagrams turn out to vanish individually.

Next we consider bosonic double bubble diagrams. The relevant quartic vertices are

$$V_{\varphi^2x^1x^1} = 16\,\varphi^2\,\left[(\partial_s - \tfrac{1}{2})\,x^1\right]^2\,, \tag{6.40}$$

$$V_{\varphi^4} = 4\,\varphi^2\left[(\partial_t\varphi)^2 + (\partial_s\varphi)^2 + \frac{1}{6}\varphi^2\right]\,, \tag{6.41}$$

$$V_{\varphi^2|z|^2} = 4\,\varphi^2\left[|\partial_t z|^2 + |\partial_s z|^2\right]\,, \tag{6.42}$$

$$V_{z^4} = \frac{1}{6}\Big[\left(\bar z_a\partial_t z^a\right)^2 + \left(\bar z_a\partial_s z^a\right)^2 + \left(z^a\partial_t\bar z_a\right)^2 + \left(z^a\partial_s\bar z_a\right)^2 - |z|^2\left(|\partial_t z|^2 + |\partial_s z|^2\right) - \left|\bar z_a\partial_t z^a\right|^2 - \left|\bar z_a\partial_s z^a\right|^2\Big]\,. \tag{6.43}$$

Despite the lengthy expressions of the vertices, the only non-vanishing contribution comes from V_{φ^4} and gives

$$\Gamma^{(2)}_{\text{bos. bubble}} = -2\,I(1)^2 \tag{6.44}$$

and cancels the divergent part of (6.39). As a result, the bosonic sector turns out to be free of divergences without the need of fermonic contributions, which was already observed in the $AdS_5 \times S^5$ case [47].

6.3.2 *Fermionic Sector*

In this section we include the diagrams arising from interactions involving fermions. The main difference between the $AdS_4 \times \mathbb{CP}^3$ spectrum in Sect. 6.2 and the one of $AdS_5 \times S^5$ [47] lies in the fermionic sector: both theories have eight fermionic degrees of freedom, but in $AdS_4 \times \mathbb{CP}^3$ they are split into six massive and two massless excitations, which interact non-trivially among themselves.

We begin with the diagrams involving at least one massless fermion. The relevant cubic vertices are (ψ denotes collectively the fermions η and θ)

$$\begin{aligned} V_{z\eta_a\eta_4} &= -2\,\partial_t z^a \eta_a \eta_4 + \text{h.c.}\,, & V_{z\eta_a\theta_4} &= 2\,\partial_s z^a \eta_a \theta_4 - \text{h.c.}\,, \\ V_{\varphi\eta_4\bar\theta^4} &= -2\,i\,\varphi\,(\bar\theta^4 \partial_s \eta_4 - \partial_s \bar\theta^4 \eta_4) - \text{h.c.}\,, & V_{x^1\bar\psi^4\psi_4} &= -2\,i\,(\bar\eta^4 \eta_4 + \bar\theta^4 \theta_4)(\partial_s - \tfrac{1}{2}) x^1\,. \end{aligned} \tag{6.45}$$

The quartic interactions are either not suitable for constructing a double tadpole diagram or they produce vanishing integrals. These include vector massless tadpoles, which vanish by parity, and tensor massless tadpoles, which have power UV divergences and are set to zero. For completeness we list them in Appendix E.1.
Focussing on the Feynman graphs which can be constructed from cubic interaction we also note that the only double tadpole diagrams that can be produced using (E.6) involve tensor massless tadpole integrals and therefore vanish. In the sector with massless fermions we are therefore left with the sunset diagrams, which, thanks to the diagonal structure of the bosonic propagators, turn out to be only five

$$\Gamma^{(2)}_{\psi_4} = -\frac{1}{2} \langle V_{z\eta_a\eta_4} V_{z\eta_a\eta_4} + V_{z\eta_a\theta_4} V_{z\eta_a\theta_4} + 2\,V_{z\eta_a\eta_4} V_{z\eta_a\theta_4} + V_{\varphi\eta_4\bar\theta^4} V_{\varphi\eta_4\bar\theta^4} + V_{x^1\bar\psi^4\psi_4} V_{x^1\bar\psi^4\psi_4} \rangle\,. \tag{6.46}$$

The explicit computation of the individual contributions shows that they are all vanishing. As an example we consider

$$-\frac{1}{2} \langle V_{\varphi\eta_4\bar\theta^4} V_{\varphi\eta_4\bar\theta^4} \rangle = 4 \int \frac{d^2p\, d^2q\, d^2r}{(2\pi)^4} \frac{(p_1 - q_1)^2 (p_0 q_0 - p_1 q_1)\, \delta^{(2)}(p+q+r)}{p^2 q^2 (r^2+1)} = 0 \tag{6.47}$$

and similar cancellations happen for the other diagrams. Therefore we conclude that $\Gamma^{(2)}_{\psi_4} = 0$ and that massless fermions are effectively decoupled at two loops.
We then move to consider massive fermions, starting from their cubic coupling to bosons

$$\begin{aligned} V_{z\eta\eta} &= -\epsilon^{abc} \partial_t \bar z_a \eta_b \eta_c + \text{h.c.}\,, & V_{z\eta\theta} &= -2\,\epsilon^{abc} \bar z_a \eta_b (\partial_s - \tfrac{1}{2}) \theta_c - \text{h.c.}\,, \\ V_{\varphi\eta\theta} &= -4\,i\,\varphi\,\eta_a (\partial_s - \tfrac{1}{2}) \bar\theta^a - \text{h.c.}\,, & V_{x^1\eta\eta} &= -4\,i\,\bar\eta^a \eta_a (\partial_s - \tfrac{1}{2}) x^1\,. \end{aligned} \tag{6.48}$$

Precisely as in the massless case, this generates five possible sunset diagrams. None of them is vanishing. We present the details of a particularly relevant example, i.e. the one involving the vertex $V_{x^1\eta\eta}$. This gives

$$-\frac{1}{2} \langle V_{x^1\eta\eta} V_{x^1\eta\eta} \rangle = 24 \int \frac{d^2p\, d^2q\, d^2r}{(2\pi)^4} \frac{(p_1^2 + \frac{1}{4})\, q_0\, r_0\, \delta^{(2)}(p+q+r)}{(p^2 + \frac{1}{2})(q^2 + \frac{1}{4})(r^2 + \frac{1}{4})} = -\frac{3}{8} I\left(\tfrac{1}{2}, \tfrac{1}{4}, \tfrac{1}{4}\right) + \frac{3}{4} I\left(\tfrac{1}{4}\right)^2. \tag{6.49}$$

We note the appearance of another integral in the class (6.37). The coefficient in front of this integral depends on the degrees of freedom of the theory and is thoroughly discussed in Sect. 6.3.3. The partial results of the remaining sunset diagrams are

$$
\begin{aligned}
-\frac{1}{2}\langle (V_{z\eta\eta} + V_{z\eta\theta})(V_{z\eta\eta} + V_{z\eta\theta})\rangle &= 3\, I\left(\tfrac{1}{4}\right)^2 - 6\, I\left(\tfrac{1}{4}\right)\, I(0)\,, \\
-\frac{1}{2}\langle V_{\varphi\eta\theta} V_{\varphi\eta\theta}\rangle_{\mathrm{1PI}} &= 6\, I\left(\tfrac{1}{4}\right)\, I(1) + \frac{3}{4}\, I\left(\tfrac{1}{4}\right)^2 .
\end{aligned} \tag{6.50}
$$

The latter vertices can be contracted also in a non-1PI manner

$$
-\frac{1}{2}\langle V_{\varphi\eta\theta} V_{\varphi\eta\theta}\rangle_{\text{non-1PI}} = -\frac{1}{2}\, G_{\varphi\varphi}(0) \times 2^6 \times 3^2 \times \int \frac{d^2p}{(2\pi)^2}\, \frac{p_1^2 + \frac{1}{4}}{p^2 + \frac{1}{4}} = -\frac{9}{2}\, I\left(\tfrac{1}{4}\right)^2 \tag{6.51}
$$

where the factor in front of the integrals comes from the expression of the vertex and from counting the degrees of freedoms that can run in the loops. As in [47], the divergent contribution proportional to $I\left(\frac{1}{4}\right)^2$ cancels exactly those coming from (6.49) and (6.50).

The total cubic fermionic part reads

$$
\Gamma^{(2)}_{\text{ferm. cubic}} = -\frac{3}{8}\, I\left(\tfrac{1}{2}, \tfrac{1}{4}, \tfrac{1}{4}\right) + 6\, I\left(\tfrac{1}{4}\right)\, I(1) - 6\, I\left(\tfrac{1}{4}\right)\, I(0)\,. \tag{6.52}
$$

Finally we consider the fermionic double bubble diagrams. These involve the fermionic quartic vertices. However, most of the vertices appearing in the Lagrangian cannot contribute to the partition function either because the bosonic propagators are diagonal or because they would produce vanishing integrals. We present the whole list of quartic vertices in Appendix E.1 and we spell out here only the relevant ones for our computation

$$
V_{\varphi^2\eta\theta} = 8\, i\, \varphi^2\, \eta_a (\partial_s - \tfrac{1}{2})\bar\theta^a - \text{h.c.} \qquad V_{zz\eta\theta} = -2\, i\, \left[|z|^2 \eta_a (\partial_s - \tfrac{1}{2})\bar\theta^a - \bar z_b z^a \eta_a (\partial_s - \tfrac{1}{2})\bar\theta^b \right] - \text{h.c.}\,. \tag{6.53}
$$

Although we can build a diagram with V_{η^4}, fermion propagators carry one component of the loop momentum in the numerator and produce vector tadpole integrals, which vanish by parity. We conclude that the contribution from fermionic double bubble graphs is

$$
\Gamma^{(2)}_{\text{ferm. bubbles}} = -6\, I\left(\tfrac{1}{4}\right)\, I(1) + 6\, I\left(\tfrac{1}{4}\right)\, I(0)\,. \tag{6.54}
$$

Summing all the partial results and reinstating the dependence on the string tension and the volume, we obtain

$$
\Gamma^{(2)} = \frac{V_2}{T}\left[\frac{1}{2}\, I\left(1, \tfrac{1}{2}, \tfrac{1}{2}\right) - \frac{3}{8} I\left(\tfrac{1}{2}, \tfrac{1}{4}, \tfrac{1}{4}\right)\right] = -\frac{1}{4}\frac{V_2}{T} I\left(1, \tfrac{1}{2}, \tfrac{1}{2}\right) = -\frac{K}{16\,\pi^2}\frac{V_2}{T} \tag{6.55}
$$

where T is defined in (1.9). Finally we can plug this expression into Eq. (6.22) and read out the second order of the strong coupling expansion (6.23) of the ABJM cusp anomalous dimension

$$a_2 = -\frac{K}{4\pi^2}\,. \tag{6.56}$$

6.3.3 Comparison with the $AdS_5 \times S^5$ Scaling Function at Two Loops

In this section we point out similarities and differences between the calculation we performed and its $AdS_5 \times S^5$ analogue [47]. The starting points, i.e. the Lagrangians in AdS light-cone gauge, look rather different. Yet the final results of the two-loop computations are strikingly similar. More precisely, when written in terms of the string tension, the two expressions have exactly the same structure up to the numerical coefficients in front of the integrals. Indeed the AdS_5 computation gives

$$\Gamma^{(2)}_{AdS_5} = \frac{V_2}{T}\left[\frac{1}{4}\,I\left(1,\tfrac{1}{2},\tfrac{1}{2}\right) - \frac{1}{4}\,I\left(\tfrac{1}{2},\tfrac{1}{4},\tfrac{1}{4}\right)\right], \tag{6.57}$$

which looks very similar in structure to (6.55). Furthermore, using (6.37), both combinations sum up to

$$\Gamma^{(2)} = -\frac{V_2}{T}\,\frac{1}{4}\,I\left(1,\tfrac{1}{2},\tfrac{1}{2}\right) \tag{6.58}$$

and only the different relation between the string tension and the 't Hooft couplings distinguishes the final results. It is easy to trace the origin of the integrals and their coefficients back in the vertices of the Lagrangian and to understand their meaning. In particular in both computations only the sunset diagrams involving the interactions $V_{\varphi xx}$ and $V_{x\psi\psi}$ (with massive fermions) seem to effectively contribute. All other terms are also important, but just serve to cancel divergences. Hence we can now focus on the relevant interactions and point out the differences between the AdS_5 and the AdS_4 cases.

We start from the bosonic sectors. The two theories differ for the number of scalar degrees of freedom with given masses. Focussing on massive fluctuations, after gauge-fixing we have one scalar with $m^2 = 1$ associated to the radial coordinate of AdS_{d+1} and $(d-2)$ real scalars with $m^2 = \frac{1}{2}$. In the metric we chose for the $AdS_4 \times \mathbb{CP}^3$ background, the size of the AdS_4 part is rescaled by a factor of $r^2 = 4$. We have compensated this, parametrizing the radial coordinate as $w = e^{r\varphi}$ and introducing a factor r in the fluctuation of x^1, so as to have the same normalization for their kinetic terms as in $AdS_5 \times S^5$. This causes some factors r to appear in interaction vertices in our Lagrangian. Apart from this, the relevant interaction vertices are exactly the same. Then, the number of x fields $(d-2)$ and this factor r determine the coefficient of the integral $I\left(1,\frac{1}{2},\frac{1}{2}\right)$ appearing in equations (6.55) and (6.57).

Turning to fermions, the first striking difference between the AdS_5 and AdS_4 cases is the presence of massless ones. As pointed out at the beginning of Sect. 6.3.2 their contribution is effectively vanishing at two loops (though they do contribute at first order). Focussing on massive fermions, the relevant cubic interactions giving rise to $I\left(\frac{1}{2},\frac{1}{4},\frac{1}{4}\right)$ look again similar in the AdS_4 and AdS_5 cases. The difference is given once more by the ratio of the radii r (through the normalization of φ and x coordinates) and the number n_f of massive fermions in the spectrum ($n_f = 8$ for $AdS_5 \times S^5$ and $n_f = 6$ for $AdS_4 \times \mathbb{CP}^3$).

The final results (6.55) and (6.57) can be re-expressed in the general form

$$\begin{aligned}\Gamma^{(2)}_{(AdS_{d+1})} &= \frac{V_2}{T}\frac{(d-2)r^2}{8}\left[I\left(1,\tfrac{1}{2},\tfrac{1}{2}\right) - \frac{n_f}{8}\,I\left(\tfrac{1}{2},\tfrac{1}{4},\tfrac{1}{4}\right)\right] \\ &= \frac{V_2}{T}\frac{(d-2)r^2}{8}\left(1-\frac{n_f}{4}\right) I\left(1,\tfrac{1}{2},\tfrac{1}{2}\right), \qquad d=3,4, \qquad (6.59)\end{aligned}$$

where the cases at hand are $d = 4$, $n_f = 8$, $r = 1$ for $\mathcal{N} = 4$ SYM and $d = 3$, $n_f = 6$, $r = 2$ for ABJM.

References

1. J. Minahan, K. Zarembo, The Bethe ansatz for superconformal Chern-Simons. JHEP **0809**, 040 (2008). arXiv:0806.3951
2. N. Gromov, P. Vieira, The all loop AdS_4/CFT_3 Bethe ansatz. JHEP **0901**, 016 (2009). arXiv:0807.0777
3. T. Klose, Review of AdS/CFT integrability, chapter IV.3: $\mathcal{N} = 6$ Chern-Simons and strings on $AdS_4 \times \mathbb{CP}^3$. Lett. Math. Phys. **99**, 401 (2012). arXiv:1012.3999
4. A. Cavaglia, D. Fioravanti, N. Gromov, R. Tateo, The quantum spectral curve of the ABJM theory. Phys. Rev. Lett. **113**, 021601 (2014). arXiv:1403.1859
5. O. Bergman, S. Hirano, Anomalous radius shift in AdS_4/CFT_3. JHEP **0907**, 016 (2009). arXiv:0902.1743
6. N. Beisert, The SU(2|2) dynamic S-matrix. Adv. Theor. Math. Phys. **12**, 945 (2008). arXiv:hep-th/0511082
7. D. Correa, J. Henn, J. Maldacena, A. Sever, An exact formula for the radiation of a moving quark in $\mathcal{N} = 4$ super Yang Mills. JHEP **1206**, 048 (2012). arXiv:1202.4455
8. D. Correa, J. Maldacena, A. Sever, The quark anti-quark potential and the cusp anomalous dimension from a TBA equation. JHEP **1208**, 134 (2012). arXiv:1203.1913
9. N. Gromov, A. Sever, Analytic solution of Bremsstrahlung TBA. JHEP **1211**, 075 (2012). arXiv:1207.5489
10. D.M. Hofman, J.M. Maldacena, Giant magnons. J. Phys. A **39**, 13095 (2006). arXiv:hep-th/0604135
11. T. Klose, T. McLoughlin, J.A. Minahan, K. Zarembo, World-sheet scattering in $AdS_5 \times S^5$ at two loops. JHEP **0708**, 051 (2007). arXiv:0704.3891
12. D. Berenstein, D. Trancanelli, S-duality and the giant magnon dispersion relation. Eur. Phys. J. C **74**, 2925 (2014). arXiv:0904.0444
13. D. Gaiotto, S. Giombi, X. Yin, Spin chains in $\mathcal{N} = 6$ superconformal Chern-Simons-Matter theory. JHEP **0904**, 066 (2009). arXiv:0806.4589
14. G. Grignani, T. Harmark, M. Orselli, The $SU(2) \times SU(2)$ sector in the string dual of $\mathcal{N} = 6$ superconformal Chern-Simons theory. Nucl. Phys. B **810**, 115 (2009). arXiv:0806.4959

15. T. Nishioka, T. Takayanagi, On type IIA Penrose limit and $\mathcal{N} = 6$ Chern-Simons theories. JHEP **0808**, 001 (2008). arXiv:0806.3391
16. J.A. Minahan, O.O. Sax, C. Sieg, Magnon dispersion to four loops in the ABJM and ABJ models. J. Phys. A **43**, 275402 (2010). arXiv:0908.2463
17. J.A. Minahan, O.O. Sax, C. Sieg, Anomalous dimensions at four loops in $\mathcal{N} = 6$ superconformal Chern-Simons theories. Nucl. Phys. B **846**, 542 (2011). arXiv:0912.3460
18. M. Leoni, A. Mauri, J. Minahan, O. Ohlsson Sax, A. Santambrogio et al., Superspace calculation of the four-loop spectrum in $\mathcal{N} = 6$ supersymmetric Chern-Simons theories. JHEP **1012**, 074 (2010). arXiv:1010.1756
19. T. McLoughlin, R. Roiban, A.A. Tseytlin, Quantum spinning strings in $AdS_4 \times \mathbb{CP}^3$: testing the Bethe Ansatz proposal. JHEP **0811**, 069 (2008). arXiv:0809.4038
20. M.C. Abbott, I. Aniceto, D. Bombardelli, Quantum strings and the AdS_4/CFT_3 interpolating function. JHEP **1012**, 040 (2010). arXiv:1006.2174
21. C. Lopez-Arcos, H. Nastase, Eliminating ambiguities for quantum corrections to strings moving in $AdS_4 \times \mathbb{CP}^3$. Int. J. Mod. Phys. A **28**, 1350058 (2013). arXiv:1203.4777
22. N. Gromov, G. Sizov, Exact slope and interpolating functions in $\mathcal{N} = 6$ Supersymmetric Chern-Simons theory. Phys. Rev. Lett. **113**, 121601 (2014). arXiv:1403.1894
23. B. Basso, *An Exact Slope for AdS/CFT*. arXiv:1109.3154
24. A. Kapustin, B. Willett, I. Yaakov, Exact results for wilson loops in superconformal Chern-Simons theories with matter. JHEP **1003**, 089 (2010). arXiv:0909.4559
25. M. Marino, P. Putrov, Exact results in ABJM theory from topological strings. JHEP **1006**, 011 (2010). arXiv:0912.3074
26. N. Drukker, M. Marino, P. Putrov, From weak to strong coupling in ABJM theory. Commun. Math. Phys. **306**, 511 (2011). arXiv:1007.3837
27. M. Preti, Studies on Wilson loops, correlators and localization in supersymmetric quantum field theories. Ph.D. thesis, https://inspirehep.net/record/1592204/files/Thesis_main.pdf
28. L. Griguolo, D. Marmiroli, G. Martelloni, D. Seminara, The generalized cusp in ABJ(M) $\mathcal{N} = 6$ Super Chern-Simons theories. JHEP **1305**, 113 (2013). arXiv:1208.5766
29. A. Lewkowycz, J. Maldacena, Exact results for the entanglement entropy and the energy radiated by a quark. JHEP **1405**, 025 (2014). arXiv:1312.5682
30. M.S. Bianchi, L. Griguolo, M. Leoni, S. Penati, D. Seminara, BPS Wilson loops and Bremsstrahlung function in ABJ(M): a two loop analysis. JHEP **1406**, 123 (2014). arXiv:1402.4128
31. D.H. Correa, J. Aguilera-Damia, G.A. Silva, Strings in $AdS_4 \times \mathbb{CP}^3$ Wilson loops in $\mathcal{N} = 6$ super Chern-Simons-matter and bremsstrahlung functions. JHEP **1406**, 139 (2014). arXiv:1405.1396
32. N. Beisert, B. Eden, M. Staudacher, Transcendentality and crossing. J. Stat. Mech. **0701**, P01021 (2007). arXiv:hep-th/0610251
33. O. Aharony, O. Bergman, D.L. Jafferis, J. Maldacena, $\mathcal{N} = 6$ superconformal Chern-Simons-matter theories, M2-branes and their gravity duals. JHEP **0810**, 091 (2008). arXiv:0806.1218
34. I. Shenderovich, *Giant magnons in AdS_4/CFT_3: Dispersion, quantization and finite-size corrections*. arXiv:0807.2861
35. T. McLoughlin, R. Roiban, Spinning strings at one-loop in $AdS_4 \times \mathbb{CP}^3$. JHEP **0812**, 101 (2008). arXiv:0807.3965
36. L.F. Alday, G. Arutyunov, D. Bykov, Semiclassical quantization of spinning strings in $AdS_4 \times \mathbb{CP}^3$. JHEP **0811**, 089 (2008). arXiv:0807.4400
37. C. Krishnan, AdS_4/CFT_3 at one loop. JHEP **0809**, 092 (2008). arXiv:0807.4561
38. N. Gromov, V. Mikhaylov, Comment on the scaling function in $AdS_4 \times \mathbb{CP}^3$. JHEP **0904**, 083 (2009). arXiv:0807.4897
39. D. Astolfi, V.G.M. Puletti, G. Grignani, T. Harmark, M. Orselli, Finite-size corrections in the $SU(2) \times SU(2)$ sector of type IIA string theory on $AdS_4 \times \mathbb{CP}^3$. Nucl. Phys. B **810**, 150 (2009). arXiv:0807.1527
40. M.A. Bandres, A.E. Lipstein, One-loop corrections to type IIA string theory in $AdS_4 \times \mathbb{CP}^3$. JHEP **1004**, 059 (2010). arXiv:0911.4061

41. M.C. Abbott, P. Sundin, The near-flat-space and BMN limits for strings in $AdS_4 \times \mathbb{CP}^3$ at one loop. J. Phys. A **45**, 025401 (2012). arXiv:1106.0737
42. D. Astolfi, V.G.M. Puletti, G. Grignani, T. Harmark, M. Orselli, Finite-size corrections for quantum strings on $AdS_4 \times \mathbb{CP}^3$. JHEP **1105**, 128 (2011). arXiv:1101.0004
43. D. Astolfi, G. Grignani, E. Ser-Giacomi, A. Zayakin, Strings in $AdS_4 \times \mathbb{CP}^3$: finite size spectrum versus Bethe Ansatz. JHEP **1204**, 005 (2012). arXiv:1111.6628
44. V. Forini, V.G.M. Puletti, O. Ohlsson, The generalized cusp in $AdS_4 \times \mathbb{CP}^3$ and more one-loop results from semiclassical strings. J. Phys. A **46**, 115402 (2013). arXiv:1204.3302
45. M. Kruczenski, A note on twist two operators in $\mathcal{N} = 4$ SYM and Wilson loops in Minkowski signature. JHEP **0212**, 024 (2002). arXiv:hep-th/0210115
46. M. Kruczenski, R. Roiban, A. Tirziu, A.A. Tseytlin, Strong-coupling expansion of cusp anomaly and gluon amplitudes from quantum open strings in $AdS_5 \times S^5$. Nucl. Phys. B **791**, 93 (2008). arXiv:0707.4254
47. S. Giombi, R. Ricci, R. Roiban, A. Tseytlin, C. Vergu, Quantum $AdS_5 \times S^5$ superstring in the AdS light-cone gauge. JHEP **1003**, 003 (2010). arXiv:0912.5105
48. J.B. Stefanski, Green-Schwarz action for type IIA strings on $AdS_4 \times \mathbb{CP}^3$, Nucl. Phys. B **808**, 80 (2009). arXiv:0806.4948
49. G. Arutyunov, S. Frolov, Superstrings on $AdS_4 \times \mathbb{CP}^3$ as a coset sigma-model. JHEP **0809**, 129 (2008). arXiv:0806.4940
50. J. Gomis, D. Sorokin, L. Wulff, The complete $AdS_4 \times \mathbb{CP}^3$ superspace for the type IIA superstring and D-branes. JHEP **0903**, 015 (2009). arXiv:0811.1566
51. D. Uvarov, $AdS_4 \times \mathbb{CP}^3$ superstring in the light-cone gauge. Nucl. Phys. B **826**, 294 (2010). arXiv:0906.4699
52. D. Uvarov, Light-cone gauge Hamiltonian for $AdS_4 \times \mathbb{CP}^3$ superstring. Mod. Phys. Lett. A **25**, 1251 (2010). arXiv:0912.1044
53. R. Metsaev, A.A. Tseytlin, Superstring action in $AdS_5 \times S^5$. Kappa symmetry light cone gauge. Phys. Rev. D **63**, 046002 (2001). arXiv:hep-th/0007036
54. S. Giombi, R. Ricci, R. Roiban, A.A. Tseytlin, C. Vergu, Generalized scaling function from light-cone gauge $AdS_5 \times S^5$ superstring. JHEP **1006**, 060 (2010). arXiv:1002.0018
55. S. Giombi, R. Ricci, R. Roiban, A. Tseytlin, Two-loop $AdS_5 \times S^5$ superstring: testing asymptotic Bethe ansatz and finite size corrections. J. Phys. A **44**, 045402 (2011). arXiv:1010.4594
56. J.M. Maldacena, Wilson loops in large N field theories. Phys. Rev. Lett. **80**, 4859 (1998). arXiv:hep-th/9803002
57. S.-J. Rey, J.-T. Yee, Macroscopic strings as heavy quarks in large N gauge theory and anti-de Sitter supergravity. Eur. Phys. J. C **22**, 379 (2001). arXiv:hep-th/9803001
58. S.S. Gubser, I.R. Klebanov, A.M. Polyakov, A semi-classical limit of the gauge/string correspondence. Nucl. Phys. B **636**, 99 (2002). arXiv:hep-th/0204051
59. R. Roiban, A. Tirziu, A.A. Tseytlin, Two-loop world-sheet corrections in $AdS_5 \times S^5$ superstring. JHEP **0707**, 056 (2007). arXiv:0704.3638
60. R. Roiban, A.A. Tseytlin, Strong-coupling expansion of cusp anomaly from quantum superstring. JHEP **0711**, 016 (2007). arXiv:0709.0681

Chapter 7
$AdS_5 \times S^5$ Superstring on the Lattice

In the last chapters we investigated the $AdS_5 \times S^5$ superstring form the perspective of perturbation theory at large 't Hooft coupling. Here we want to present a natural, genuinely field-theoretical route to study the finite-coupling region by discretizing the two-dimensional sigma-model and proceed with numerical methods for the lattice field theory so defined.

In Sect. 1.5 we recalled that lattice discretizations of the two-dimensional theory of the $AdS_5 \times S^5$ sigma-model was inaugurated in [1] with the study of the *universal scaling function* $f(\lambda)$ associated to a Maldacena-Wilson loop with path on two light-like semi-infinite lines intersecting at a cusp point. In the dual string theory this quantity is measured by the path-integral of an open string moving in $AdS_5 \times S^5$ and bounded by the null cusp at the AdS boundary, which is where $\mathcal{N} = 4$ SYM lives:

$$\langle \mathcal{W}_{\text{cusp}} \rangle = Z = \int \mathcal{D}\delta X\, \mathcal{D}\delta\Psi\, e^{-S_{\text{cusp}}[X_{\text{cl}}+\delta X,\, \delta\Psi]} \equiv e^{-\Gamma(\lambda)} = e^{-\frac{1}{4}\frac{f(\lambda)}{2} V_2}\,, \; V_2 \equiv \int_{-\infty}^{\infty} dt \int_{-\infty}^{\infty} ds\,. \tag{7.1}$$

As explained in full details in Sect. 1.3, to write this expression one has to consider the classical solution $X_{\text{cl}} = X_{\text{cl}}(t, s)$ of the string equations of motion describing the world surface of an open string ending on a null cusp, where (t, s) is a pair of convenient coordinates spanning the classical worldsheet. The classical solution X_{cl} is the *null cusp background* [2] and it is related to the *Gubser-Klebanov-Polyakov (GKP)* string vacuum [3]. The latter is of crucial importance in AdS/CFT, as holographic dual to several fundamental observables in the gauge theory [4], which can be studied exploiting the underlying integrability of this AdS/CFT system, e.g. in [5–9]. The action $S_{\text{cusp}}[X + \delta X, \delta\Psi]$ is the $AdS_5 \times S^5$ supercoset action 2.33 expanded in the fluctuation fields around this vacuum and it is reported below in (7.15) in terms of the bosonic and fermionic degrees of freedom remaining after gauge-fixing.
In (7.1) the scaling function is proportional to the free energy $\Gamma \equiv -\log Z$. The (infinite) worldsheet volume V_2 simply factorizes out, since the fluctuation Lagrangian has constant coefficients, the normalization factor $1/4$ follows from the conventions

E. Vescovi, *Perturbative and Non-perturbative Approaches to String Sigma-Models in AdS/CFT*, Springer Theses, DOI 10.1007/978-3-319-63420-3_7

of [2] and an additional 1/2 takes into account that $f(\lambda)$ equals twice the so-called *universal* (or *null*) *cusp anomalous dimension* associated to the UV divergency of a light-like Wilson cusp, anticipating what we will review in the next section.

Bethe ansatz techniques based on the assumed quantum integrability of $\mathcal{N}=4$ SYM allow to write the *Beisert-Eden-Staudacher (BES)* integral equation [10] to compute the scaling function numerically[1] at finite λ or in a perturbative series at weak/strong coupling. The starting point of our investigations is still (7.1), although the principles of our analysis differ from [1]. We measure the vacuum expectation value of the fluctuation Lagrangian defined as above

$$\langle S_{\mathrm{cusp}}\rangle \equiv \frac{\int \mathcal{D}\delta X\, \mathcal{D}\delta\Psi\;\, S\, e^{-S}}{\int \mathcal{D}\delta X\, \mathcal{D}\delta\Psi\;\, e^{-S}} = -g\,\frac{d\ln Z}{dg} = g\,\frac{V_2}{8}\, f'(g) \tag{7.2}$$

and we want to obtain information on the derivative of the scaling function. We shall use

$$g \equiv \frac{T}{2} = \frac{\sqrt{\lambda}}{4\pi} \tag{7.3}$$

as new coupling constant on the lattice. Note that the large-g (or large-λ) region is referred to as *strong-coupling* regime in the context of AdS/CFT. However, the sigma-model is *weakly-coupled* because it admits a perturbation theory in powers of $1/g$.

Even though our research is still in its infancy, our work has readdressed the study of the derivative of the scaling function and extended the previous analysis of [1] to measure the mass of two bosonic AdS fluctuation fields, corresponding to the transverse directions to the null cusp classical solution. These two observables and the AdS light-cone gauge-fixed action in the continuum are reviewed in Sects. 7.1 and 7.2.

The field content of the worldsheet theory in (7.2) contains anti-commuting scalars, also referred to as fermions for short. Certain difficulties arise when they are treated by means of lattice field theory. Following [1], we need to introduce a set of auxiliary fields to linearize four-point fermionic interactions and then perform a Gaussian integration of the quadratic ones, as shown in Sect. 7.3. In this way fermions can be integrated out and the resulting determinant becomes part of the definition of a purely bosonic path-integral. Secondly, we need to eliminate unphysical states that naturally arise in the lattice discretization of the fermionic operator. We consider two possible discretizations breaking different subgroups of the global symmetry for the sigma-model: the $SO(6)$-preserving discretization in Sect. 7.4 for the simulations in the chapter and the $SO(6)$-breaking one in Appendix F.6. So far our results seem not to be sensitive to the discretization adopted.

Lattice simulations are performed employing a Rational Hybrid Monte Carlo algorithm similar to the one of [1]. The prescription given in Sect. 7.5 to approach the continuum limit demands physical masses to be kept constant, which in the case

[1] We thank Dmytro Volin for sharing a Mathematica script providing the numerical solution.

of finite mass renormalization requires no tuning of the "bare" mass parameter of the theory (here, the light-cone momentum p^+). The study of the correlator and physical mass of the bosonic fields supports this hypothesis.

We reached a good agreement for both the bosonic mass (Sect. 7.6.1) and the scaling function (Sect. 7.6.2) at large 't Hooft coupling, namely the perturbative regime around the null cusp vacuum of the sigma-model. On the other hand, for smaller values of g, the expectation value of the action, unlike the bosonic correlator, exhibits a deviation compatible with the presence of quadratic divergences of order a^{-2} when the lattice spacing $a \to 0$. It is certainly possible that our way to take the continuum limit might be subject to a change once all field two-point functions are investigated in the future, although such divergences are expected in lattice regularization. Indeed in the continuum perturbation theory, power-divergences arise in [2] as well as in Chap. 6, but dimensional regularization sets them equal to zero. The problem of renormalization in presence of power-like divergences is generally a non-trivial issue. Here, they will be non-perturbative subtracted before taking the continuum limit.

In Sect. 7.6.3 we discuss the presence of a complex phase in the determinant of the fermionic operator, resulting from the chosen fermionic linearization, that leads at small g to a sign problem not treatable via the standard reweighting method. Additional future work to cure this problem is outlined in Sect. 7.7.

We have tried to keep the chapter focused on presenting our results, but the interested reader can also benefit from a basic introduction to lattice methods in Appendix F.

7.1 The Cusp Anomaly of $\mathcal{N} = 4$ SYM and the Light-Like Limit

The *cusp anomaly* was originally introduced [11] as the coefficient of the logarithmic divergence in the expectation value of a Wilson loop whose path makes a sudden turn by a Euclidean angle ϕ (Fig. 7.1, left panel). Analogously, in planar $\mathcal{N} = 4$ SYM we define $\Gamma(\lambda, \phi)$ as the coefficient of the divergence of a cusped Wilson loop[2] defined by a generalized connection that couples to a fixed combination of scalar fields:

$$\mathcal{W}_{\rm cusp} = \frac{1}{N}{\rm tr}\,\mathcal{P}\exp\left[\int \left(i A_\mu \dot{x}^\mu + |\dot{x}|\, n^I \phi_I\right) d\tau\right], \qquad \langle \mathcal{W}_{\rm cusp}\rangle \sim e^{-\Gamma(\lambda,\phi)\,\ln\frac{\epsilon}{L}}\,, \tag{7.4}$$

where n^I is any constant unit vector ($I = 1, \dots 6$) and the cutoffs L and ϵ screen respectively the IR divergence arising from the loop points at infinity and the UV divergence close to the singular point. One can also analytically continue $\phi = i\varphi$, so

[2]The presence of the cusp makes the operator only *locally* supersymmetric as defined around 5.2.

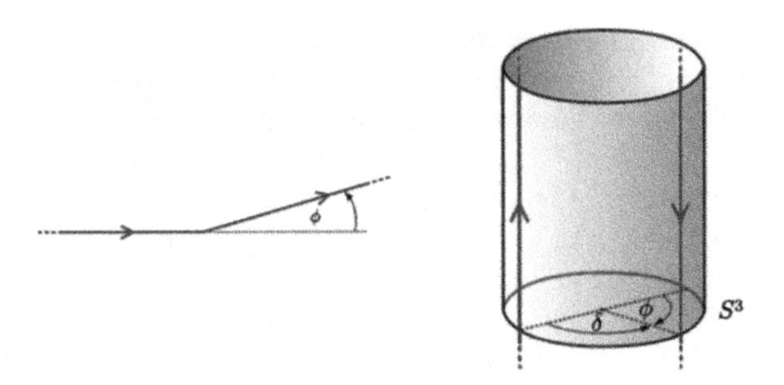

Fig. 7.1 Two situations where the *angle-dependent* cusp anomaly $\Gamma(\lambda, \phi)$ appears in $\mathcal{N} = 4$ SYM: a Wilson line with a singular point (*left panel*, [17]) and a quark-antiquark pair sitting on a three-sphere and extended along the time direction (*right panel*, [19])

that now φ is a boost angle in Lorentzian signature in $\mathbb{R}^{1,3}$. The cusp anomaly[3] is an ubiquitous quantity related to a wide range of physical observables, as summarized below.

- It governs the IR divergences of scattering amplitudes and form factors, where φ is the angle between a pair of external momenta, as in a supersymmetric generalization of QCD [13, 14].
- Four-point massive scattering amplitudes on the Coulomb branch [15] are also governed by the same function [16]. In the Regge limit $t \gg m^2, s$ the planar amplitude, normalized by its tree-level value, scales as $\log\left(\frac{\mathcal{A}}{\mathcal{A}_{\text{tree-level}}}\right) \sim \Gamma(\lambda, \phi) \log t$.
- The *Bremsstrahlung function* [17] measures the first deviation of a Wilson cusp expectation value from the BPS-protected straight Wilson line $\Gamma(\lambda, \phi) = -B(\lambda)\phi^2 + O(\phi^4)$ (see end of Sect. 5.2 and Eq. 5.17 for references). In [17, 18] the same function appears also in the power radiated by a slow-moving quark, in the two-point function of the displacement operator evaluated on the Wilson line and in the stress tensor expectation value in the presence of a Wilson line.
- After conformally mapping $\mathbb{R}^4$ to $\mathbb{R} \times S^3$, the cusp anomaly gives the static potential between a quark and an anti-quark separated by an angle $\delta = \pi - \phi$ on the spatial three-sphere (Fig. 7.1, right panel). The cusp anomaly coincides with the Coulomb potential in flat space when $\Gamma(\lambda, \phi) \sim \frac{V(\lambda)}{\delta}$ for small distances $\delta \to 0$.

The observable that will be of central importance in this chapter is the *universal scaling function* $f(\lambda)$, defined as twice the coefficient $\Gamma(\lambda)$ [20–22] that multiplies the linear divergence $\Gamma(\lambda, \phi = i\varphi) \sim \varphi\, \Gamma(\lambda)$ in the light-like (or null) limit $\varphi \to \infty$. Due to its relevance, we typically omit "universal" or "light-like" and understand, also here in the chapter, that we always refer to the angle-independent $f(\lambda)$. Another interesting occurrence is in the conformal dimension $\Delta(\lambda)$ of twist-two operators for large spin S [4, 22, 23]

$$\mathcal{O} = \text{tr}(Z D_+^S Z)\,, \qquad \Delta(\lambda) = S + f(\lambda) \log S + O(\log S/S) \text{for } \; S \gg 1\,. \quad (7.5)$$

[3]Let us also mention that a further generalization $\Gamma(\lambda, \phi, \theta)$ of the cusp anomaly includes an "angle" θ in the coupling to the scalars [12].

On the string theory side these operators are represented by fast rotating *closed* strings with energy corrections dictated by the scaling function $f(\lambda)$. Support to this identification arrived from the *open*-string picture in [24] (also [25]), where the observation that $f(\lambda)$ is related to a null cusp Wilson loop [22, 23] suggested to extract $f(\lambda)$ from the classical worldsheet bounded by two light-like segments using the AdS/CFT correspondence (1.11). The equivalence of the scaling function computed in the open/closed approaches was argued to extend to all-loop orders by relating the corresponding open/closed minimal surfaces through a conformal transformation and an analytic continuation [26]. Semiclassical analysis around these two string vacua allowed to compute the scaling function up to two loops at strong coupling [2, 3, 26–31]

$$f(g)=4\,g-\frac{3\log 2}{\pi}-\frac{K}{4\,\pi^2\,g}+O(g^{-2}) \tag{7.6}$$

with $K\approx 0.916$ being the Catalan constant. These first few orders are also reproduced by the BES equation.

7.2 The Continuum Action and Its Symmetries

7.2.1 The Action in the AdS Light-Cone Gauge

In this section we briefly cover the main steps leading from the κ-symmetry gauge-fixed action (2.33) to the fully gauge-fixed one (7.15) following [2]. The $AdS_5\times S^5$ background metric (3.6), setting the radii of AdS_5 and S^5 to unity, can be put into the form

$$\begin{aligned} ds^2 &= z^{-2}\,(dx^m\,dx_m+dz^M\,dz^M)=z^{-2}(dx^m\,dx_m+dz^2)+du^M du^M\,,\\ x^m x_m &= x^+x^-+x^*x\,,\qquad x^{\pm}=x^3\pm x^0\,,\qquad x=x^1+ix^2\,,\\ z^M &= z\,u^M\,,\qquad u^M\,u^M=1\,,\qquad z=\sqrt{z^M z^M}\,. \end{aligned} \tag{7.7}$$

Above, $x^{\pm}$ are the light-cone directions, built out of two of the four coordinates x^m ($m=1,\ldots 4$) parametrizing the four-dimensional boundary of AdS_5. The radial coordinate z is the modulus of z^M ($M=1\ldots 6$).

The *AdS light-cone gauge* [32, 33] is defined by fixing the local symmetries of the superstring action, bosonic diffeomorphisms and κ-symmetry, via a sort of "non-conformal" gauge[4]

$$\sqrt{-g}g^{ij}=\mathrm{diag}(-z^2,z^{-2})\,,\qquad x^+=p^+\tau\,, \tag{7.8}$$

[4] As in the standard conformal gauge, the choice $x^+=p^+\tau$ is allowed by residual diffeomorphisms after the choice for the auxiliary metric g_{ij}.

and a more standard light-cone gauge on the two Majorana-Weyl fermions ($I = 1, 2$) of the type IIB superstring action

$$\Gamma^+ \theta^I = 0\,. \tag{7.9}$$

The resulting action—entering the path-integral with weight e^{iS}—is given by

$$\begin{aligned}
S &= g \int d\tau d\sigma\, \mathcal{L} \\
\mathcal{L} &= \dot{x}^*\dot{x} + (\dot{z}^M + \mathrm{i}p^+ z^{-2} z^N \eta_i \rho^{MN\,i}{}_j \eta^j)^2 + \mathrm{i}p^+(\theta^i\dot{\theta}_i + \eta^i\dot{\eta}_i + \theta_i\dot{\theta}^i + \eta_i\dot{\eta}^i) + \\
&\quad -(p^+)^2 z^{-2}(\eta^i\eta_i)^2 - z^{-4}(x'^*x' + z'^M z'^M) \\
&\quad -2\Big[\, p^+ z^{-3}\eta^i \rho^M_{ij} z^M(\theta'^j - \mathrm{i}z^{-1}\eta^j x') + p^+ z^{-3}\eta^i(\rho^\dagger_M)^{ij} z^M(\theta'^j + \mathrm{i}z^{-1}\eta^j x'^*)\Big] \\
&\equiv \dot{x}^*\dot{x} + (\dot{z}^M + \mathrm{i}p^+ z^{-2} z^N \eta_i \rho^{MN\,i}{}_j \eta^j)^2 + \mathrm{i}p^+(\theta^i\dot{\theta}_i + \eta^i\dot{\eta}_i - \text{h.c.}) - (p^+)^2 z^{-2}(\eta^i\eta_i)^2 \\
&\quad -z^{-4}(x'^*x' + z'^M z'^M) - 2\Big[\, p^+ z^{-3}\eta^i \rho^M_{ij} z^M(\theta'^j - \mathrm{i}z^{-1}\eta^j x') + \text{h.c.}\Big]\,.
\end{aligned} \tag{7.10}$$

The six 4×4 matrices ρ^M are collected in Appendix F.1. The fields θ_i, η_i ($i = 1, 2, 3, 4$) are $4 + 4$ complex Grassmann-odd variables for which $\theta^i = (\theta_i)^\dagger$, $\eta^i = (\eta_i)^\dagger$. They transform in the fundamental representation of the $SU(4)$ R-symmetry, which can be seen as a flavour symmetry, and do not carry (Lorentz) spinor indices. Note that the AdS light-cone gauge delivers an action that is *at most quartic* in these anti-commuting variables. Note also that we often name them *fermions* referring to the fact that they are degrees of freedom descending from the type IIB Majorana-Weyl spinors in ten dimensions.

Wick-rotating $\tau \to -\mathrm{i}\tau$, $p^+ \to \mathrm{i}p^+$, and setting the light-cone momentum $p^+ = 1$ as standard in literature, one gets a factor e^{-S_E} in the path-integral with

$$\begin{aligned}
S_E &= g \int d\tau d\sigma\, \mathcal{L}_E \\
\mathcal{L}_E &= \dot{x}^*\dot{x} + (\dot{z}^M + i\, z^{-2} z_N \eta_i (\rho^{MN})^i{}_j \eta^j)^2 + i(\theta^i\dot{\theta}_i + \eta^i\dot{\eta}_i - \text{h.c.}) - z^{-2}\left(\eta^2\right)^2 \\
&\quad + z^{-4}(x'^*x' + z'^M z'^M) + 2i\Big[z^{-3} z^M \eta^i \rho^M{}_{ij}(\theta'^j - i\, z^{-1}\eta^j x') + \text{h.c.}\Big]\,.
\end{aligned} \tag{7.11}$$

The *null cusp background* [2]

$$x^+ = \tau\,,\ \ x^- = -\frac{1}{2\sigma}\,,\ \ x = x^* = 0\,,\ \ z = \sqrt{-2x^+x^-} = \sqrt{\frac{\tau}{\sigma}}\,,\ \ \tau, \sigma > 0\,, \tag{7.12}$$

is the classical solution of (7.11) that describes a Euclidean open string surface ending at $z = 0$ on the path

$$(x^+, x^-) = \begin{cases} (0, -u) & \text{for}\, u \le 0 \\ (u, 0) & \text{for}\ u \ge 0 \end{cases}\,, \qquad\qquad x = x^* = 0 \tag{7.13}$$

that supports the Wilson cusp in $\mathcal{N}=4$ SYM. We can then choose a field parametrization

$$x=\sqrt{\frac{\tau}{\sigma}}\,\tilde{x}\,,\quad z^M=\sqrt{\frac{\tau}{\sigma}}\,\tilde{z}^M\,,\quad z=\sqrt{z^M z^M}\,,\quad \theta=\frac{1}{\sqrt{\sigma}}\tilde{\theta}\,,\quad \eta=\frac{1}{\sqrt{\sigma}}\tilde{\eta} \tag{7.14}$$

and absorb powers of the worldsheet coordinates by posing $t=\log\tau$ and $s=\log\sigma$. Dropping tildes over the fields for simplicity, we arrive to the Euclidean gauge-fixed Lagrangian $\mathcal{L}_{\text{cusp}}$ with *constant* coefficients

$$S_{\text{cusp}}=g\int dt\,ds\,\mathcal{L}_{\text{cusp}} \tag{7.15}$$

$$\begin{aligned}
\mathcal{L}_{\text{cusp}}=&\left|\partial_t x+\tfrac{1}{2}x\right|^2+\frac{1}{z^4}\left|\partial_s x-\tfrac{1}{2}x\right|^2+\left(\partial_t z^M+\frac{1}{2}z^M+\frac{i}{z^2}z_N\eta_i\left(\rho^{MN}\right)^i{}_j\,\eta^j\right)^2\\
&+\frac{1}{z^4}\left(\partial_s z^M-\tfrac{1}{2}z^M\right)^2+i\left(\theta^i\partial_t\theta_i+\eta^i\partial_t\eta_i+\theta_i\partial_t\theta^i+\eta_i\partial_t\eta^i\right)-\tfrac{1}{z^2}\left(\eta^i\eta_i\right)^2\\
&+\frac{2i}{z^3}z^M\eta^i\left(\rho^M\right)_{ij}\left(\partial_s\theta^j-\frac{1}{2}\theta^j-\frac{i}{z}\eta^j\left(\partial_s x-\frac{1}{2}x\right)\right)\\
&+\frac{2i}{z^3}z^M\eta_i(\rho_M^\dagger)^{ij}\left(\partial_s\theta_j-\frac{1}{2}\theta_j+\frac{i}{z}\eta_j\left(\partial_s x-\frac{1}{2}x\right)^*\right).
\end{aligned}$$

We remark that it has been obtained through the redefinitions (7.14) without operating any truncation of high-order interactions—as we would do in semiclassical approximation at $g\gg 1$—so it constitutes a valid starting point for the exploration of non-perturbative physics of the string sigma-model and of the dual Wilson loop.

7.2.2 The Mass Spectrum

This string vacuum (7.12) is $SO(6)$-degenerate because any rotation of the six-dimensional vector z^M leaves the norm of the vector unchanged. The fluctuation spectrum of the solution can be easily found if we break the symmetry of the vacuum by fixing a direction in the z^M-space, e.g. the unit vector $u^M=(0,\,0,\,0,\,0,\,0,\,1)$, and use it to define new fluctuation fields that vanish on the vacuum ($\tilde{\phi}=y^a=0$)

$$\tilde{z}^M=e^{\tilde{\phi}}\tilde{u}^M\,,\quad \tilde{z}=e^{\tilde{\phi}}\,, \tag{7.16}$$

$$\tilde{u}^a=\frac{y^a}{1+\frac{1}{4}y^2}\,,\quad \tilde{u}^6=\frac{1-\frac{1}{4}y^2}{1+\frac{1}{4}y^2}\,,\qquad y^2\equiv\sum_{a=1}^{5}(y^a)^2\,,\quad a=1,\ldots,5\,.$$

The classical value $\frac{g}{2} V_2$ of the action on the vacuum gives the leading term 4 g in (7.6) through (7.1). If we truncate the Lagrangian at quadratic order in the fluctuations fields and drop tildes again, we have

$$\begin{aligned}\mathcal{L}^{(2)} &= (\partial_t \phi)^2 + (\partial_s \phi)^2 + \phi^2 + |\partial_t x|^2 + |\partial_s x|^2 + \frac{1}{2}\,|x|^2 + (\partial_t y^a)^2 + (\partial_s y^a)^2 \\ &\quad + 2i\,(\theta^i \partial_t \theta_i + \eta^i \partial_t \eta_i) + 2i\,\eta^i (\rho^6)_{ij} (\partial_s \theta^j - \theta^j) + 2i\,\eta_i (\rho_6^\dagger)^{ij} (\partial_s \theta_j - \theta_j) \qquad (7.17)\\ &= (\partial_t \phi)^2 + (\partial_s \phi)^2 + \phi^2 + |\partial_t x|^2 + |\partial_s x|^2 + \frac{1}{2}\,|x|^2 + (\partial_t y^a)^2 + (\partial_s y^a)^2 + \psi^T\, K_F\, \psi\,, \qquad (7.18)\end{aligned}$$

where we called $\psi \equiv (\theta^i, \theta_i, \eta^i, \eta_i)^T$ and the 16×16 free fermionic operator reads

$$K_F = \begin{pmatrix} 0 & i\mathbb{I}_4 \partial_t & -i(\partial_s + \frac{1}{2})\rho^6 & 0 \\ i\mathbb{I}_4 \partial_t & 0 & 0 & -i(\partial_s + \frac{1}{2})\rho_6^\dagger \\ i(\partial_s - \frac{1}{2})\rho^6 & 0 & 0 & i\mathbb{I}_4 \partial_t \\ 0 & i(\partial_s - \frac{1}{2})\rho_6^\dagger & i\mathbb{I}_4 \partial_t & 0 \end{pmatrix}. \qquad (7.19)$$

It is easy to see that the bosonic excitation spectrum consists of one field (ϕ) with $m^2 = 1$, two fields (x, x^*) with $m^2 = \frac{1}{2}$ and five fields (y^a) with $m^2 = 0$ [2]. Going to Fourier space ($\partial_\mu \to i p_\mu$ with $\mu = 0, 1 = t, s$) and adding the fermionic determinant $\text{Det}\, K_F = \left(p_0^2 + p_1^2 + \frac{1}{4}\right)^8$, the one-loop free energy $\Gamma^{(1)} = -\log Z^{(1)}$ evaluates to

$$\Gamma^{(1)} = \frac{V_2}{2} \int \frac{dp_0 dp_1}{(2\pi)^2} \log\left[\frac{(p_0^2 + p_1^2 + 1)(p_0^2 + p_1^2 + \frac{1}{2})^2 (p_0^2 + p_1^2)^5}{(p_0^2 + p_1^2 + \frac{1}{4})^8}\right] = -\frac{3 \log 2}{8\pi}\, V_2 \qquad (7.20)$$

and it is in agreement with the one-loop constant term $\frac{-3 \log 2}{\pi}$ in (7.6) (Fig. 7.2).

The mass spectrum extracted from (7.17) is renormalized at finite g. The all-loop prediction was obtained via asymptotic Bethe ansatz [34] and later confirmed by semiclassical string theory around the folded closed string in AdS_5 in the large spin

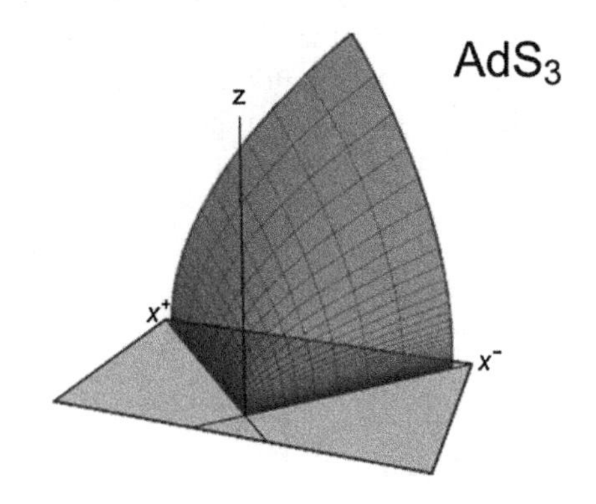

Fig. 7.2 The projection of the string vacuum (7.12) into AdS_3. The open string has support on the light-cone directions $x^\pm$ at $z = 0$, where the Wilson loop is located. The mesh grid is made of the same lines of constant t and s that draw the discrete grid in Fig. F.1

limit [35]. Notice that the mass spectrum in light-cone gauge coincides with the one in conformal gauge up to a factor of 4 [35].

The corrections to the mass of the bosonic field x [35] read

$$m_x^2(g) = \frac{m^2}{2}\left(1 - \frac{1}{8\,g} + O(g^{-2})\right). \tag{7.21}$$

The expression (7.21) (and also the $O(g^{-1})$ coefficient in (7.6)) are results obtained in a dimensional regularization scheme in which power divergent contributions are set to zero. In Sect. 7.6.1 we will compute the lattice correlators of the fields x, x^* so to study whether our discretization changes the renormalization pattern above.

In (7.21) and in what follows, we already make explicit the presence of a massive parameter m since it will be crucial to keep track of dimensionful quantities, in principle subject to renormalization, in the perspective of lattice field theory. We will also use the dimensionless combination $M = a\,m$ with the lattice spacing a. The parameter m and the (dimensionless) g are the only "bare" parameters characterizing the model in the continuum.

7.2.3 Global Symmetries of the Action

Two important global symmetries are explicitly realized in the gauge-fixed action (7.15).

- The first one is the $SU(4) \sim SO(6)$ symmetry originating from the isometries of S^5, which is unaffected by the gauge-fixing. Under this symmetry the fields z^M change in the **6** representation (vector representation), the fermions $\{\eta_i, \theta_i\}$ and $\{\eta^i, \theta^i\}$ transform in the **4** and $\bar{\mathbf{4}}$ (fundamental and anti-fundamental) respectively and the fields x and x^* are neutral.
- The second global symmetry is a $U(1) \sim SO(2)$ arising from the rotational symmetry in the two AdS_5 directions orthogonal to AdS_3 (i.e. transverse to the classical solution). Therefore, contrary to the previous case, the fields x and x^* are charged (with charges 1 and -1 respectively) while the z^M are neutral. The invariance of the action simply requires the fermions η_i and θ^i to have charge $\frac{1}{2}$ and consequently η^i and θ_i acquire charge $-\frac{1}{2}$.

An optimal discretization should preserve the full global symmetry of the continuum model. In Sect. 7.4 we will see that this is not possible in the case of the $SO(2)$ symmetry.

7.3 Linearization of the Anti-commuting Scalars Interactions

While the bosonic part of (7.15) can be easily discretized and simulated, the Lagrangian (7.15) is not in a suitable form to be simulated on a calculator because we have to take into account the Fermi statistics of the θ- and η-fields. Grassmann-odd fields can be either ignored (*quenched approximation*)[5] or formally integrated out. In the latter case, the fermionic contribution becomes part—via exponentiation in terms of *pseudo-fermions* in (7.29) below—of the Boltzmann weight of each configuration in the statistical ensemble. In the case of interactions at most quartic in fermions as in (7.15), this is possible via the introduction of a certain number of auxiliary fields. One can introduce 7 real auxiliary fields [1], one scalar ϕ and a $SO(6)$ vector ϕ^M, with the *Hubbard-Stratonovich transformation* [36, 37]

$$\begin{aligned} &\exp\Big\{-g\int dtds\Big[-\tfrac{1}{z^2}\left(\eta^i\eta_i\right)^2+\left(\tfrac{i}{z^2}z_N\eta_i\rho^{MN\,i}{}_j\eta^j\right)^2\Big]\Big\} \\ &\quad\propto\int\mathcal{D}\phi\mathcal{D}\phi^M\,\exp\Big\{-g\int dtds\,[\tfrac{1}{2}\phi^2+\tfrac{\sqrt{2}}{z}\phi\,\eta^2+\tfrac{1}{2}(\phi_M)^2-i\,\tfrac{\sqrt{2}}{z^2}\phi^M\left(\tfrac{i}{z^2}z_N\eta_i\rho^{MN\,i}{}_j\eta^j\right)]\Big\}\,. \end{aligned} \tag{7.22}$$

In the second line we have written the Lagrangian for the auxiliary field ϕ^M in a manner to emphasize that it has an imaginary part: it is easy to show that the bilinear form in round brackets is hermitian

$$\left(i\,\eta_i\rho^{MN\,i}{}_j\eta^j\right)^\dagger=i\eta_j\,\rho^{MN\,j}{}_i\,\eta^i \tag{7.23}$$

with the properties of the ρ-matrices and the flipping formulas in Appendix F.1. Since ϕ^M takes real values, its Yukawa term above sets a phase problem a priori,[6] the only question being whether the phase is treatable via standard *reweighting*. In Sect. 7.6.3 we will see that this method is not possible for small values of g, suggesting that an alternative linearization should be found to explore the full nonperturbative region, see [38, 39] and Chap. 8 (Fig. 7.3).

After the transformation (7.22), the Lagrangian (7.15) reads

$$\begin{aligned} \mathcal{L}=&\left|\partial_t x+\frac{m}{2}x\right|^2+\frac{1}{z^4}\left|\partial_s x-\frac{m}{2}x\right|^2+\left(\partial_t z^M+\frac{m}{2}z^M\right)^2+\frac{1}{z^4}\left(\partial_s z^M-\frac{m}{2}z^M\right)^2 \\ &+\frac{1}{2}\phi^2+\frac{1}{2}(\phi_M)^2+\psi^T O_F\psi\,, \end{aligned} \tag{7.24}$$

[5] This strategy is not promising in this context because fermionic contributions are expected to be of the same order of magnitude of the bosonic ones.

[6] The second term in the first line of (7.22) is the square of an hermitian object and comes in the exponent in second line as a "repulsive" potential. This has the final effect of an imaginary part in the auxiliary Lagrangian, precisely as the $i\,b\,x$ in $e^{-\frac{b^2}{4a}}\propto\int dx\,e^{-ax^2+ibx}$ with $b\in\mathbb{R}$.

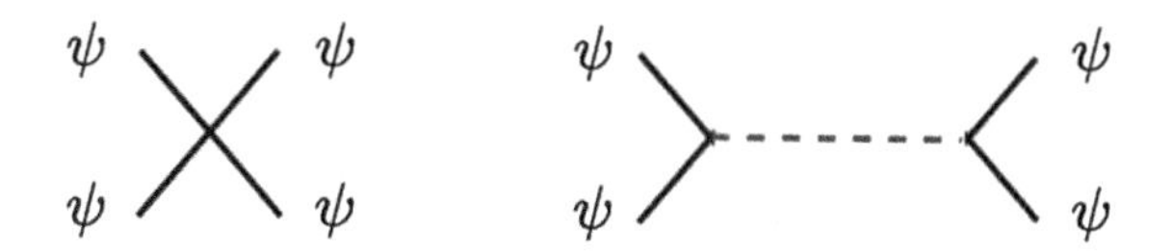

Fig. 7.3 The Hubbard-Stratonovich transformation "splits" quartic fermionic interactions (*left*) into Yukawa vertices with two fermionic lines (*right*)

with the 16-component vector $\psi \equiv (\theta^i, \theta_i, \eta^i, \eta_i)^T$ acted upon by the fermionic operator

$$\mathcal{O}_F = \begin{pmatrix} 0 & i\mathbb{I}_4\partial_t & -i\rho^M\left(\partial_s+\frac{m}{2}\right)\frac{z^M}{z^3} & 0 \\ i\mathbb{I}_4\partial_t & 0 & 0 & -i\rho_M^\dagger\left(\partial_s+\frac{m}{2}\right)\frac{z^M}{z^3} \\ i\frac{z^M}{z^3}\rho^M\left(\partial_s-\frac{m}{2}\right) & 0 & 2\frac{z^M}{z^4}\rho^M\left(\partial_s x-m\frac{x}{2}\right) & i\mathbb{I}_4\partial_t-A^T \\ 0 & i\frac{z^M}{z^3}\rho_M^\dagger\left(\partial_s-\frac{m}{2}\right) & i\mathbb{I}_4\partial_t+A & -2\frac{z^M}{z^4}\rho_M^\dagger\left(\partial_s x^*-m\frac{x}{2}^*\right) \end{pmatrix}$$

$$A = \frac{1}{\sqrt{2}z^2}\phi_M\rho^{MN}z_N - \frac{1}{\sqrt{2}z}\mathbb{I}_4\phi + i\,\frac{z_N}{z^2}\rho^{MN}\,\partial_t z^M\,. \tag{7.25}$$

It is convenient to separate the bosonic $\mathcal{L}_B$ and the fermionic part $\mathcal{L}_F = \psi^T\mathcal{O}_F\psi$ of the Lagrangian $\mathcal{L} = \mathcal{L}_B + \mathcal{L}_F$ to write the path-integral

$$Z = \int \mathcal{D}x\,\mathcal{D}x^*\,\mathcal{D}z^M\,\mathcal{D}\phi\,\mathcal{D}\phi^M\mathcal{D}\psi\,e^{-S}, \qquad S = g\int dt\,ds\,\mathcal{L} \equiv S_B + S_F\,. \tag{7.26}$$

We dropped the label "cusp" in passing from (7.15) to the equivalent formulation (7.24) only to signal that we enlarged the field content using the integral identity (7.22). The integration measure in (7.26) involves only the field ψ but not its complex conjugate,[7] thus formally integrating fermions out generates a Pfaffian Pf $\mathcal{O}_F$

$$\int D\psi\,e^{-\int dtds\,\psi^T\mathcal{O}_F\psi} = \text{Pf}\,\mathcal{O}_F \equiv \left[\text{Det}\left(\mathcal{O}_F\,\mathcal{O}_F^\dagger\right)\right]^{\frac{1}{4}}\,. \tag{7.27}$$

It is important to note that Pf $\mathcal{O}_F$ must be real and nonnegative in order to enter the Boltzmann weight and later be interpreted as a probability in Monte Carlo simulations (Appendix F.4.). Therefore one proceeds to replace it with a determinant in the second equality, where we do not keep track of potential phases or anomalies. We will comment further on this point in Sect. 7.6.3.

[7]The vector ψ in (7.24) collects the 8+8 *complex* θ^i and η^i in a formally "redundant" way which includes also their complex conjugates. Making real and imaginary parts of θ, η explicit, it is easy to see that the fermionic contribution coming from the 16×16 operator $\mathcal{O}_F$ coincides then to the one of 16 *real* anti-commuting degrees of freedom.

At this stage we have reduced the path-integral (7.26) as in [1] to the form

$$Z = \int \mathcal{D}x\, \mathcal{D}x^*\, \mathcal{D}z^M\, \mathcal{D}\phi\, \mathcal{D}\phi^M \left[\text{Det}\left(\mathcal{O}_F\, \mathcal{O}_F^\dagger\right)\right]^{\frac{1}{4}} e^{-S_B}\,. \tag{7.28}$$

Formula (7.28) is our final form of the action in the remnant of the chapter, but we anticipate that a consistent lattice discretization of fermions will replace $\mathcal{O}_F$ (7.25) with the operator $\hat{\mathcal{O}}_F$ (7.40) and (7.41) below, as it will be justified in Sect. 7.4. Secondly, still taking this remark into account, we want to emphasise that the numerical implementation works with the form

$$Z = \int \mathcal{D}x\, \mathcal{D}x^*\, \mathcal{D}z^M\, \mathcal{D}\phi\, \mathcal{D}\phi^M\, \mathcal{D}\zeta_k\, \mathcal{D}\zeta_k^\dagger\; e^{-S_B - S_\zeta}\,, \quad S_\zeta \equiv \int dt ds\, \zeta^\dagger (\mathcal{O}_F \mathcal{O}_F^\dagger)^{-\frac{1}{4}} \zeta\,. \tag{7.29}$$

Here as in [1], we brought the fermionic determinant to the exponent with an additional functional integration over the commuting complex scalars $\zeta = (\zeta_k)$ with $k = 1, \dots 16$, which we name *pseudo-fermions* to stick to the lattice QCD terminology. Further details on the actual simulation of (7.29) are in Appendix F.4.

7.4 Discretization and Lattice Perturbation Theory

In order to investigate the lattice model corresponding to (7.24), we proceed with the discretization of the 2d spacetime and the action as described at length in Appendix F.2. While this works well for the bosonic sector, a *naive* lattice formulation of fermions is known to suffer from the *fermion doubling problem*, which consists in the appearance of spurious states (*doublers*) in the spectrum of fermionic two-point functions. As clear from Appendix F.3, this is a consequence of the standard Dirac operator being of the first order, therefore it comes as no surprise that it also affects the θ- and η-*scalars* with first-order kinetic operator spelt out in (7.30) below. To remove those lattice artifacts we will follow the Wilson approach: the fermionic Lagrangian is deformed by the addition of an operator that shifts the doublers mass by a term of order $1/a$, so that in the continuum limit they will become infinitely heavy and decouple from the theory.

7.4.1 Wilson-Like Term for Free Fermions

For simplicity we work in the continuum model (Sect. 7.2.1) and we denote with u^M a particular unit vector defining the vacuum around which we expand the operator (7.25) perturbatively, for instance the vector $u^M = (0, 0, 0, 0, 0, 1)$ as above (7.16). The matrix-valued differential operator (7.25) incapsulates the free Lagrangian of θ-

and η-fermions and their Yukawa interactions with physical and auxiliary bosons. The free part in Fourier space

$$K_F = \begin{pmatrix} 0 & -p_0\mathbb{I}_4 & (p_1 - i\frac{m}{2})\rho^M u_M & 0 \\ -p_0\mathbb{I}_4 & 0 & 0 & (p_1 - i\frac{m}{2})\rho_M^\dagger u^M \\ -(p_1 + i\frac{m}{2})\rho^M u_M & 0 & 0 & -p_0\mathbb{I}_4 \\ 0 & -(p_1 + i\frac{m}{2})\rho_M^\dagger u^M & -p_0\mathbb{I}_4 & 0 \end{pmatrix}, \tag{7.30}$$

has determinant

$$\text{Det}\, K_F = \left(p_0^2 + p_1^2 + \frac{m^2}{4}\right)^8 . \tag{7.31}$$

Fermionic propagators are proportional to the relevant entries of the inverse of the fermionic kinetic operator (7.30). It is immediate to realize that the naive discretization (see formula F.23)

$$p_\mu \to \mathring{p}_\mu \equiv \frac{1}{a}\sin(p_\mu a) \tag{7.32}$$

would give rise to the phenomenon of fermion doublers.

In principle there are many possible ways of introducing a Wilson-like operator due to the rather non-standard structure of the Dirac-like operator (7.25). An optimal discretization should

- preserve all symmetries of the continuum action, including the global $U(1)$ symmetry $x \to e^{i\alpha}\, x$, $\theta^i \to e^{i\alpha/2}\,\theta^i$, $\eta^i \to e^{-i\alpha/2}\,\eta^i$ with $\alpha \in \mathbb{R}$, discussed in Sect. 7.2.3,
- reproduce the result (7.6) in the $a \to 0$ limit,
- not induce complex phases in the fermionic determinant in addition to the one already implicit in the Hubbard-Stratonovich procedure adopted, see (7.23), in order not to obstruct Monte Carlo simulations.

Notice that the vanishing entries in (7.30) are set to zero by the $U(1)$ symmetry because they couple fermions with the same charge, and therefore a $U(1)$-preserving discretization should not alter them. Furthermore, the $SO(6)$ symmetry fixes completely the structure of the matrix (7.30) so that the only Wilson term preserving all the symmetries would be of the form $p_0 \to p_0 + a_i$ and $p_1 \to p_1 + b_i$ for different a_i and b_i in the four entries where p_0 and p_1 appear in (7.30). After implementing such a shift and computing the determinant of the fermionic operator, one immediately finds that this would not yield the perturbative result (7.6) for any value of a_i and b_i. For this reason, we choose to give up the global $U(1)$ symmetry and introduce the *(free) Wilson-like lattice operator*

$$\hat{K}_F = \begin{pmatrix} W_+ & -\mathring{p}_0\mathbb{I}_4 & (\mathring{p}_1 - i\frac{m}{2})\rho^M u_M & 0 \\ -\mathring{p}_0\mathbb{I}_4 & -W_+^\dagger & 0 & (\mathring{p}_1 - i\frac{m}{2})\rho_M^\dagger u^M \\ -(\mathring{p}_1 + i\frac{m}{2})\rho^M u_M & 0 & W_- & -\mathring{p}_0\mathbb{I}_4 \\ 0 & -(\mathring{p}_1 + i\frac{m}{2})\rho_M^\dagger u^M & -\mathring{p}_0\mathbb{I}_4 & -W_-^\dagger \end{pmatrix}. \tag{7.33}$$

where

$$W_{\pm} = \frac{r}{2}\left(\hat{p}_0^2 \pm i\,\hat{p}_1^2\right)\rho^M u_M\,, \qquad |r| = 1 \tag{7.34}$$

and, from (F.25),

$$\hat{p}_\mu \equiv \frac{2}{a}\sin\frac{p_\mu a}{2}\,. \tag{7.35}$$

The analogue of (7.31) reads now

$$\mathrm{Det}\hat{K}_F = \left(\mathring{p}_0^2 + \mathring{p}_1^2 + \frac{r^2}{4}\left(\hat{p}_0^4 + \hat{p}_1^4\right) + \frac{M^2}{4}\right)^8 \tag{7.36}$$

and can be used together with its bosonic counterpart—obtained via the naive replacement $p_\mu \to \hat{p}_\mu$ in the numerator of the ratio (7.20)—to define in this discretized setting the one-loop free energy

$$\Gamma^{(1)}_{\mathrm{LAT}} = -\log Z^{(1)}_{\mathrm{LAT}} = \mathcal{I}(a)\,, \tag{7.37}$$

where set $r = 1$ and used (F.18) for an infinite-volume lattice

$$\begin{aligned}\mathcal{I}(a) = \frac{V_2}{2a^2}\int_{-\pi}^{\pi}\frac{d^2p}{(2\pi)^2}\Big\{&5\log\Big[4\Big(\sin^2\frac{p_0}{2}+\sin^2\frac{p_1}{2}\Big)\Big]+2\log\Big[4\Big(\sin^2\frac{p_0}{2}+\sin^2\frac{p_1}{2}+\frac{M^2}{8}\Big)\Big]\\ &+\log\Big[4\Big(\sin^2\frac{p_0}{2}+\sin^2\frac{p_1}{2}+\frac{M^2}{4}\Big)\Big]-8\log\Big[4\sin^4\frac{p_0}{2}+\sin^2 p_0+4\sin^4\frac{p_1}{2}+\sin^2 p_1+\frac{M^2}{4}\Big]\Big\}\,.\end{aligned} \tag{7.38}$$

A consistent discretization makes (7.37)-(7.38) converge to the value in the continuum (7.20) in the $a \to 0$ limit. The numerical integration of (7.38) yields

$$\Gamma^{(1)} = -\log Z^{(1)} = \lim_{a\to 0}\mathcal{I}(a) = -\frac{3\log 2}{8\pi}\left(2N^2\right)M^2\,, \tag{7.39}$$

where we used that $V_2 = T\,L = 2(Na)^2$ from (F.13). If we plug $M = m\,a$ and expand the integrand in (7.38) for $a \to 0$, the $O(a^0)$ and $O(a^1)$ terms vanish and ensure the cancellation of quadratic $O(a^{-2})$ and linear $O(a^{-1})$ divergences in (7.39). In the continuum model this is a consequence of the equal number of fermionic and bosonic degrees of freedom and of the mass-squared sum rule. The $O(a^2)$ part of the integrand finally gives the expected finite part (7.39).

7.4.2 *Promoting the Wilson-Like Term to the Interacting Case*

Given the structure of the Wilson term evaluated on the vacuum (7.33), it is quite natural to find a suitable generalization to the interacting case. We shall promote (7.25) to the following operator

$$\hat{O}_F = \begin{pmatrix} W_+ & -\mathring{p}_0\mathbb{I}_4 & (\mathring{p}_1 - i\,\frac{m}{2})\rho^M\,\frac{z^M}{z^3} & 0 \\ -\mathring{p}_0\mathbb{I}_4 & -W_+^\dagger & 0 & \rho_M^\dagger(\mathring{p}_1 - i\,\frac{m}{2})\,\frac{z^M}{z^3} \\ -(\mathring{p}_1 + i\,\frac{m}{2})\rho^M\,\frac{z^M}{z^3} & 0 & 2\frac{z^M}{z^4}\rho^M\left(\partial_s x - m\,\frac{x}{2}\right) + W_- & -\mathring{p}_0\mathbb{I}_4 - A^T \\ 0 & -\rho_M^\dagger(\mathring{p}_1 + i\,\frac{m}{2})\,\frac{z^M}{z^3} & -\mathring{p}_0\mathbb{I}_4 + A & -2\frac{z^M}{z^4}\rho_M^\dagger\left(\partial_s x^* - m\,\frac{x}{2}^*\right) - W_-^\dagger \end{pmatrix} \qquad (7.40)$$

with

$$W_\pm = \frac{r}{2\,z^2}\left(\hat{p}_0^2 \pm i\,\hat{p}_1^2\right)\rho^M z_M\,. \qquad (7.41)$$

In the expression above we added the factor $1/z^2$, which becomes invisible to (7.34) because $z = 1$ on the vacuum, since it improves the stability of the simulations.[8]

It is worth pausing to appreciate that (7.40)–(7.41) meets the requirements below (7.32), save for the broken $U(1)$ symmetry that is not possible to conciliate with the other demands.

- The operator is manifestly $SO(6)$-invariant because the index M is fully contracted.
- The operator evaluated on the vacuum is (7.33), which is known to reproduce the correct, finite number in (7.39) in combination with the bosonic contribution.
- To see that the discretization does not induce (additional) complex phases, let us begin with the fermionic operator (7.25) obtained by setting to zero the auxiliary fields ϕ^M that bring a Yukawa term responsible for the phase problem. It is easy to check that it satisfies the properties (antisymmetry and a constraint reminiscent of the γ_5-hermiticity in lattice QCD [40])

$$\left(\hat{O}_F|_{\phi^M=0}\right)^T = -\,\hat{O}_F|_{\phi^M=0}\,, \qquad \left(\hat{O}_F|_{\phi^M=0}\right)^\dagger = \Gamma_5\left(\hat{O}_F|_{\phi^M=0}\right)\Gamma_5\,, \qquad (7.42)$$

where for us Γ_5 is the 16×16 unitary, antihermitian matrix

$$\Gamma_5 = \begin{pmatrix} 0 & \mathbb{I}_4 & 0 & 0 \\ -\mathbb{I}_4 & 0 & 0 & 0 \\ 0 & 0 & 0 & \mathbb{I}_4 \\ 0 & 0 & -\mathbb{I}_4 & 0 \end{pmatrix}, \qquad \Gamma_5^\dagger\Gamma_5 = \mathbb{I}_4 \qquad \Gamma_5^\dagger = -\Gamma_5\,. \qquad (7.43)$$

[8]A two-loop calculation in lattice perturbation theory would help to clarify the proposed structure (7.40)–(7.41).

The properties (7.42) are enough to ensure that $\text{Det}\,\hat{\mathcal{O}}_F|_{\phi^M=0}$ is *real* and *non-negative*.
Requiring that the addition of a Wilson term in the discretization of the full fermionic operator should preserve (7.42) is one of the criteria leading to $\hat{\mathcal{O}}_F$ in (7.40) and (7.41). This is indeed what happens, as can be checked both numerically and analytically, confirming that the phase problem described in Sect. 7.6.3 is only due to the Hubbard-Stratonovich transformation.

In Appendix F.6 we present simulations obtained with another fermionic discretization (F.46) and (F.47) consistent only with lattice perturbation theory around a vacuum chosen in one of the six directions u^M (see (7.16)) with all vanishing entries but one. It breaks the $U(1)$ symmetry like (7.40) and (7.41) and also the $SO(6)$ invariance of the model. We will show that in the range of the couplings explored, the measurements of the x-mass, the derivative of the scaling function and the occurrence of a complex phase in the fermion determinant appear not to be sensitive to the different discretization.

7.5 Continuum Limit

The parameter space of the continuum model consists of two "bare" variables: the dimensionless coupling $g = \frac{\sqrt{\lambda}}{4\pi}$ and the mass scale m. The lattice regularization introduces additional dimensionless parameters built out of the lattice spacing a: the combination $M = am$ and the lattice sizes $N_T = T/a$ and $N_L = L/a$, which are always chosen to be $N_T = 2N_L \equiv 2N$ to improve the study of the two-point function $\langle xx^* \rangle$ in Sect. 7.6.1.

The discrete nature of the lattice spacetime provides a natural regulator for all observables measured in the simulations, but we are eventually interested to obtain results for the underlying continuum model. Therefore, one wants to remove the lattice cutoffs using a scheme tailored to the model itself. The guideline is to adjust the simulation parameters in order not to alter the physics of the discretized model for decreasing a. The dimensionless quantities that it is natural to keep constant when $a \to 0$ are the physical masses of the field excitations rescaled by L, the spatial lattice extent. In our investigation we have focused only on the x-mass (7.21) so far. This means that our prescription for the *continuum limit* is to send $a \to 0$ along a *line of constant physics* characterized by

$$L^2 m_x^2 = \text{constant}\,, \qquad \text{leading to} \qquad L^2 m^2 \equiv (NM)^2 = \text{constant}\,. \tag{7.44}$$

The second equation in (7.44) relies first on the hypothesis that g is not (infinitely) renormalized.[9] Secondly, one should also investigate whether the relation (7.21), and the analogue ones for the other fields of the model, are still true in the discretized

[9]This supposition is somewhat supported, a posteriori, by our analysis of the (derivative of the) scaling function, which can be used as a definition of the renormalized coupling. As discussed in

model—i.e. the physical masses undergo only a finite renormalization. In this case, at given g, fixing $L^2\, m^2$ constant would be enough to keep the rescaled physical masses constant, namely no tuning of the "bare" parameter m would be necessary.

In the present study, we start by considering the example of bosonic x, x^* correlators, where indeed we find that the dimensionless ratio m_x^2/m^2 is free of $1/a$-divergences and approaches the expected continuum value $1/2$ (7.21) for large g, as shown in Sect. 7.6.1 below. Having this as hint, and because with the proposed discretization we have recovered in perturbation theory the one-loop cusp anomaly (7.39), we assume that in the discretized model no further scale but the lattice spacing a is present.

Any expectation value of a functional of lattice fields $\langle F_{\rm LAT}\rangle$ is therefore a function of the triplet of input (dimensionless) "bare" parameters (g, N, M) and in the limit discussed above we expect to extrapolate its value $\langle F\rangle$ in the continuum model

$$\langle F_{\rm LAT}(g, N, M)\rangle = \langle F(g)\rangle + O\Big(N^{-1}\Big) + O\left(e^{-MN}\right) . \qquad (7.45)$$

Lattice spacing effects are N^{-1} corrections, while finite volume effects lead to exponential corrections $e^{-M\,N}$.[10] At fixed coupling g and fixed large $M\,N = L\,m$, the observable $\langle F_{\rm LAT}\rangle$ is evaluated for different values of N and extrapolated to infinite N to obtain its continuum limit $\langle F\rangle$.

Table F.1 summarizes the parameters employed in our simulation runs. While most of them are done at $L\,m = 4$, for one value of the coupling ($g = 30$) we perform simulations at a larger value $L\,m = 6$ to explicitly check finite volume effects. For the physical observables under investigation, Figs. 7.6 (right panel) and 7.10, we find these effects to be very small and within the present statistical errors. They appear to play a role only in the case of the coefficient of the divergences which must be subtracted non-perturbatively in order to define the cusp action, see Sect. 7.6.2, as in Fig. 7.7 (right panel).

Each value of $\langle F_{\rm LAT}(g, N, M)\rangle$ is the output of a Monte Carlo evolution that uses the standard RHMC algorithm to compute quantum expectation values as time averages over a fictitious lattice dynamics (Appendix F.4.). We show two examples of Monte Carlo histories for the correlator $\langle x x^*\rangle$ and the action $\langle S\rangle$ in Fig. 7.4. We determined auto-correlation times of the observables and included their effect in the error analysis [41]. Multiple points at the same value of g and N in Figs. 7.6 (left panel), 7.8 and 7.9—and similarly in Figs. F.6 (left panel), F.8 and F.9—indicate multiple replica.

(Footnote 9 continued)
Sect. 7.6.2, occurring divergences in $S_{\rm LAT}$ can be consistently subtracted showing an agreement with the continuum expectation, at least for the region of lattice spacings and couplings that we explore.

[10]The combination $MN = (\frac{m}{a})(La)$ can be understood as the number of typical Compton wavelengths m^{-1} that fits into the lattice box. Finite volume effects $e^{-L\,m}$ are suppressed when interactions around the spatial direction are negligible for $m^{-1} \ll L$. Similarly, in QCD the mass of the lightest particle would give the largest error.

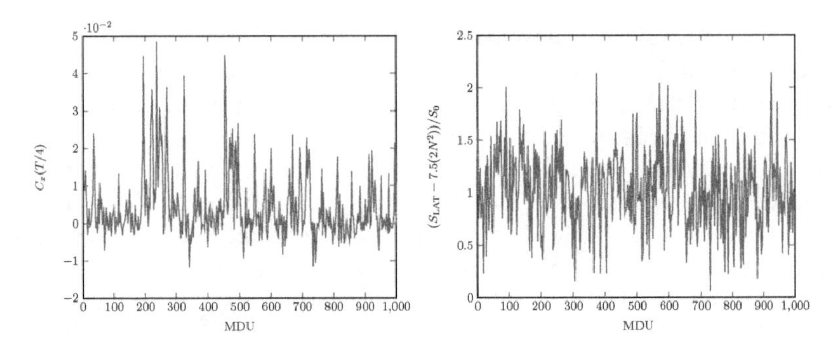

Fig. 7.4 Monte Carlo histories for the correlator $\langle x\, x^* \rangle$ at time separation $T/4$ and for $\langle S_{\mathrm{cusp}} \rangle$, at $g = 10$ and $L/a = 16$, in terms of Molecular Dynamic Units (MDU). The HMC produces a series of bosonic field configurations, on each of them the observable is evaluated and plotted here for the same series at the given parameters. The fact that successive configurations produced by the RHMC are statically correlated might lead to strong so-called auto-correlations in the data, which would appear in these plots as fluctuations with long periods. As one can see, the histories presented here do not suffer from such long fluctuations, and sample well the observables under investigation

7.6 Measuring the Observables

7.6.1 The $\langle X\, x^* \rangle$ Correlator

To motivate the line of constant physics (7.44), we first investigate the physical mass of the bosonic fluctuation field x around the string vacuum (7.12) as determined from the $\langle x\, x^* \rangle$ correlator. The mass of the bosonic field x, defined as the value of energy at vanishing momentum, in (7.15) can be read off at leading order from the expansion of the quadratic fluctuation Lagrangian (7.17). The leading quantum correction to its dispersion relation has been computed in [35], leading to the expression (7.21).

One can estimate the dependence of the physical mass on the coupling constant by measuring the connected two-point correlation function of the discretized x-field on the lattice (see for example [40]). In configuration space one defines the two-point function

$$G_x(t_1,\, s_1;\, t_2,\, s_2) = \langle x(t_1,\, s_1) x^*(t_2,\, s_2) \rangle \tag{7.46}$$

and Fourier-transforms over the spatial direction (cf. (F.19)) to define the *lattice timeslice correlator*

$$C_x(t;\, k) = \sum_{s_1,\, s_2=0}^{N_L-1} e^{-ik(s_1-s_2)} G_x(t,\, s_1;\, 0,\, s_2)\,. \tag{7.47}$$

The latter admits a spectral decomposition over propagating states of different energies, given spatial momentum k and amplitude c_n

$$C_x(t;\ k) = \sum_n |c_n|^2 e^{-t E_x(k;\, n)} \tag{7.48}$$

which is dominated by the state of lowest energy for sufficiently large temporal distance t. This effectively single asymptotic exponential decay corresponds to a one-particle state with energy equal—for vanishing spatial momentum—to the physical mass of the x-field

$$C_x(t;\ 0) \sim e^{-t\, m_{x\text{LAT}}}, \quad t \gg 1, \qquad m_{x\text{LAT}} = E_x(k=0)\,. \tag{7.49}$$

Corrections to the asymptotic behaviour of $C_x(t;\ 0)$ are proportional to $e^{-\Delta E\, t}$, where ΔE is the energy splitting with the nearest excited state. The physical mass on the lattice $m_{x\text{LAT}}$ is usefully obtained as a limit of an *effective mass* m_x^{eff}, defined at a given temporal extension T of the lattice and fixed pair of neighbouring points at time t and $t+a$ by the discretized logarithmic derivative of the timeslice correlation function (7.47) at zero momentum

$$m_{x\text{LAT}} = \lim_{T,t\to\infty} m_x^{\text{eff}} \equiv \lim_{T,t\to\infty,} \frac{1}{a} \log \frac{C_x(t;\ 0)}{C_x(t+a;\ 0)}\,. \tag{7.50}$$

Figure 7.5 (left) shows the effective mass measured from (7.50) as a function of the time separation t in units of $m_{x\text{LAT}}$ for different g and lattice sizes. To reduce uncertainty about the saturation of the lowest-energy state in the asymptotics of the

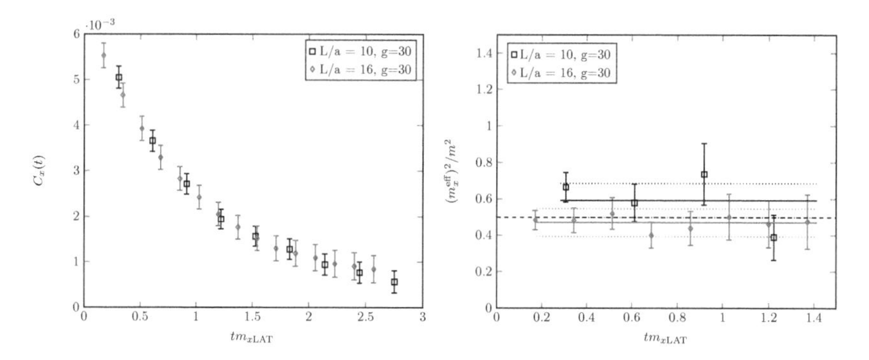

Fig. 7.5 Correlator $C_x(t;0) = \sum_{s_1,s_2}\langle x(t,s_1)x^*(0,s_2)\rangle$ of bosonic fields x, x^* (*left panel*) and corresponding effective mass $m_x^{\text{eff}} = \frac{1}{a}\log\frac{C_x(t)}{C_x(t+a)}$ normalized by m^2 (*right panel*), plotted as functions of the (dimensionless) time separation $t\, m_{x\text{LAT}}$ for different g and lattice sizes. The flatness of the effective mass indicates that the lowest-energy state in the two-point function saturates the correlation function, and allows for a reliable extraction of the mass of the x-excitation. Data points are masked by large errorbars for time scales greater than unity because the signal of the correlator degrades exponentially compared with the statistical noise

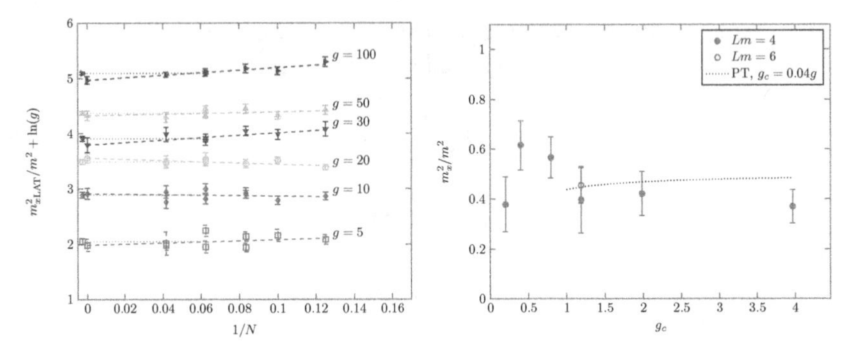

Fig. 7.6 Plot of $m^2_{x\text{LAT}}(N, g)/m^2 = m_x(g) + O(1/N)$ (left panel), as from *plateaux average* of results, which for $g = 30$ are shown in Fig. 7.5 (*right panel*). To ensure better visibility of the fits at different g values, $\log g$ has been added. Dashed lines represent a linear fit to all the data points for one value of g, while for *dotted lines* the fit is to a constant and only includes the two smallest lattice spacings. Multiple points at the same value of g and N indicate multiple replica. Continuum extrapolation (*right panel*) corresponding to the linear fits in the *left* panel. The simulations represented by the orange point ($L\,m = 6$) are used for a check of the finite volume effects, that appear to be within statistical errors. The extrapolation is plotted as a function of the continuum coupling $g_c = 0.04\,g$ to facilitate the comparison with the prediction coming from the sigma-model perturbation theory expectation (PT) (7.21), and uses the matching procedure performed for the observable action. The latter is described at the end of Sect. 7.6.2

correlation function (7.49), we work in a lattice of size $N_T \times N_L$ with temporal extent $N_T = T/a$ always twice the spatial extent $N_L = L/a$. The flatness of the effective mass in Fig. 7.5 (right) indicates that t and T are large enough to wash out the excited states in (7.50) and the estimate of the x-mass through (7.50) is reliable. We extract the value $m_x^2/m^2 = \frac{1}{2}$, which appears to be consistent with the classical, large g prediction (7.21). We do not see a clear signal yet for the expected bending down at smaller g. Decreasing the coupling drives up the numerical cost of the simulations and parallel computing would be necessary (Fig. 7.6).

The most important corollary of the analysis of the $\langle xx^* \rangle$ correlator is the following. As it happens in the continuum, also in the discretized setting there appears to be no infinite renormalization occurring for (7.21), and thus no need of tuning the bare parameter m to adjust for it. This corroborates the choice of (7.44) as the line of constant physics along which a continuum limit can be taken.

7.6.2 The Bosonic Action and the Scaling Function

In measuring the action (7.2) on the lattice, exploring first the "weak coupling" (large g) region we are supposed to recover the following general linear behavior in g

$$\langle S_{\text{LAT}} \rangle \equiv \frac{c}{2}(2N^2) + S_0 \,, \qquad g \gg 1, \qquad \text{where } S_0 \equiv \frac{1}{2}\,(2N^2)\,M^2\,g\,. \quad (7.51)$$

In the formula above we reinstated the mass scale m, used $V_2 = T\,L = a^2\,(2N^2)$ (F.13) and recalled the leading classical behavior $f(g) = 4\,g$ in (7.6). We also introduced the value of the (discretized) classical action S_0, which is fixed in each simulation for given g and NM. In (7.51) we also added a shift $\frac{c}{2}N^2$ constant in g and quadratically divergent in N^2 in the continuum limit.

The constant c can be extrapolated for very large values of g with a fit linear in N^{-2} of the relation $\frac{\langle S\rangle}{2N^2} = \frac{c}{2} + \frac{S_0}{2N^2}$. Data points for $g = 100, 50, 30$—red, green and violet fits respectively in the left panel of Fig. 7.7[11] –yield a value $c/2 = 7.5(1)$ consistent with the number $15 = 8 + 7$ of bosonic fields appearing in the path-integral. In other words, we expected that this (field-independent and proportional to the lattice volume) contribution to the expectation value $\langle S\rangle = -\partial \log Z/\partial g$ in (7.51) simply counts the number of degrees of freedom which appear quadratically in the action and carry a power of the coupling constant g. Indeed the theory is quadratic in bosons for very large g[12] and equipartition holds, namely integration over the bosonic variables yields a factor proportional to $g^{-\frac{(2N^2)}{2}}$ for each bosonic field species.[13]

Having determined with good precision the coefficient of the divergence, we can proceed first fixing it to be exactly $c = 15$ and subtracting from $\langle S_{\rm LAT}\rangle$ the corresponding contribution. In the finite g region we perform simulations to determine the ratio

$$\frac{\langle S_{LAT}\rangle - \frac{c}{2}\,(2N^2)}{S_0} \equiv \frac{f'_{\rm LAT}(g)}{4}\,. \tag{7.52}$$

On the right hand side we consistently restored the general definition (7.2) valid for any coupling. The plots at $g = 100, 50, 30, 20$ in Fig. 7.8 show a good agreement with the leading order prediction in (7.6) for which $f'(g) = 4$. For lower values of g—orange and light blue data points in Fig. 7.8—we observe deviations that obstruct the continuum limit and signal the presence of further quadratic divergences in N^2. They are compatible with an ansatz for $\langle S_{\rm LAT}\rangle$ in which the "constant" c contribution multiplying $2N^2$ in (7.51) and (7.52) is actually g-dependent. It seems natural to relate these power divergences to those arising in continuum perturbation theory, where they are usually set to zero using dimensional regularization [35]. From the perspective of a hard cut-off regularization like the lattice one, this is related to the emergence in the continuum limit of power divergences—quadratic, in the present two-dimensional case—induced by mixing of the (scalar) Lagrangian with the iden-

---

[11] Recall that in Fig. 7.7 $\log g$ has been added to ensure better visibility of the fits at different g values.

[12] In lattice codes and in (7.29) here, it is conventional to omit the coupling form the pseudo-fermionic part of the action, since this is quadratic in the fields and hence its contribution in g can be evaluated by a simple scaling argument.

[13] It is interesting to mention that in theories with exact supersymmetry this constant contribution of the bosonic action (this time on the trivial vacuum) is valid at all orders in g, due to the coupling constant independence of the free energy. For twisted $\mathcal{N} = 4$ SYM this is the origin of the supersymmetry Ward identity $S_{\rm bos} = 9N^2/2$ per lattice site, one of the observables used to measure soft supersymmetry breaking, see [42]. We thank David Schaich and Andreas Wipf for pointing this out to us.

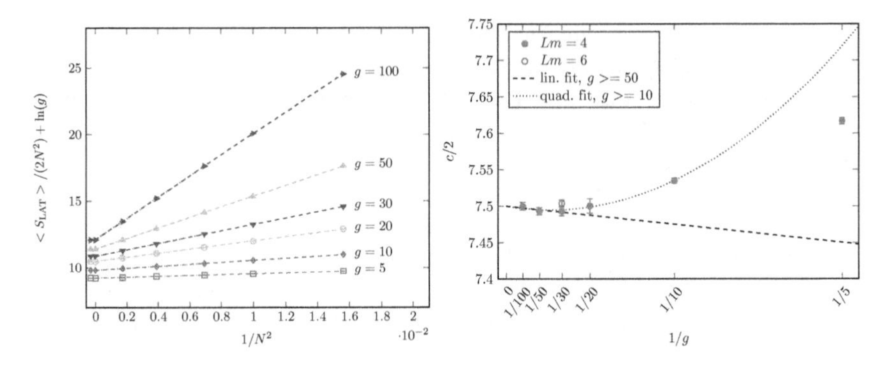

Fig. 7.7 In the *left* panel, we plot $\frac{\langle S_{\text{LAT}}\rangle}{2N^2}$ with linear fits in $1/N^2$ (*dashed lines*). To ensure better visibility of the fits at different g values, $\log g$ has been added. The extrapolation to the continuum limit (symbol at infinite N) determines the coefficient $c/2$ of the divergent ($\sim N^2$) contribution in (7.51) and (7.52) and is represented in the diagram of the right of this figure. In the *right panel*, data points estimate the continuum value of $c/2$ as from the extrapolations of the linear fits in the *left panel*. The simulations at $g = 30$, $L\,m = 6$ (*orange point*) are used for a check of the finite volume effects, which appear here to be visible. *Dashed* and *dotted lines* are the results of, respectively, a linear fit in $1/g$ and a fit to a polynomial of degree two

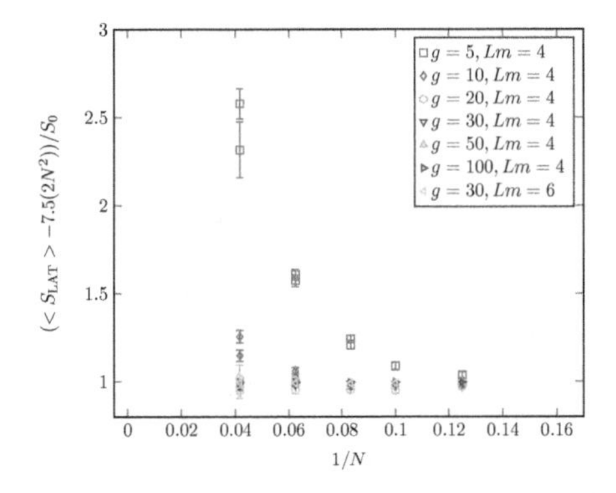

Fig. 7.8 Plot of the ratio $\frac{\langle S_{LAT}\rangle - \frac{c}{2}(2N^2)}{S_0} \equiv \frac{f'(g)}{4}$, where the coefficient of the divergent contribution c has been here *fixed* to the exact value $c = 15$ and $S_0 = \frac{1}{2}M^2\,(2N^2)\,g$. For very large g, there is agreement with the continuum prediction $f'(g) = 4$ in (7.6). For smaller values ($g = 10, 5$, *orange* and *light blue* data points) strong deviations appear, compatible with quadratic divergences

tity operator under UV renormalization. Additional contributions to these deviations might be due to the (possibly wrong) way the continuum limit is taken, i.e. they could be related to a possible infinite renormalization occurring in those field correlators and corresponding physical masses which have been not investigated here (fermionic and z excitations).

While such points should be investigated in the future to shed light on the issue, here we proceed with a non-perturbative subtraction of these divergences. Namely, from the data of Fig. 7.8 we subtract the continuum extrapolation of $\frac{c}{2}$ (multiplied

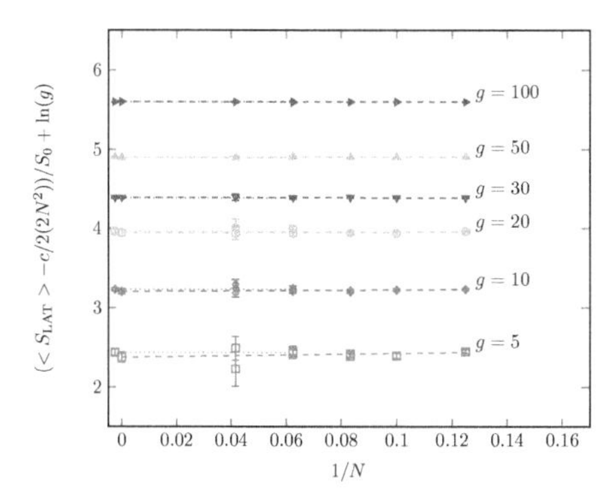

Fig. 7.9 Plots for the ratio $\frac{\langle S_{LAT}\rangle - \frac{c}{2}(2N^2)}{S_0} + \log g$ as a function of $1/N$, where the divergent contribution $c\,N^2/2$ is now the continuum extrapolation determined in Fig. 7.7. To ensure better visibility of the fits at different g values, $\log g$ has been added. *Dashed lines* represent a linear fit to all the data points for one value of g, while for *dotted lines* the fit is to a constant and only includes the two smallest lattice spacings. Symbols at zero (infinite N) are extrapolations from the fit constant in $1/N$

by the number of lattice points $2N^2$), as determined in Fig. 7.7 (right panel), for the full range of the coupling explored. The result is shown in Fig. 7.9. The divergences appear to be completely subtracted, confirming their purely quadratic nature. The flatness of data points, which can be fitted by a constant, indicates very small lattice artifacts. At least in the region of lattice spacings explored from our simulations errors are small, and do not diverge as one approaches the $N \to \infty$ limit. We can thus use the extrapolations at infinite N of Fig. 7.9 to show the continuum limit for the left hand side of (7.52) in Fig. 7.10. This is our measure for $f'(g)/4$, and in principle it allows a direct comparison with the perturbative series (dashed line) and with the prediction obtained via the integrability of the model (continuous line, representing the first derivative of the cusp as obtained from a numerical solution of the BES equation [10]).

To compare our extrapolations with the continuum expectation, we match the lattice point for the observable $f'(g)$ at $g = 10$—as determined from the $N \to \infty$ limit of $f'(g)_{\rm LAT}$ (7.52)—with the continuum value for the observable $f_c'(g_c)$ as determined from the integrability prediction, i.e. as obtained from a numerical solution of the BES equation. This is where in Fig. 7.10 the lattice point lies exactly on the integrability curve. The value $g = 10$ has been chosen as a reference point since it is far enough from both the region where the observable is substantially flat and proportional to one (which ensure a better matching procedure) and the region of higher errors (also, where the sign problem plays no role yet, see Sect. 7.6.3). Assuming that a simple finite rescaling relates the lattice bare coupling g and the (bare) continuum one g_c, from $f'(g) = f_c'(g_c)$ we then derive that $g_c = 0.04g$. Figure 7.10 shows that in the perturbative region our analysis and the related assumption for the finite rescaling of the coupling yield a good qualitative agreement with the integrability prediction. About direct comparison with the perturbative series (7.6), since we are

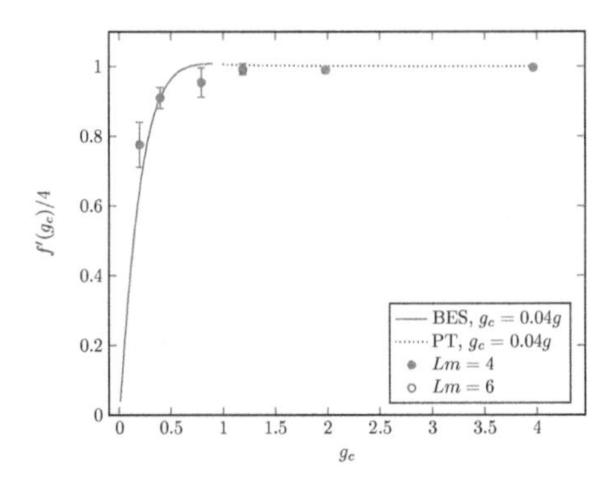

Fig. 7.10 Plot for $f'(g)/4$ as determined from the $N \to \infty$ extrapolation of (7.52), i.e. from the extrapolations of the fits in Fig. 7.9, and plotted as a function of the (bare) continuum coupling g_c under the hypothesis that the latter is just a finite rescaling of the lattice bare coupling g ($g_c = 0.04\, g$), see discussion at the end of this section. The *dashed line* represents the first few terms in the perturbative series (7.6), the continuous line is obtained from a numerical solution of the BES equation and represents therefore the prediction from the integrability of the model. The simulations at $g = 30$, $L\, m = 6$ (*orange point*) are used for a check of the finite volume effects, that appear to be within statistical errors

considering the derivative of (7.6) the first correction to the expected large g behavior $f'(g)/4 \sim 1$ is positive and proportional to the Catalan constant K. The plot in Fig. 7.10 does not catch the upward trend of such a first correction (which is too small, about 2%, if compared to the statistical error). Notice that, again under the assumption that such simple relation between the couplings exists—something that within our error bars cannot be excluded—the nonperturbative regime beginning with $g_c = 1$ would start at $g = 25$, implying that our simulations at $g = 10, 5$ would already test a fully non-perturbative regime of the string sigma-model under investigation. The mild discrepancy observed in that point of this region ($g = 5$ or $g_c = 0.2$) which is not fixed by definition via the "matching" procedure discussed above could be the effect of several contributing causes. Among them, systematic factors as the ones related to the complex phase—and its omission from the measurements, see next section—as well as finite volume effects with related errors in the non-perturbative subtraction of divergences. We emphasize that the relation between the lattice and continuum bare couplings might well be *not* just a finite rescaling. To shed light on this point, the matching procedure should use points at further smaller values of g. In the next section we discuss in more detail one of the most relevant issues which inhibits measurements at the interesting, small values of g.

7.6.3 The Pfaffian Phase

After the linearization realized via the Hubbard-Stratonovich transformation (7.22), the formal integration over the fermionic components leads to a Pfaffian in (7.27).

For any given bosonic configuration, the latter is manifestly *not* real. As discussed in Sect. 7.3, the Yukawa term (7.23) introduces a phase, so that $\mathrm{Pf}\,\hat{\mathcal{O}}_F \equiv \sqrt{\left|\mathrm{Det}\hat{\mathcal{O}}_F\right|}\, e^{i\varphi} = \left[\mathrm{Det}\left(\hat{\mathcal{O}}_F\hat{\mathcal{O}}_F^\dagger\right)\right]^{\frac{1}{4}} e^{i\varphi}$. The standard way to bypass this problem is to perform *phase-quenched* simulations, omitting $e^{i\varphi}$ from the integration measure

$$\langle F\rangle \equiv \frac{\int \mathcal{D}x\mathcal{D}x^*\mathcal{D}z^M\mathcal{D}\phi\mathcal{D}\phi^M\ \mathrm{Pf}\hat{\mathcal{O}}_F\ F\ e^{-S_B}}{\int \mathcal{D}x\mathcal{D}x^*\mathcal{D}z^M\mathcal{D}\phi\mathcal{D}\phi^M\ \mathrm{Pf}\hat{\mathcal{O}}_F\ e^{-S_B}} \tag{7.53}$$

$$\rightarrow\ \langle F\rangle_{\text{phase-quenching}} \equiv \frac{\int \mathcal{D}x\mathcal{D}x^*\mathcal{D}z^M\mathcal{D}\phi\mathcal{D}\phi^M\ \left[\mathrm{Det}\left(\hat{\mathcal{O}}_F\hat{\mathcal{O}}_F^\dagger\right)\right]^{\frac{1}{4}}\ F\ e^{-S_B}}{\int \mathcal{D}x\mathcal{D}x^*\mathcal{D}z^M\mathcal{D}\phi\mathcal{D}\phi^M\ \left[\mathrm{Det}\left(\hat{\mathcal{O}}_F\hat{\mathcal{O}}_F^\dagger\right)\right]^{\frac{1}{4}}\ e^{-S_B}}\,.$$

Such a procedure ensures drastic computational simplifications and still can deliver the true expectation values via the *phase reweighting method*, which prescribes to redefine the measure by incorporating the non-positive part of the Boltzmann weight (here the complex phase) into the observable

$$\langle F\rangle \equiv \frac{\int \mathcal{D}x\mathcal{D}x^*\mathcal{D}z^M\mathcal{D}\phi\mathcal{D}\phi^M\ \mathrm{Pf}\hat{\mathcal{O}}_F\ F\ e^{-S_B}}{\int \mathcal{D}x\mathcal{D}x^*\mathcal{D}z^M\mathcal{D}\phi\mathcal{D}\phi^M\ \mathrm{Pf}\hat{\mathcal{O}}_F\ e^{-S_B}} \tag{7.54}$$

$$= \frac{\langle F e^{i\varphi}\rangle_{\text{reweighting}}}{\langle e^{i\varphi}\rangle_{\text{reweighting}}} \equiv \frac{\int \mathcal{D}x\mathcal{D}x^*\mathcal{D}z^M\mathcal{D}\phi\mathcal{D}\phi^M\ \left[\mathrm{Det}\left(\hat{\mathcal{O}}_F\hat{\mathcal{O}}_F^\dagger\right)\right]^{\frac{1}{4}}\ \left(F e^{i\varphi}\right)\ e^{-S_B}}{\int \mathcal{D}x\mathcal{D}x^*\mathcal{D}z^M\mathcal{D}\phi\mathcal{D}\phi^M\ \left[\mathrm{Det}\left(\hat{\mathcal{O}}_F\hat{\mathcal{O}}_F^\dagger\right)\right]^{\frac{1}{4}}\ \left(e^{i\varphi}\right)\ e^{-S_B}}\,.$$

However, the phase averages to zero if it displays a highly oscillatory behaviour far from zero in a significant part of the bosonic field configurations. In this case the reweighting procedure breaks down because numerator and denominator evaluate to a small number, whose exact value is masked by the stochastic noise of the Monte Carlo sampling process. The severity of *sign problem* is quantified by the smallness of the (ensamble-averaged) phase

$$\int \mathcal{D}x\mathcal{D}x^*\mathcal{D}z^M\mathcal{D}\phi\mathcal{D}\phi^M\ \left[\mathrm{Det}\left(\hat{\mathcal{O}}_F\hat{\mathcal{O}}_F^\dagger\right)\right]^{\frac{1}{4}}\ \left(e^{i\varphi}\right)\ e^{-S_B}\,. \tag{7.55}$$

We have explicitly computed the reweighting (phase) factor for smaller lattices, up to $L/a = 12$, and observed that the reweighting has no effect on the central value of the two observables that we study. Thus, in the analysis presented in the previous sections and in Appendix F.6 we omit the phase from the simulations in order to be able to consistently take the continuum limit. In absence of data for the phase factor in the case of larger lattices, we do not assess the possible systematic error related to this procedure.

To explore the possibility of a sign problem in simulations, we have then studied the relative frequency for the real part (the imaginary part is zero within errors, as

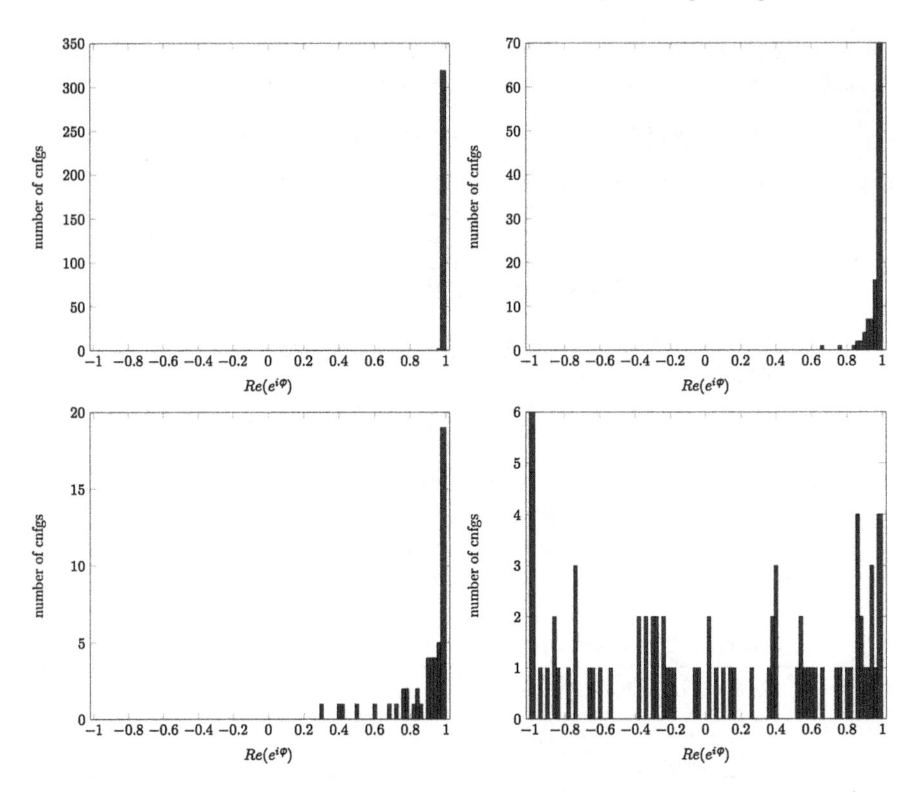

Fig. 7.11 Histograms for the frequency of the real part of the reweighting phase factor $e^{i\varphi}$ of the Pfaffian $\text{Pf}\,\hat{\mathcal{O}}_F \equiv \sqrt{\left|\text{Det}\hat{\mathcal{O}}_F\right|}\,e^{i\varphi}$, based on the ensembles generated at $g = 30, 10, 5, 1$ (from *left* to *right*, *top* to *down*) on a lattice with $N_L = 8$. The plots use the discretization (F.46)–(F.47), however we found no substantial difference between this analysis and the one performed with the discretization (7.40)–(7.41)

predicted from the reality of the observables studied) of the Pfaffian phase $e^{i\varphi}$ at $g = 30, 10, 5, 1$ (from left to right, top to down) in Fig. 7.11. At $g = 1$ (right bottom histogram) the observed $\langle e^{i\varphi}\rangle$ is consistent with zero, thus preventing the use of standard reweighting. As explained above, the analysis that we present here is also limited to the values $g = 100, 50, 30, 20, 10, 5$ (and with the further parameters listed in Tables F.1 and F.2). Therefore, a severe sign problem is appearing precisely for values of the coupling referring to a fully non-perturbative regime (corresponding to weakly-coupled $\mathcal{N} = 4$ SYM). We conclude that alternative algorithms or settings (in terms of a different, phase-free linearization) should be considered in order to investigate this interesting region of g.

7.7 New Insight into the Pfaffian Phase

One important question that we are currently addressing [38, 39] is whether one can engineer a Hubbard-Stratonovich transformation for a new set of auxiliary fields that leads to a *real* and *positive* Pfaffian in the path-integral.[14] We are modelling a new proposal upon the linearization of the $SO(4)$-invariant four-fermion interactions in [44]. After some lengthy manipulations, the relevant global symmetry $SO(6)$ suggests to write the integral identity

$$\begin{aligned}
&\exp\Big\{-g\int dtds\Big[-\tfrac{1}{z^2}\left(\eta^i\eta_i\right)^2+\left(\tfrac{i}{z^2}z_N\eta_i\rho^{MN^i}{}_j\eta^j\right)^2\Big]\Big\}\\
&\propto\int \mathcal{D}\phi\,\mathcal{D}\phi^i_j\,\exp\left\{-g\int dtds\,\Big[6\phi^2+\frac{12}{z}\left(\eta^i\eta_i\right)^2\phi\right.\\
&\left.\quad+\phi^i_j\phi^j_i+\frac{2}{z}\eta_j\phi^j_i\eta^i+\frac{2}{z^3}(\rho^N)^{ik}z_N\eta_k\phi^j_i(\rho^L)_{jl}z_L\eta^l\Big]\right\}.
\end{aligned}\tag{7.56}$$

where now we integrate over 17 real auxiliary degrees of freedom, split between a real field ϕ and a set of 16 fields ϕ^i_j with the property $\left(\phi^i_j\right)^*=\phi^j_i$ $(i,j=1,\ldots 4)$. The nice feature is that each term in the linearized Lagrangian is now hermitian (cf. (7.24)–(7.25)):

$$\begin{aligned}
\mathcal{L}^{\text{new}}&=\left|\partial_t x+\frac{m}{2}x\right|^2+\frac{1}{z^4}\left|\partial_s x-\frac{m}{2}x\right|^2+\left(\partial_t z^M+\frac{m}{2}z^M\right)^2+\frac{1}{z^4}\left(\partial_s z^M-\frac{m}{2}z^M\right)^2\\
&\quad+6\phi^2+\phi^i_j\phi^j_i+\psi^T\mathcal{O}_F^{\text{new}}\psi
\end{aligned}\tag{7.57}$$

with

$$\mathcal{O}_F^{\text{new}}=\begin{pmatrix}
0 & i\mathbb{I}_4\partial_t & -i\rho^M\left(\partial_s+\frac{m}{2}\right)\frac{z^M}{z^3} & 0\\
i\mathbb{I}_4\partial_t & 0 & 0 & -i\rho^\dagger_M\left(\partial_s+\frac{m}{2}\right)\frac{z^M}{z^3}\\
i\frac{z^M}{z^3}\rho^M\left(\partial_s-\frac{m}{2}\right) & 0 & 2\frac{z^M}{z^4}\rho^M\left(\partial_s x-m\frac{x}{2}\right) & i\mathbb{I}_4\partial_t-A^T_{\text{new}}\\
0 & i\frac{z^M}{z^3}\rho^\dagger_M\left(\partial_s-\frac{m}{2}\right) & i\mathbb{I}_4\partial_t+A_{\text{new}} & -2\frac{z^M}{z^4}\rho^\dagger_M\left(\partial_s x^*-m\frac{x}{2}^*\right)
\end{pmatrix},$$

$$A_{\text{new}}=-\frac{6}{z}\phi+\frac{1}{z}\tilde{\phi}+\frac{1}{z^3}\rho^*_N\tilde{\phi}^T\rho^L z^N z^L+i\frac{z^N}{z^2}\rho^{MN}\partial_t z^M\tag{7.58}$$

and $\tilde{\phi}$ being a 4×4 matrix whose (i,j)-entry is ϕ^i_j. We perform again a fermion discretization that replaces $\mathcal{O}_F^{\text{new}}$ with a double-free $\hat{\mathcal{O}}_F^{\text{new}}$ while preserving the reality of the operator. This eliminates the presence of a complex phase in the new Pfaffian Pf $\hat{\mathcal{O}}_F^{\text{new}}\equiv\sqrt{\left|\text{Det}\,\hat{\mathcal{O}}_F^{\text{new}}\right|}\,e^{i\varphi^{\text{new}}}$, leaving only the two possibilities $\varphi^{\text{new}}=0,\pi$. The only question left to answer is whether the Pfaffian in also *positive* ($\varphi^{\text{new}}=0$).

[14]This question is legitimate because the problem of the complex phase *depends* on the discretization adopted, so it can be avoided by rewriting the action in a suitable form, for instance as done in [43].

Preliminary studies suggest that such sign problem is not ruled out yet because the simulation visits configurations with (field-dependent) alternating sign for the Pfaffian $\text{Pf}\,\hat{O}_F^{\text{new}} = \pm\sqrt{\left|\text{Det}\,\hat{O}_F^{\text{new}}\right|}$, especially when we depart from the weakly-coupled regime of the theory. Further work is also needed to assess whether the argument of [44] can be traduced to our case in a way different from (7.56) and directly circumvent the sign problem in (7.58).

References

1. R. W. McKeown, R. Roiban, The quantum $AdS_5 \times S^5$ superstring at finite coupling, arXiv:1308.4875
2. S. Giombi, R. Ricci, R. Roiban, A. Tseytlin, C. Vergu, Quantum $AdS_5 \times S^5$ superstring in the AdS light-cone gauge. JHEP **1003**, 003 (2010), arXiv:0912.5105
3. S. S. Gubser, I. R. Klebanov, A. M. Polyakov, A semi-classical limit of the gauge/string correspondence, Nucl. Phys. B **636**, 99 (2002), arXiv:hep-th/0204051
4. L. F. Alday and J. M. Maldacena, Comments on operators with large spin. JHEP **0711**, 019 (2007), arXiv:0708.0672
5. L. F. Alday, D. Gaiotto, J. Maldacena, A. Sever, P. Vieira, An operator product expansion for polygonal null Wilson loops. JHEP **1104**, 088 (2011), arXiv:1006.2788
6. B. Basso, A. Sever, P. Vieira, Spacetime and flux tube S-matrices at finite coupling for $\mathcal{N} = 4$ supersymmetric yang-mills theory. Phys. Rev. Lett. **111**, 091602 (2013), arXiv:1303.1396
7. B. Basso, A. Sever, P. Vieira, Space-time S-matrix and flux tube S-matrix II. Extracting and matching data. JHEP **1401**, 008 (2014), arXiv:1306.2058
8. D. Fioravanti, S. Piscaglia, M. Rossi, Asymptotic Bethe Ansatz on the GKP vacuum as a defect spin chain: scattering, particles and minimal area Wilson loops. Nucl. Phys. B **898**, 301 (2015)
9. A. Bonini, D. Fioravanti, S. Piscaglia, M. Rossi, Strong Wilson polygons from the lodge of free and bound mesons. JHEP **1604**, 029 (2016), arXiv:1511.05851
10. N. Beisert, B. Eden and M. Staudacher, Transcendentality and crossing. J. Stat. Mech. **0701**, P01021 (2007), arXiv:hep-th/0610251
11. A. M. Polyakov, Gauge fields as rings of glue. Nucl. Phys. B **164**, 171 (1980)
12. N. Drukker, D. J. Gross and H. Ooguri, Wilson loops and minimal surfaces. Phys. Rev. D **60**, 125006 (1999), arXiv:hep-th/9904191
13. G. P. Korchemsky and A. V. Radyushkin, Infrared factorization, Wilson lines and the heavy quark limit. Phys. Lett. B **279**, 359 (1992), arXiv:hep-ph/9203222
14. T. Becher, M. Neubert, Infrared singularities of QCD amplitudes with massive partons. Phys. Rev. D **79**, 125004 (2009), arXiv:0904.1021, [Erratum: Phys. Rev. D **80**, 109901 (2009)]
15. L. F. Alday, J. M. Henn, J. Plefka, T. Schuster, Scattering into the fifth dimension of $\mathcal{N} = 4$ super Yang-Mills. JHEP **1001**, 077 (2010), arXiv:0908.0684
16. J. M. Henn, S. G. Naculich, H. J. Schnitzer, M. Spradlin, Higgs-regularized three-loop four-gluon amplitude in $\mathcal{N} = 4$ SYM: exponentiation and Regge limits. JHEP **1004**, 038 (2010), arXiv:1001.1358
17. D. Correa, J. Henn, J. Maldacena, A. Sever, An exact formula for the radiation of a moving quark in $\mathcal{N} = 4$ super Yang Mills. JHEP **1206**, 048 (2012), arXiv:1202.4455
18. B. Fiol, B. Garolera, A. Lewkowycz, Exact results for static and radiative fields of a quark in $\mathcal{N} = 4$ super Yang-Mills, JHEP **1205**, 093 (2012), arXiv:1202.5292
19. D. Correa, J. Henn, J. Maldacena, A. Sever, The cusp anomalous dimension at three loops and beyond. JHEP **1205**, 098 (2012), arXiv:1203.1019
20. G. P. Korchemsky, A. V. Radyushkin, Loop space formalism and renormalization group for the infrared asymptotics of QCD. Phys. Lett. B **171**, 459 (1986), in *International Seminar: Quarks 86 Tbilisi, USSR*, April 15–17, 1986, pp. 459–467

21. G. P. Korchemsky, A. V. Radyushkin, Renormalization of the Wilson loops beyond the leading order. Nucl. Phys. B **283**, 342 (1987)
22. G. P. Korchemsky, G. Marchesini, Structure function for large x and renormalization of Wilson loop. Nucl. Phys. B **406**, 225 (1993), arXiv:hep-ph/9210281
23. G. P. Korchemsky, Asymptotics of the Altarelli-Parisi-Lipatov evolution kernels of parton distributions. Mod. Phys. Lett. A **4**, 1257 (1989)
24. M. Kruczenski, A Note on twist two operators in $\mathcal{N} = 4$ SYM and Wilson loops in Minkowski signature. JHEP **0212**, 024 (2002), arXiv:hep-th/0210115
25. Y. Makeenko, Light cone Wilson loops and the string/gauge correspondence. JHEP **0301**, 007 (2003), arXiv:hep-th/0210256
26. M. Kruczenski, R. Roiban, A. Tirziu, A. A. Tseytlin, Strong-coupling expansion of cusp anomaly and gluon amplitudes from quantum open strings in $AdS_5 \times S^5$. Nucl. Phys. B **791**, 93 (2008), arXiv:0707.4254
27. S. Frolov, A. A. Tseytlin, Semiclassical quantization of rotating superstring in $AdS_5 \times S^5$. JHEP **0206**, 007 (2002), arXiv:hep-th/0204226
28. S. Frolov, A. Tirziu, A. A. Tseytlin, Logarithmic corrections to higher twist scaling at strong coupling from AdS/CFT. Nucl. Phys. B **766**, 232 (2007), arXiv:hep-th/0611269
29. R. Roiban, A. A. Tseytlin, Strong-coupling expansion of cusp anomaly from quantum superstring. JHEP **0711**, 016 (2007), arXiv:0709.0681
30. R. Roiban, A. Tirziu, A. A. Tseytlin, Two-loop world-sheet corrections in $AdS_5 \times S^5$ superstring. JHEP **0707**, 056 (2007), arXiv:0704.3638
31. R. Roiban, A. A. Tseytlin, Spinning superstrings at two loops: strong-coupling corrections to dimensions of large-twist SYM operators. Phys. Rev. D **77**, 066006 (2008), arXiv:0712.2479
32. R. Metsaev, A. A. Tseytlin, Superstring action in $AdS_5 \times S^5$. Kappa symmetry light cone gauge. Phys. Rev. D **63**, 046002 (2001), arXiv:hep-th/0007036
33. R. Metsaev, C. B. Thorn, A. A. Tseytlin, Light cone superstring in AdS space-time. Nucl. Phys. B **596**, 151 (2001), arXiv:hep-th/0009171
34. B. Basso, Exciting the GKP string at any coupling. Nucl. Phys. B **857**, 254 (2012), arXiv:1010.5237
35. S. Giombi, R. Ricci, R. Roiban, A. Tseytlin, Quantum dispersion relations for excitations of long folded spinning superstring in $AdS_5 \times S^5$. JHEP **1101**, 128 (2011), arXiv:1011.2755
36. R. L. Stratonovich, On a method of calculating quantum distribution functions. Soviet Phys. Doklady **2**, 416 (1957)
37. J. Hubbard, Calculation of partition functions. Phys. Rev. Lett. **3**, 77 (1959)
38. V. Forini, L. Bianchi, B. Leder, P. Toepfer, E. Vescovi, Strings on the lattice and AdS/CFT. PoS LATTICE2016, in *Proceedings, 34th International Symposium on Lattice Field Theory (Lattice 2016)*, vol. 206, Southampton, UK, July 24–30 2016, p. 206, arXiv:1702.02005
39. V. Forini, L. Bianchi, B. Leder, P. Toepfer, E. Vescovi, in preparation
40. I. Montvay, G. Muenster, *Quantum Fields on a Lattice* (Cambridge University Press, Cambridge, 1994)
41. ALPHA Collaboration, U. Wolff, Monte Carlo errors with less errors. Comput. Phys. Commun. **156**, 143 (2004), arXiv:hep-lat/0306017, [Erratum: Comput. Phys. Commun. **176**, 383 (2007)]
42. S. Catterall, First results from simulations of supersymmetric lattices. JHEP **0901**, 040 (2009), arXiv:0811.1203
43. C. Gattringer, T. Kloiber, V. Sazonov, Solving the sign problems of the massless lattice Schwinger model with a dual formulation. Nucl. Phys. B **897**, 732 (2015), arXiv:1502.05479
44. S. Catterall, Fermion mass without symmetry breaking. JHEP **1601**, 121 (2016), arXiv:1510.04153

Chapter 8
Conclusion and Outlook

In this thesis we have reviewed the construction of superstring theory in two background spaces—$AdS_5 \times S^5$ and $AdS_4 \times \mathbb{CP}^3$—relevant for the study of the respective AdS/CFT integrable systems. We performed perturbative calculations at large't Hooft coupling in order to compute quantum corrections to non-trivial classical solutions of the superstring action. They correspond to solitonic solutions (related to certain local gauge-theory operators) and open string solutions (here associated to BPS and non-BPS Wilson loops). On a parallel route, we have investigated a promising non-perturbative approach that relies on purely lattice field theory methods to obtain information on observables defined in the AdS_5/CFT_4 duality at finite 't Hooft coupling.
The motivation behind all these analyses is twofold. The main interest is to collect complementary perspectives on string sigma-models that provide support to predictions based on integrability and supersymmetric localization. Our results offer also a direct demonstration of the quantum consistency of superstring actions in perturbative quantization around particular string vacua embedded in curved spacetimes.

8.1 Summary of the Main Results

In Chap. 3 we made the first step in the direction of computing one-loop quantum corrections to *arbitrary* classical solutions in $AdS_5 \times S^5$ by deriving the quadratic action for the small fluctuations. While it would be virtually impossible to address regularization procedures at this general level, we pedagogically illustrated a *manifestly-covariant* algorithm to derive the structure of differential operators entering the fluctuation Lagrangian. The appropriate bosonic and fermionic gauge-fixings, the diffeomorphism-ghost operator, the decoupling of bosonic longitudinal modes, the gauge connections associated to the normal bundle and the reduction of type IIB

E. Vescovi, *Perturbative and Non-perturbative Approaches to String Sigma-Models in AdS/CFT*, Springer Theses, DOI 10.1007/978-3-319-63420-3_8

Green-Schwarz fermions to 2d Dirac spinors are the principal results scattered in literature [1, 2] that we have organized and generalized here.

In particular, a novelty of our formulas is the treatment of the fermionic fluctuation Lagrangian and the general expressions for bosonic and fermionic masses (as in all cases previously analyzed, simplifications occur due to the flatness of the normal bundle). The sum rules for the masses find a nice application in reviewing the argument for the cancellation of the one-loop conformal anomaly in the $AdS_5 \times S^5$ superstring [2].

In Chap. 4 we analytically solved the matrix-valued fluctuations determinant for two non-trivial string configurations in $AdS_5 \times S^5$. We evaluated exactly the one-loop partition function for the quantum Landau-Lifshitz model on the $SU(2)$ folded string solution of [3]. The relevant differential operator can be written in terms of a fourth-order differential operator with coefficients being meromorphic, doubly periodic functions (combinations of Jacobi elliptic functions) on the complex plane.

The same procedure allows the diagonalization of the operator in conformal gauge that governs the bosonic excitations around the two-spin folded string solution of [4] in $AdS_5 \times S^5$. Here the differential equations governing the fermionic spectrum do not satisfy the conditions that allowed us to solve the bosonic sector and we did not find the necessary generalization of the tools we have developed in appendix C. However, our analysis still proves to be useful to analytically demonstrate the equivalence between the full exact one-loop partition function for the one-spin folded string in conformal and static gauge, which is a non-trivial statement verified only numerically in [5].

The purpose of Chap. 5 is to re-examine a delicate issue [6] in the semiclassical computation of the partition function of $AdS_5 \times S^5$ superstring that represents a 1/2-BPS circular Wilson loop in $\mathcal{N} = 4$ SYM, where the one-loop string correction does not match the expansion of the localization result for the gauge-theory observable [7]. To avoid any measure-related ambiguity in the string path-integral, we calculated the ratio between the one-loop partition functions for two classical solutions with the same topology: a generic 1/4-BPS latitude Wilson loop and the 1/2-BPS circular loop. By comparing with the gauge-theory prediction for such ratio, we addressed the question whether such procedure could eliminate the measure-related ambiguity under the assumption that they only depend on the worldsheet topology. Our answer is that the standard setup to compute one-loop corrections (infinite sum of 1d determinants obtained via Gel'fand-Yaglom theorem and cutoff regularization) fails to restore the agreement with gauge theory. In Sect. 5.7 we have summarized the current status of the mismatch after a similar attempt was carried out in [8] and we devoted Sect. 5.8 to point out how our work suggests possible continuations.

In Chap. 6 we extended the computation of the cusp anomalous dimension of ABJM theory to second order in its strong coupling expansion. This result has been determined considering the κ-symmetry gauge-fixed action of [9, 10] in $AdS_4 \times \mathbb{CP}^3$. We studied its fluctuations about the null cusp background and the AdS light-cone gauge makes the explicit evaluation of the free energy rather manageable, as in the $AdS_5 \times S^5$ counterpart of this calculation [11].

While at one loop we confirm the result of [12–14], the two-loop correction provides a new important information that, once combined with a proposal based on the Bethe Ansatz [15] and the correction to the effective string tension [16], gives information on the two-loop coefficient in the expansion of the interpolating function $h(\lambda)$ of the AdS_4/CFT_3 duality. At this perturbative order, our calculation gives a test of the validity of the exact expression for $h(\lambda)$ conjectured in [17] and also indirect evidence of quantum integrability for the type IIA superstring in the AdS light-cone gauge.

The cancellation of UV divergences gives strong evidences of the quantum consistency of the non-supercoset action of [9, 10] and shows that it can be used for other non-trivial strong-coupling computations in $AdS_4 \times \mathbb{CP}^3$ [18].

In Chap. 7 we investigate a path that has led us out of the context of perturbation theory. We proposed possible lattice discretizations of the two-dimensional $AdS_5 \times S^5$ supercoset sigma-model in which local symmetries are fixed in the AdS light-cone gauge.

We measured numerically the derivative of the light-like cusp anomaly of $\mathcal{N} = 4$ SYM (from the value of the bosonic action) and the mass of the two bosonic AdS worldsheet excitations (from their two-point function) that are transverse to the relevant classical solution. Lattice simulations are performed with a Rational Hybrid Monte Carlo algorithm and two Wilson-like fermion discretizations breaking different subgroups of the global symmetry of the action. They agree within errors in continuum extrapolations of the observables, in which no tuning of the "bare" mass parameter (the light-cone momentum p^+) seems to be required.

Both observables are in good agreement (qualitative for the mass and quantitative for the action) with integrability predictions in the weakly-coupled regime at large't Hooft coupling. For smaller values, the action develops quadratic divergences that we non-perturbatively subtracted before taking the continuum limit. Our choice for the linearization of fermionic interactions leads to a complex phase in the fermionic determinant. In the non-perturbative regime this causes a severe sign problem that we can not be treated via standard reweighting.

8.2 Future Directions

All this work has opened up a number of interesting directions that is worth exploring.

In the perspective adopted in Chap. 3 there is no explicit reference to the classical integrability of the $AdS_5 \times S^5$ superstring [19]. In a number of semiclassical studies [5, 20] the underlying integrable structure of the $AdS_5 \times S^5$ background emerges in the appearance of certain "solvable" differential operators [5], whose determinants can be calculated explicitly and result in closed form expressions for the one-loop partition functions. The question of a deeper relation between our geometric approach to fluctuations and the integrability of the sigma-model of interest should become more manifest within the algebraic curve approach to semiclassical quantization [21–23], likely on the lines of [24].

The observations above apply also to the spectral problem encountered in Chap. 4. For the two-spin string of [4] it is arguably more urgent to complete the solution of the spectral problem in the fermionic sector, which might require a non-trivial field redefinition for the corresponding Lagrangian, or equivalently a modification of the ansatz for the solution of the related differential operator. Perfecting the analysis in this direction should enlarge the range of the class of problems that can be solved analytically.

An immediate follow-up [25] inspired by the work in Chap. 5 is the development of heat kernel techniques to evaluate one-loop determinants around minimal surfaces that possess a geometry "infinitesimally close" to the hyperbolic space H^2. This occurs whenever the worldsheet is actually a family of classical surfaces, controlled by a certain deformation parameter α, that reaches such special configuration for $\alpha = 0$. The power of our method relies on the knowledge of the heat kernels of Laplace and Dirac operators on the maximally symmetric H^2 [26–29]. The fundamental idea is to approximate the heat kernels of the fluctuation operators on the (non-maximally-symmetric) worldsheet as a perturbative series in small α around the known heat kernels at $\alpha = 0$. The fluctuation determinants can be eventually extracted from this spectral information via zeta-function regularization. The principles of the algorithm have strong resemblance to the perturbative solution of a Schrödinger-like equation (heat equation in formula (1.15)), where one adds a perturbation parameter α to an exactly solvable potential (maximally-symmetric worldsheet metric) and then tries to determine the small corrections to a certain wavefunction (heat kernel). Our method reproduces [25] the small-angle coefficient in the one-loop partition function (here $\alpha = k^2$)

$$\log \frac{Z^{(1)}(k=0)}{Z^{(1)}(k)} \equiv \Gamma^{(1)}(k) - \Gamma^{(1)}(k=0) = \frac{3T}{8}k^2 + O(k^2) \tag{8.1}$$

dual to a Wilson loop made of two nearly-antipodal lines on $\mathbb{R} \times S^3$ [20].[1] We have started to address the case of latitude Wilson loops of Chap. 5 for small values of the angular parameter θ_0, so in a configuration very close to the 1/2-BPS circular loop (left panel in Fig. 5.2, with $\phi_0 \equiv \pi - \theta_0 \approx \pi$). Preliminary results indicate that we can recover (here $\alpha = \theta_0^2$)

$$\log \frac{Z^{(1)}(\theta_0)}{Z^{(1)}(\theta_0=0)} \equiv \Gamma^{(1)}(\theta_0=0) - \Gamma^{(1)}(\theta_0) = \frac{3}{4}\theta_0^2 + O(\theta_0^4) \tag{8.2}$$

in *agreement* with the small-angle limit of $-\frac{3}{2}\log\cos\theta_0$ in (5.22). A rather necessary step will be of course to explain why the Gel'fand-Yaglom method applied to the antiparallel lines of [20] agrees with thermodynamic Bethe ansatz [30] and quantum

[1] See formula (D.48) therein, where $\mathcal{T} = \int dt + O(k^2)$ and $T \equiv \int dt$ is the AdS time cutoff on the temporal extension of the Wilson lines in Fig. 1. In the notation of the paper, we are considering a loop coupled to a fixed scalar (setting $\theta = 0$ in (2.1)) and made of two lines separated by an angle $\pi - \phi$ along a big circle on S^3, where in first approximation $\phi \approx \pi k$ in (B.10) for $k \to 0$.

spectral curve [31], whereas it fails to match localization for the latitude loops in Chap. 5 and [8]. On a related note, one should be able to understand why the Gel'fand-Yaglom approach seems nevertheless to capture the expected localization result once some contributions are artificially excluded in [8] and here in (5.93). The authors of this paper speculated that this *ad hoc* prescription might hint at the existence of some localization mechanism in string theory that would reproduce the full partition function by integrating over only some zero-modes, in the same way as localization writes the matrix model of the circular Wilson loop in gauge theory [7]. Without endeavouring to prove such speculative statement, it is certainly worthwhile to see [32] whether the prescription of [8] still applies to other DGRT Wilson loops [33] and correlators with local operators, as they are known to localize on a matrix model in $\mathcal{N}=4$ SYM [34].

The two-loop computation in Chap. 6 provided support to the quantum consistency of the $AdS_4 \times \mathbb{CP}^3$ superstring action [9, 10]. Beyond the possibility of pushing the perturbative check of the all-loop proposal for the ABJM interpolating function (6.4) to three-loop level, one intriguing direction would be to see whether this action can be discretized on a lattice as we did for the $AdS_5 \times S^5$ sigma-model action in Chap. 7. In this way we would be able to obtain information on the ABJM scaling function in terms of the coupling constant, for any value thereof. By comparison with its integrability prediction in terms of the ABJM interpolating function $h(\lambda)$, one could then provide numerical values of $h(\lambda)$ at some finite values of the 't Hooft coupling, which could then be contrasted with (6.4).

Another topic of high relevance is the non-perturbative investigation of the $AdS_5 \times S^5$ sigma-model set up in Chap. 7. The next step currently underway is to extend the analysis to all field correlators and proceed with the comparison of their physical masses—obtained from the exponential decay of the two-point lattice correlators—with their predictions from semiclassical string theory [35] or at smaller coupling via asymptotic Bethe ansatz [36].[2] A rather stringent test of our computational setup would be to see that fermions remain massless at any coupling [38].

Current data in Chap. 7 has already clearly exposed many numerical and conceptual challenges that we are addressing in parallel. The action at small values of g appears to be quadratically divergent in the inverse lattice spacing $\sim a^{-2}$. It is possible that the reasoning leading to the proposed line of constant physics might be subject to change once all fields correlators are investigated. At the same time, such divergences are expected in our lattice regularization because in the continuum theory in this [11] and analogue models (Chap. 6) power-like divergences are set to zero in dimensional regularization. While we have proceeded to non-perturbative subtractions of those divergences so far, in general this procedure leads to potentially severe ambiguities, with errors diverging in the continuum limit. We remark however that for the other physical observable here investigated, the $\langle x\, x^* \rangle$ correlator, we encountered no problems in proceeding to the continuum limit.

---

[2]We thank Benjamin Basso and Pedro Vieira for proving a `Mathematica` script solving for the mass of the x excitation based on [37].

The comparison for both observable under investigation with the integrability predictions (Figs. 7.6 and 7.10) is carried out by matching at a given coupling the corresponding values for the continuum extrapolation of $f'(g)_{\rm LAT}$ and the integrability prediction $f'(g_c)_c$. Assuming that a simple finite rescaling relates the lattice bare coupling g and the (bare) continuum one g_c, we have derived the relation $g_c = 0.04g$ and proceeded with the comparison of further data points. Nevertheless, it is possible that further data at smaller values of g might prove that the assumption is not justified. A non-trivial relation between g and g_c would take away any predictivity from the lattice measurements for the (derivative of the) cusp. To proceed, one could then define the continuum (BES) prediction as the point where to study the theory and tune accordingly the lattice bare coupling, i.e. numerically determine such non-trivial interpolating function of the bare couplings. This could then be used as an input for the—this time fully predictive—measurements of other physical observables (like the mass m_x^2 here).

Our simulations have also detected a phase in the fermionic determinant that descends from the linearization of fermionic interactions taken from [39]. The phase undergoes strong fluctuations, signalling a severe sign problem, when we approach the non-perturbative regime at small values of g. Current numerical simulations are now employing the new proposal in Sect. 7.7 and they will furnish a non-trivial benchmark for the new fermionic discretization. Current efforts are also focussing on an implementation of these improvements in a parallel software.

References

1. C. G. Callan, Jr., L. Thorlacius, Sigma models and string theory, in: *"In *Providence 1988, Proceedings, Particles, strings and supernovae*, vol. 2*, pp. 795–878
2. N. Drukker, D.J. Gross, A.A. Tseytlin, Green-Schwarz string in $AdS_5 \times S^5$: Semiclassical partition function. JHEP **0004**, 021 (2000). hep-th/0001204
3. J. Minahan, A. Tirziu, A.A. Tseytlin, $1/J$ corrections to semiclassical AdS/CFT states from quantum Landau-Lifshitz model. Nucl. Phys. B **735**, 127 (2006). hep-th/0509071
4. S. Frolov, A.A. Tseytlin, Semiclassical quantization of rotating superstring in $AdS_5 \times S^5$. JHEP **0206**, 007 (2002). hep-th/0204226
5. M. Beccaria, G.V. Dunne, V. Forini, M. Pawellek, A.A. Tseytlin, Exact computation of one-loop correction to energy of spinning folded string in $AdS_5 \times S^5$. J. Phys. A **43**, 165402 (2010). arXiv:1001.4018
6. M. Kruczenski, A. Tirziu, Matching the circular Wilson loop with dual open string solution at 1-loop in strong coupling. JHEP **0805**, 064 (2008). arXiv:0803.0315
7. V. Pestun, Localization of gauge theory on a four-sphere and supersymmetric Wilson loops. Commun. Math. Phys. **313**, 71 (2012). arXiv:0712.2824
8. A. Faraggi, L. A. Pando Zayas, G. A. Silva, D. Trancanelli, Toward precision holography with supersymmetric Wilson loops. JHEP **1604**, 053 (2016), arxiv:1601.04708
9. D. Uvarov, $AdS_4 \times \mathbb{CP}^3$ superstring in the light-cone gauge. Nucl. Phys. B **826**, 294 (2010). arXiv:0906.4699
10. D. Uvarov, Light-cone gauge Hamiltonian for $AdS_4 \times \mathbb{CP}^3$ superstring. Mod. Phys. Lett. A **25**, 1251 (2010). arXiv:0912.1044
11. S. Giombi, R. Ricci, R. Roiban, A. Tseytlin, C. Vergu, Quantum $AdS_5 \times S^5$ superstring in the AdS light-cone gauge. JHEP **1003**, 003 (2010). arXiv:0912.5105

12. T. McLoughlin, R. Roiban, A.A. Tseytlin, Quantum spinning strings in $AdS_4 \times \mathbb{CP}^3$: Testing the Bethe Ansatz proposal. JHEP **0811**, 069 (2008). arXiv:0809.4038
13. M.C. Abbott, I. Aniceto, D. Bombardelli, Quantum strings and the AdS_4/CFT_3 interpolating function. JHEP **1012**, 040 (2010). arXiv:1006.2174
14. C. Lopez-Arcos, H. Nastase, Eliminating ambiguities for quantum corrections to strings moving in $AdS_4 \times \mathbb{CP}^3$. Int. J. Mod. Phys. A **28**, 1350058 (2013). arXiv:1203.4777
15. N. Gromov, P. Vieira, The all loop AdS_4/CFT_3 Bethe ansatz. JHEP **0901**, 016 (2009). arXiv:0807.0777
16. O. Bergman, S. Hirano, Anomalous radius shift in AdS_4/CFT_3. JHEP **0907**, 016 (2009). arXiv:0902.1743
17. N. Gromov, G. Sizov, Exact slope and interpolating functions in $\mathcal{N} = 6$ supersymmetric Chern-Simons theory. Phys. Rev. Lett. **113**, 121601 (2014). arXiv:1403.1894
18. L. Bianchim, M. S. Bianchi, Quantum dispersion relations for the $AdS_4 \times \mathbb{CP}^3$ GKP string. JHEP **1511**, 031 (2015), arXiv:1505.00783
19. I. Bena, J. Polchinski, R. Roiban, Hidden symmetries of the $AdS_5 \times S^5$ superstring. Phys. Rev. D **69**, 046002 (2004). hep-th/0305116
20. N. Drukker, V. Forini, Generalized quark-antiquark potential at weak and strong coupling. JHEP **1106**, 131 (2011). arxiv:1105.5144
21. N. Beisert, L. Freyhult, Fluctuations and energy shifts in the Bethe ansatz. Phys. Lett. B **622**, 343 (2005). hep-th/0506243
22. N. Gromov, Integrability in AdS/CFT correspondence: Quasi-classical analysis. J. Phys. A **42**, 254004 (2009)
23. B. Vicedo, Semiclassical quantisation of finite-gap strings. JHEP **0806**, 086 (2008). arXiv:0803.1605
24. R. Ishizeki, M. Kruczenski, S. Ziama, Notes on Euclidean Wilson loops and Riemann Theta functions. Phys. Rev. D **85**, 106004 (2012). arXiv:1104.3567
25. V. Forini, A. A. Tseytlin, E. Vescovi, Perturbative computation of string one-loop corrections to Wilson loop minimal surfaces in $AdS_5 \times S^5$". JHEP **1703**, 003 (2017), arXiv:1702.02164
26. R. Camporesi, Harmonic analysis and propagators on homogeneous spaces. Phys. Rept. **196**, 1 (1990)
27. R. Camporesi, The Spinor heat kernel in maximally symmetric spaces. Commun. Math. Phys. **148**, 283 (1992)
28. R. Camporesi, A. Higuchi, Spectral functions and zeta functions in hyperbolic spaces. J. Math. Phys. **35**, 4217 (1994)
29. R. Camporesi, A. Higuchi, On the eigenfunctions of the Dirac operator on spheres and real hyperbolic spaces. J. Geom. Phys. **20**, 1 (1996), gr-qc/9505009
30. N. Gromov, A. Sever, Analytic solution of Bremsstrahlung TBA. JHEP **1211**, 075 (2012). arXiv:1207.5489
31. N. Gromov, F. Levkovich-Maslyuk, Quantum spectral curve for a cusped Wilson line in $\mathcal{N} = 4$ SYM. JHEP **1604**, 134 (2016), arXiv:1510.02098
32. J. Aguilera-Damia, A. Faraggi, L. A. Pando Zayas, V. Rathee, G. A. Silva, D. Trancanelli, E. Vescovi, in preparation
33. N. Drukker, S. Giombi, R. Ricci, D. Trancanelli, Supersymmetric Wilson loops on S^3. JHEP **0805**, 017 (2008). arXiv:0711.3226
34. S. Giombi, V. Pestun, Correlators of local operators and 1/8 BPS Wilson loops on S^2 from 2d YM and matrix models. JHEP **1010**, 033 (2010). arXiv:0906.1572
35. S. Giombi, R. Ricci, R. Roiban, A. Tseytlin, Quantum dispersion relations for excitations of long folded spinning superstring in $AdS_5 \times S^5$. JHEP **1101**, 128 (2011). arXiv:1011.2755
36. B. Basso, Exciting the GKP string at any coupling. Nucl. Phys. B **857**, 254 (2012). arXiv:1010.5237
37. B. Basso, A. Sever, P. Vieira, Space-time S-matrix and flux tube S-matrix II. Extracting and matching data. JHEP **1401**, 008 (2014). arXiv:1306.2058
38. L.F. Alday, J.M. Maldacena, Comments on operators with large spin. JHEP **0711**, 019 (2007). arXiv:0708.0672
39. R. W. McKeown, R. Roiban, The quantum $AdS_5 \times S^5$ superstring at finite coupling, arXiv:1308.4875

Appendix A
Jacobi Elliptic Functions

In the course of proving the results of Chap. 4 and Appendix C we need the following relations for the Jacobi elliptic functions, mainly taken from Appendices A and C of [1]. Two more complete resources are [2–4]. We adopt the standard notation of [5], which is the same of the `Mathematica` built-in functions.

The *incomplete elliptic integrals* of the first, second and third kind are respectively defined through

$$
\begin{aligned}
F(z|k^2) &\equiv \int_0^z \frac{d\theta}{\sqrt{1-k^2\sin^2\theta}} = \int_0^{\sin z} \frac{dt}{\sqrt{(1-t^2)(1-k^2t^2)}}\,,\\
E(z|k^2) &\equiv \int_0^z d\theta\sqrt{1-k^2\sin^2\theta} = \int_0^{\sin z} dt\sqrt{\frac{1-k^2t^2}{1-t^2}}\,,\\
\Pi(\ell^2;z|k^2) &\equiv \int_0^z \frac{d\theta}{(1-\ell^2\sin^2\theta)\sqrt{1-k^2\sin^2\theta}} = \int_0^{\sin z} \frac{dt}{(1-\ell^2t^2)\sqrt{(1-t^2)(1-k^2t^2)}}\,,
\end{aligned}
\tag{A.1}
$$

where k^2 is the *elliptic modulus* and ℓ^2 is the *characteristic*. The modulus can take any complex values, but it shall be restricted to the interval $0 \leq k < 1$ for the applications in this thesis.

We obtain the *complete elliptic integrals* by setting $z = \frac{\pi}{2}$

$$
\begin{aligned}
&\mathbb{K} \equiv \mathbb{K}(\urcorner^2) \equiv F(k^2) = F(\tfrac{\pi}{2}|k^2)\,, && \mathbb{K}' \equiv \mathbb{K}'(k^2) \equiv \mathbb{K}(k')\,,\\
&\mathbb{E}(k^2) \equiv E(\tfrac{\pi}{2}|k^2)\,, && \Pi(\ell^2|k^2) \equiv \Pi(\ell^2;\tfrac{\pi}{2}|k^2)\,,
\end{aligned}
\tag{A.2}
$$

with $k' \equiv \sqrt{1-k^2}$ being the *complementary modulus*.

Defining the *Jacobi amplitude* by

$$
\vartheta \equiv \mathrm{am}(u|k^2) \qquad \text{with} \qquad u = F(\vartheta|k^2)\,,
\tag{A.3}
$$

the *sine, cosine* and *delta amplitudes* are given by

$$
\mathrm{sn}(u|k^2) = \sin\vartheta\,, \qquad \mathrm{cn}(u|k^2) = \cos\vartheta\,, \qquad \mathrm{dn}(u|k^2) = \sqrt{1-k^2\sin^2\vartheta}\,.
\tag{A.4}
$$

E. Vescovi, *Perturbative and Non-perturbative Approaches to String Sigma-Models in AdS/CFT*, Springer Theses, DOI 10.1007/978-3-319-63420-3

They are doubly periodic functions of the complex variable u. Like the trigonometric functions, they have a real-valued period ($2\mathbb{K}$ for dn and $4\mathbb{K}$ for sn, cn) and, like the hyperbolic ones, a purely-imaginary period ($2i\mathbb{K}'$ for sn and $4i\mathbb{K}'$ for cn, dn). By convention, $\mathbb{K}$ is called *real quarter period* and $i\mathbb{K}'$ *imaginary quarter period*. The *fundamental parallelogram* of the elliptic functions is a rectangle on the complex plane with corners at $(0,\ 4\mathbb{K},\ 4i\mathbb{K}',\ 4\mathbb{K}+4i\mathbb{K}')$.

One can also define nine minor Jacobi elliptic functions built from the three above. The *quotients* and *reciprocals* are designated in *Glaisher's notation* by

$$\begin{aligned}
&\operatorname{ns} u \equiv \frac{1}{\operatorname{sn} u}\,, && \operatorname{tn} u \equiv \operatorname{sc} u \equiv \frac{\operatorname{sn} u}{\operatorname{cn} u}\,, && \operatorname{sd} u \equiv \frac{\operatorname{sn} u}{\operatorname{dn} u}\,, \\
&\operatorname{nc} u \equiv \frac{1}{\operatorname{cn} u}\,, && \frac{1}{\operatorname{tn} u} \equiv \operatorname{cs} u \equiv \frac{\operatorname{cn} u}{\operatorname{sn} u}\,, && \operatorname{cd} u \equiv \frac{\operatorname{cn} u}{\operatorname{dn} u}\,, \\
&\operatorname{nd} u \equiv \frac{1}{\operatorname{dn} u}\,, && \operatorname{ds} u \equiv \frac{\operatorname{dn} u}{\operatorname{sn} u}\,, && \operatorname{dc} u \equiv \frac{\operatorname{dn} u}{\operatorname{cn} u}\,,
\end{aligned} \tag{A.5}$$

where we suppressed the modulus k^2 in order not to clutter formulas.[1]

The functions satisfy the algebraic relations

$$\begin{aligned}
-\operatorname{dn}^2(u|k^2) + k'^2 &= -k^2 \operatorname{cn}^2(u|k^2) = k^2 \operatorname{sn}^2(u|k^2) - k^2\,, \\
-k'^2 \operatorname{nd}(u|k^2) + k'^2 &= -k^2 k'^2 \operatorname{sd}^2(u|k^2) = k^2 \operatorname{cd}(u|k^2) - k^2\,.
\end{aligned} \tag{A.6}$$

The inverse functions of (A.4) possess an integral representation in terms of elliptic integrals

$$\operatorname{sn}^{-1}(u|k^2) = \int_u^1 \frac{dt}{\sqrt{(1-t^2)(1-k^2t^2)}} = F(\arcsin\, u|k^2)\,, \tag{A.7}$$

$$\operatorname{cn}^{-1}(u|k^2) = \int_u^1 \frac{dt}{\sqrt{(1-t^2)(k'^2+k^2t^2)}} = F(\arcsin\sqrt{1-u^2}|k^2)\,, \tag{A.8}$$

$$\operatorname{dn}^{-1}(u|k^2) = \int_u^1 \frac{dt}{\sqrt{(1-t^2)(t^2-k'^2)}} = F(\arcsin\sqrt{\frac{u^2}{1-u^2}}|k^2) \tag{A.9}$$

and similar expressions for (A.5). The derivatives of the three basic Jacobi functions are

$$\frac{d}{dz}\operatorname{sn}(u|k^2) = \operatorname{cn}(u|k^2)\operatorname{dn}(u|k^2)\,, \tag{A.10}$$

$$\frac{d}{dz}\operatorname{cn}(u|k^2) = -\operatorname{sn}(u|k^2)\operatorname{dn}(u|k^2)\,, \tag{A.11}$$

$$\frac{d}{dz}\operatorname{dn}(u|k^2) = -k^2\operatorname{sn}(u|k^2)\operatorname{cn}(u|k^2)\,,. \tag{A.12}$$

[1] We drop the dependence on k^2 and k'^2 also in Chap. 4.

The Jacobi H, Θ and Z functions are given by the Jacobi θ functions with *nome* $q = q(k^2) = \exp(-\pi\frac{\mathbb{K}'}{\mathbb{K}})^2$

$$H(u|k^2) = \vartheta_1\left(\frac{\pi u}{2\,\mathbb{K}}; q\right), \quad \Theta(u|k^2) = \theta_4\left(\frac{\pi u}{2\,\mathbb{K}}; q\right), \quad Z(u|k^2) = \frac{\pi}{2\,\mathbb{K}}\,\frac{\theta_4'(\frac{\pi u}{2\,\mathbb{K}}; q)}{\theta_4(\frac{\pi u}{2\,\mathbb{K}}; q)}. \tag{A.13}$$

Useful integral representations are

$$Z(\mathrm{sn}^{-1}(u|k^2)|k^2) = \int_0^u \mathrm{d}t\left[\sqrt{\frac{1-k^2t^2}{1-t^2}} - \frac{\mathbb{E}(k^2)}{\mathbb{K}(k^2)}\frac{1}{\sqrt{(1-t^2)(1-k^2t^2)}}\right], \tag{A.14}$$

$$\mathbb{Z}(u|k^2) = \int_0^u dv\,\mathrm{dn}^2(v|k^2) - \frac{\mathbb{E}(k^2)}{\mathbb{K}(k^2)}\,u\,, \qquad 0 < u < \mathbb{K}\,. \tag{A.15}$$

Bibliography

1. M. Beccaria, G.V. Dunne, V. Forini, M. Pawellek, A.A. Tseytlin, Exact computation of one-loop correction to energy of spinning folded string in $AdS_5 \times S^5$, J. Phys. A **43**, 165402 (2010), arXiv:1001.4018.
2. G.N. Watson, E.T. Whittaker, in *A Course of Modern Analysis* (Cambridge University Press, 1902).
3. P. Byrd, M. Friedman, Handbook of elliptic integrals for engineers and scientists (Springer, 1971).
4. I. Gradshteyn, I. Ryzhik, in *Table of Integrals, Series, and Products* (Academic Press, 1980).
5. M. Abramowitz, I.A. Stegun, in *Handbook of Mathematical Functions with Formulas, Graphs, and Mathematical Tables* (Courier Corporation, 1970).

[2] See [5] for further discussion.

Appendix B
Methods for Functional Determinants in One Dimension

The theory of functional determinants of partial differential operators is an important subject in many areas of mathematical physics.[3] In this thesis we shall concentrate on the problem of computing one-loop corrections to string effective actions around minimal-area surfaces, which is a task requiring the knowledge of the determinant of highly non-trivial matrix operators in two variables. This appendix continues the discussion started in Sect. 1.4.1 for fluctuation operators that are translationally invariant in one worldsheet direction.

B.1 Gel'fand-Yaglom Method

The method originally developed by Gel'fand and Yaglom [2] provides a simple way to evaluate *ratios* of functional determinants of ordinary differential operators, defined on compact intervals under a large class boundary conditions. Although equivalent to zeta-function regularization, this algorithm bypasses the need of knowing the full sequence of the eigenvalues. In particular, no previous information about the spectrum, e.g. bound or continuum states, is needed. The complexity of the spectral problem is reduced to the solution of an initial value problem, which can be then analytically or numerically solvable.[4]

The situation we typically encounter is finding a determinant *relative* to a fiducial one

$$\frac{\operatorname{Det}\mathcal{O}_1}{\operatorname{Det}\mathcal{O}_2}\,. \tag{B.1}$$

[3]The reader can consult [1] and references therein to find a report on recent progress.

[4]The solutions of the 1d spectral problems in this thesis (4.45), (4.63), (5.45)–(5.48), (5.65) and (5.68)–(5.70) are all known in closed form.

E. Vescovi, *Perturbative and Non-perturbative Approaches to String Sigma-Models in AdS/CFT*, Springer Theses, DOI 10.1007/978-3-319-63420-3

The linear differential operators $\mathcal{O}_1$, $\mathcal{O}_2$ are either of Dirac-type for spinor fields

$$\mathcal{O}_1 = P_0(x)\frac{d}{dx} + P_1(x), \qquad \mathcal{O}_2 = P_0(x)\frac{d}{dx} + Q_1(x), \tag{B.2}$$

or of Laplace-type for scalar fields

$$\mathcal{O}_1 = P_0(x)\frac{d^2}{dx^2} + P_1(x)\frac{d}{dx} + P_2(x), \qquad \mathcal{O}_2 = P_0(x)\frac{d^2}{dx^2} + Q_1(x)\frac{d}{dx} + Q_2(x). \tag{B.3}$$

The coefficients above are complex matrices, continuous functions of x on the finite interval $I = [a, b]$. The two operators have the same principal symbol $P_0(x)$. This assumption ensures that the behaviour of the large eigenvalues is comparable, thus the ratio is well-defined despite the fact each determinant is the product of infinitely-many eigenvalues of increasing magnitude. More formally, it guarantees the cancellation of the leading Seeley-deWitt coefficients, responsible for the divergences of the individual determinants [3–5], in the difference of the zeta-functions of $\mathcal{O}_1$ and $\mathcal{O}_2$.

In Appendix B.2 we deal with a class of spectral problems free of zero modes (i.e. vanishing eigenvalues) using some results built on the Gel'fand-Yaglom theorem. We closely follow the technology developed by Forman [6,7], who gave a prescription to work with even more general elliptic boundary value problems. We derive a compact formula for the square of first-order operators in Appendix B.3. Appendices B.4 and B.5 conclude the discussion with some corollaries.

B.2 *n*th-Order Operators

Let us consider the pair of n-order ordinary differential operators

$$\mathcal{O}_1 = P_0(x)\frac{d^n}{dx^n} + \sum_{k=0}^{n-1} P_{n-k}(x)\frac{d^k}{dx^k}, \qquad \mathcal{O}_2 = P_0(x)\frac{d^n}{dx^n} + \sum_{k=0}^{n-1} Q_{n-k}(x)\frac{d^k}{dx^k} \tag{B.4}$$

with coefficients being $r \times r$ complex matrices. The principal symbols of the two operators (proportional to the coefficient $P_0(x)$ of the highest-order derivative) are equal and invertible ($\det P_0(x) \neq 0$) on the compact interval $I = [a, b]$. We do not impose further conditions on the matrix coefficients, besides the requirement of being continuous functions on I. The operators act on the space of square-integrable r-component functions

$$\bar{f} \equiv (f_1, f_2, ..., f_r)^T \in \mathcal{L}^2(I), \tag{B.5}$$

where for our purposes one defines the Hilbert inner product ($*$ stands for complex conjugation)

$$(\bar{f}, \bar{g}) \equiv \int_a^b \sqrt{h(x)}\, dx \sum_{i=1}^{r} f_i^*(x)\, g_i(x). \tag{B.6}$$

The inclusion of the measure factor, given by the invariant volume element $\sqrt{h}$, guarantees that the worldsheet operators are self-adjoint when supplemented with appropriate boundary conditions. One has to specify the two $nr \times nr$ constant matrices M, N to implement the linear boundary conditions at the endpoints of I

$$M \begin{pmatrix} \bar{f}(a) \\ \frac{d}{dx}\bar{f}(a) \\ \vdots \\ \frac{d^{n-1}}{dx^{n-1}}\bar{f}(a) \end{pmatrix} + N \begin{pmatrix} \bar{f}(b) \\ \frac{d}{dx}\bar{f}(b) \\ \vdots \\ \frac{d^{n-1}}{dx^{n-1}}\bar{f}(b) \end{pmatrix} = \begin{pmatrix} 0 \\ 0 \\ \vdots \\ 0 \end{pmatrix}. \tag{B.7}$$

We restrict ourselves to operators without zero modes on the functions satisfying (B.7). The proof uses a particular integral representation of the zeta-function to eventually arrive to the elegant formula for the ratio (B.1)

$$\frac{\mathrm{Det}\,\mathcal{O}_1}{\mathrm{Det}\,\mathcal{O}_2} = \frac{\exp\left\{\int_a^b \mathrm{tr}\left[\mathcal{R}(x) P_1(x) P_0^{-1}(x)\right] dx\right\} \det\left[M + N Y_{\mathcal{O}_1}(b)\right]}{\exp\left\{\int_a^b \mathrm{tr}\left[\mathcal{R}(x) Q_1(x) P_0^{-1}(x)\right] dx\right\} \det\left[M + N Y_{\mathcal{O}_2}(b)\right]}. \tag{B.8}$$

The matrix $\mathcal{R}$ is defined below. This result agrees with the one obtained via zeta-function regularization for elliptic differential operators, where the divergent part of the individual determinants cancels in the ratio. Notice that M, N in (B.7) are uniquely fixed up to a constant rescaling, which leaves the ratio unaffected. When $P_1(x) = Q_1(x)$, the exponential factors cancel out. The $nr \times nr$ matrix

$$Y_{\mathcal{O}_1}(x) = \begin{pmatrix} \bar{f}_{(I)}(x) & \bar{f}_{(II)}(x) & \dots & \bar{f}_{(nr)}(x) \\ \frac{d}{dx}\bar{f}_{(I)}(x) & \frac{d}{dx}\bar{f}_{(II)}(x) & \dots & \frac{d}{dx}\bar{f}_{(nr)}(x) \\ \vdots & \vdots & \ddots & \vdots \\ \frac{d^{n-1}}{dx^{n-1}}\bar{f}_{(I)}(x) & \frac{d^{n-1}}{dx^{n-1}}\bar{f}_{(II)}(x) & \dots & \frac{d^{n-1}}{dx^{n-1}}\bar{f}_{(nr)}(x) \end{pmatrix} \tag{B.9}$$

accommodates all the independent homogeneous solutions of[5]

$$\mathcal{O}_1\, \bar{f}_{(i)}(x) = 0 \qquad i = I, II, ..., 2r \tag{B.10}$$

[5]The solutions of this *auxiliary* problem are *not* zero modes of the operator with the boundary condition (B.7).

normalized such that $Y_{\mathcal{O}_1}(a) = \mathbb{I}_{nr}$. The matrix $Y_{\mathcal{O}_2}(x)$ is similarly defined with respect to $\mathcal{O}_2$.

If we focus on even-order differential operators, then $\mathcal{R}(x) = \frac{1}{2}\mathbb{I}_{nr}$ is proportional to the identity matrix and (B.8) becomes

$$\frac{\mathrm{Det}\,\mathcal{O}_1}{\mathrm{Det}\,\mathcal{O}_2} = \frac{\exp\left\{\frac{1}{2}\int_a^b \mathrm{tr}\left[P_1(x)P_0^{-1}(x)\right]dx\right\}\det\left[M + NY_{\mathcal{O}_1}(b)\right]}{\exp\left\{\frac{1}{2}\int_a^b \mathrm{tr}\left[Q_1(x)P_0^{-1}(x)\right]dx\right\}\det\left[M + NY_{\mathcal{O}_2}(b)\right]}. \tag{B.11}$$

For odd n one gets a slightly more complicated structure, constructed as follows. Let us assume that the spectrum of the principal symbols, i.e. the matrix $(-i)^n\,P_0(x)$, has no intersection with the cone

$$C \equiv \{z \in \mathbb{C} \,:\, \theta_1 < \arg(z) < \theta_2\} \tag{B.12}$$

for some choice of θ_1, θ_2. This is to say that $\mathcal{O}_1$ and $\mathcal{O}_2$ have principal angle between θ_1 and θ_2. It also follows that no eigenvalue falls in the opposite cone

$$-C \equiv \{z \in \mathbb{C} \,:\; \theta_1 + \pi < \arg(z) < \theta_2 + \pi\} \tag{B.13}$$

for odd n. Consequently, the eigenvalues fall under two sets, depending on which sector of $\mathbb{C} \setminus (C \cup -C)$ they belong to. The matrix $\mathcal{R}(x)$ is then defined[6] as the projector onto the subspace spanned by the eigenvectors associated to all eigenvalues in one of these two subsets of the complex plane.

We stress again that the powerful, yet simple, result (B.8) is possible only for the ratio. We do not known of any simple formula when the two operators are defined on different intervals or the boundary conditions do not fall under (B.7). Another severe limitation is the equality of the principal symbols. For instance, this prevents to compare two scalar Laplacians defined on different 1d manifolds, as the coefficient of the highest derivative would depend on the metric of the space.

The results above generalise in several ways. It is possible to use a similar formalism to deal with *single* determinants of second-order operators with periodic coefficients on S^1 [8, 9] and on a line [10], as well as with operators of any order and more general boundary conditions [11]. We also emphasise that a number of physically interesting systems displays zero modes and one has to find a way to single them out from the spectrum and treat them separately [12, 13]. A regularization procedure via contour integration methods is discussed in detail in [14–18].

[6] Up to a factor $\frac{1}{n}$ according to the amendment in [7].

B.3 Square of First-Order Operators

As a corollary of the Forman's construction, we can easily compute the ratio of determinants of the square of first-order operators with reference only to the operators themselves. Consider the matrix operator of the form (B.2)[7]

$$\mathcal{O}_1 = P_0(x)\frac{d}{dx} + P_1(x) \qquad \mathcal{O}_2 = P_0(x)\frac{d}{dx} + Q_1(x) \tag{B.14}$$

and denote by $Y_{\mathcal{O}_1}(x)$ its fundamental matrix that solves the equation (here $'$ is the derivative with respect to x)

$$P_0(x)Y'_{\mathcal{O}_1}(x) + P_1(x)Y_{\mathcal{O}_1}(x) = 0, \qquad Y_{\mathcal{O}_1}(a) = \mathbb{I}_r. \tag{B.15}$$

The matrix of fundamental solutions of the square of this operator

$$\mathcal{O}_1^2 = P_0^2(x)\frac{d^2}{dx^2} + \left[P_0(x)P_0'(x) + \{P_0(x), P_1(x)\}\right]\frac{d}{dx} + P_1^2(x) + P_0(x)P_1'(x) \tag{B.16}$$

can be constructed via the method of reduction of order as

$$Y_{\mathcal{O}_1^2}(x) = \begin{pmatrix} Y_{\mathcal{O}_1}(x) - Z_{\mathcal{O}_1}(x)Y'_{\mathcal{O}_1}(a) & Z_{\mathcal{O}_1}(x) \\ Y'_{\mathcal{O}_1}(x) - Z'_{\mathcal{O}_1}(x)Y'_{\mathcal{O}_1}(a) & Z'_{\mathcal{O}_1}(x) \end{pmatrix}, \qquad Y_{\mathcal{O}_1^2}(a) = \mathbb{I}_{2r}, \tag{B.17}$$

in which

$$Z_{\mathcal{O}_1}(x) = Y_{\mathcal{O}_1}(x)\int_a^x ds\left[Y_{\mathcal{O}_1}^{-1}(s)P_0^{-1}(s)Y_{\mathcal{O}_1}(s)\right]P_0(a) \qquad Z_{\mathcal{O}_1}(a) = 0 \qquad Z'_{\mathcal{O}_1}(a) = \mathbb{I}_r. \tag{B.18}$$

encapsulates the solutions of $\mathcal{O}_1\bar{f} = 0$ and two more ones of $\mathcal{O}_1^2\bar{f} = 0$.

Suppose that the spectral problem of the squared operator is determined by the boundary condition

$$M_{\mathcal{O}^2}\bar{f}(a) + N_{\mathcal{O}^2}\bar{f}(b) = 0. \tag{B.19}$$

After some algebra, successive applications of (B.11), (B.17), (B.18) bring

$$\mathrm{Det}\,\mathcal{O}^2 = \sqrt{\frac{\det P_0(b)}{\det P_0(a)}\frac{\det\left[M_{\mathcal{O}^2} + N_{\mathcal{O}^2}Y_{\mathcal{O}_1^2}(b)\right]}{\det Y_{\mathcal{O}_1}(b)}}. \tag{B.20}$$

---

[7] We do not report the analogue formulas for $\mathcal{O}_2$.

For Dirichlet boundary conditions at the endpoints $x = a,\ b$

$$f_1(a) = f_2(a) = f_1(b) = f_2(b) = 0\,, \qquad \text{(B.21)}$$
$$M_{\mathcal{O}^2} = \begin{pmatrix} \mathbb{I}_r & 0 \\ 0 & 0 \end{pmatrix}, \qquad N_{\mathcal{O}^2} = \begin{pmatrix} 0 & 0 \\ \mathbb{I}_r & 0 \end{pmatrix},$$

then (B.20) gives

$$\left(\text{Det}\,\mathcal{O}_1^2\right)_{\text{Dirichlet b.c.}} = \sqrt{\det P_0(a)\det P_0(b)}\ \det\left[\int_a^b ds Y_{\mathcal{O}_1}^{-1}(x)\, P_0^{-1}(x)\, Y_{\mathcal{O}_1}(x)\right] \qquad \text{(B.22)}$$

which is understood to hold when the convenient normalisation with the fiducial $\mathcal{O}_2$ is taken, see (B.25).

B.4 Corollaries for Second-Order Operators

- Scalar-valued differential operators,
  Dirichlet boundary conditions $f_1(a) = f_1(b) = 0$.

$$M = \begin{pmatrix} 1 & 0 \\ 0 & 0 \end{pmatrix}, \qquad N = \begin{pmatrix} 0 & 0 \\ 1 & 0 \end{pmatrix}, \qquad \frac{\text{Det}\left[\frac{d^2}{dx^2} + P_2(x)\right]}{\text{Det}\left[\frac{d^2}{dx^2} + Q_2(x)\right]} = \frac{f_{(II)1}(b)}{g_{(II)1}(b)}\,. \qquad \text{(B.23)}$$

  The normalization of the matrix (B.9) tells that the functions $f_{(II)1}(x)$ and $g_{(II)1}(x)$ solve the initial value problems

$$f''_{(II)1}(x) + P_2(x) f_{(II)1}(x) = 0\,, \qquad f_{(II)1}(a) = 0\,, \qquad f'_{(II)1}(a) = 1\,,$$
$$\text{(B.24)}$$
$$g''_{(II)1}(x) + Q_2(x) g_{(II)1}(x) = 0\,, \qquad g_{(II)1}(a) = 0\,, \qquad g'_{(II)1}(a) = 1\,.$$

- 2×2 matrix-valued differential operators,
  Dirichlet boundary conditions $f_1(a) = f_2(a) = f_1(b) = f_2(b) = 0$.
  This is a corollary of (B.22).

$$M = \begin{pmatrix} 1 & 0 & 0 & 0 \\ 0 & 1 & 0 & 0 \\ 0 & 0 & 0 & 0 \\ 0 & 0 & 0 & 0 \end{pmatrix}, \qquad N = \begin{pmatrix} 0 & 0 & 0 & 0 \\ 0 & 0 & 0 & 0 \\ 1 & 0 & 0 & 0 \\ 0 & 1 & 0 & 0 \end{pmatrix},$$

$$Y_{P_0\partial_x+P_2}(x) = \begin{pmatrix} f_{(I)1}(x) & f_{(II)1}(x) \\ f_{(I)2}(x) & f_{(II)2}(x) \end{pmatrix}, \qquad Y_{P_0\partial_x+Q_2}(x) = \begin{pmatrix} g_{(I)1}(x) & g_{(II)1}(x) \\ g_{(I)2}(x) & g_{(II)2}(x) \end{pmatrix}, \tag{B.25}$$

$$\frac{\mathrm{Det}\left[P_0(x)\frac{d}{dx}+P_1(x)\right]^2}{\mathrm{Det}\left[P_0(x)\frac{d}{dx}+Q_1(x)\right]^2} = \frac{\int_a^b ds Y^{-1}_{P_0\partial_x+P_2}(s)P_0^{-1}(s)Y_{P_0\partial_x+P_2}(s)}{\int_a^b ds Y^{-1}_{P_0\partial_x+Q_2}(s)P_0^{-1}(s)Y_{P_0\partial_x+Q_2}(s)}$$

with

$$\begin{gathered}
P_0(x)\begin{pmatrix} f'_{(I)1}(x) \\ f'_{(I)2}(x) \end{pmatrix} + P_1(x)\begin{pmatrix} f_{(I)1}(x) \\ f_{(I)2}(x) \end{pmatrix} = \begin{pmatrix} 0 \\ 0 \end{pmatrix}, \\
f_{(I)1}(a) = 1, \qquad f_{(I)2}(a) = 0, \\
P_0(x)\begin{pmatrix} f'_{(II)1}(x) \\ f'_{(II)2}(x) \end{pmatrix} + P_1(x)\begin{pmatrix} f_{(II)1}(x) \\ f_{(II)2}(x) \end{pmatrix} = \begin{pmatrix} 0 \\ 0 \end{pmatrix}, \\
f_{(II)1}(a) = 0, \qquad f_{(II)2}(a) = 1, \\
P_0(x)\begin{pmatrix} g'_{(I)1}(x) \\ g'_{(I)2}(x) \end{pmatrix} + Q_1(x)\begin{pmatrix} g_{(I)1}(x) \\ g_{(I)2}(x) \end{pmatrix} = \begin{pmatrix} 0 \\ 0 \end{pmatrix}, \\
g_{(I)1}(a) = 1, \qquad g_{(I)2}(a) = 0, \\
P_0(x)\begin{pmatrix} g'_{(II)1}(x) \\ g'_{(II)2}(x) \end{pmatrix} + Q_1(x)\begin{pmatrix} g_{(II)1}(x) \\ g_{(II)2}(x) \end{pmatrix} = \begin{pmatrix} 0 \\ 0 \end{pmatrix}, \\
g_{(II)1}(a) = 0, \qquad g_{(II)2}(a) = 1.
\end{gathered} \tag{B.26}$$

It is straightforward to apply the first block of formulas (B.23)–(B.24) to (5.40)–(5.42) and the second one (B.25)–(B.26) to (5.53), provided that $x \in [a, b]$ is replaced by $\sigma \in [\epsilon_0, R]$ and the one-dimensional Det is understood as the Det_ω at fixed frequency. Since we square the fermionic operator (5.53), we do not make use of formulas for first-order operators in this thesis. Note that the Weyl rescaling of the operators by $\sqrt{h}$ done in Sects. 5.5.1 and 5.5.2 removes the measure from (B.6) and also the θ_0-dependence in the principal symbol of each operator, effectively making possible to normalize latitude and circle operators in Sect. (5.6).

B.5 Corollary for Fourth-Order Operators

- Scalar-valued differential operators,
  boundary conditions $\rho\, f_1\,(a) = f_1\,(b) = 0,\ \rho \in \mathbb{C}$.

$$M = -\rho\,\mathbb{I}_4\,, \qquad N = \mathbb{I}_4\,, \qquad \frac{\mathrm{Det}\left[\frac{d^4}{dx^4}+P_2(x)\frac{d^2}{dx^2}+P_3(x)\frac{d}{dx}+P_4(x)\right]}{\mathrm{Det}\left[\frac{d^4}{dx^4}+Q_2(x)\frac{d^2}{dx^2}+Q_3(x)\frac{d}{dx}+Q_4(x)\right]} =$$

$$= \frac{\det\begin{pmatrix} f_{(I)1}(x)-\rho & f_{(II)1}(x) & f_{(III)1}(x) & f_{(IV)1}(x) \\ f'_{(I)1}(x) & f'_{(II)1}(x)-\rho & f'_{(III)1}(x) & f'_{(IV)1}(x) \\ f''_{(I)1}(x) & f''_{(II)1}(x) & f''_{(III)1}(x)-\rho & f''_{(IV)1}(x) \\ f'''_{(I)1}(x) & f'''_{(II)1}(x) & f'''_{(III)1}(x) & f'''_{(IV)1}(x)-\rho \end{pmatrix}}{\det\begin{pmatrix} g_{(I)1}(x)-\rho & g_{(II)1}(x) & g_{(III)1}(x) & g_{(IV)1}(x) \\ g'_{(I)1}(x) & g'_{(II)1}(x)-\rho & g'_{(III)1}(x) & g'_{(IV)1}(x) \\ g''_{(I)1}(x) & g''_{(II)1}(x) & g''_{(III)1}(x)-\rho & g''_{(IV)1}(x) \\ g'''_{(I)1}(x) & g'''_{(II)1}(x) & g'''_{(III)1}(x) & g'''_{(IV)1}(x)-\rho \end{pmatrix}} \tag{B.27}$$

with

$$\begin{gathered} f''''_{(I)1} + P_2(x) f''_{(I)1} + P_3(x) f'_{(I)1} + P_4(x) f_{(I)1} = 0\,, \\ f'_{(I)1}\,(a) = f''_{(I)1}\,(a) = f'''_{(I)1}\,(a) = 0\,, \qquad f_{(I)1}\,(a) = 1\,, \\[1em] f''''_{(II)1} + P_2(x) f''_{(II)1} + P_3(x) f'_{(II)1} + P_4(x) f_{(II)1} = 0\,, \\ f_{(II)1}\,(a) = f''_{(II)1}\,(a) = f'''_{(II)1}\,(a) = 0\,, \qquad f'_{(II)1}\,(a) = 1\,, \\[1em] f''''_{(III)1} + P_2(x) f''_{(III)1} + P_3(x) f'_{(III)1} + P_4(x) f_{(III)1} = 0\,, \\ f_{(III)1}\,(a) = f'_{(III)1}\,(a) = f'''_{(III)1}\,(a) = 0\,, \qquad f''_{(III)1}\,(a) = 1\,, \\[1em] f''''_{(IV)1} + P_2(x) f''_{(IV)1} + P_3(x) f'_{(IV)1} + P_4(x) f_{(IV)1} = 0\,, \\ f_{(IV)1}\,(a) = f'_{(IV)1}\,(a) = f''_{(IV)1}\,(a) = 0\,, \qquad f'''_{(IV)1}\,(a) = 1 \end{gathered} \tag{B.28}$$

and similar relations for the reference operator upon replacing $P \to Q$ and $f \to g$.

The case encompasses periodic ($\rho = 1$) and anti-periodic boundary conditions ($\rho = -1$) for differential operators (C.1). The formulas above will find application in (C.14)–(C.16).

Bibliography

1. G.V. Dunne, Functional determinants in quantum field theory J. Phys. A **41**, 304006 (2008), arXiv:0711.1178 (in *Proceedings, 5th International Symposium on Quantum Theory and Symmetries (QTS5)*, p. 304006).
2. I.M. Gelfand, A.M. Yaglom, Integration in functional spaces and it applications in quantum physics. J. Math. Phys. **1**, 48 (1960).

3. P.B. Gilkey, in *Invariance Theory, the Heat Equation and the Atiyah-Singer Index Theorem* (CRC Press, Boca Raton 1995).
4. D.V. Vassilevich, Heat kernel expansion: User's manual. Phys. Rept. **388**, 279 (2003), arXiv:hep-th/0306138.
5. D. Fursaev, D. Vassilevich, in *Operators, Geometry and Quanta: Methods of Spectral Geometry in Quantum Field Theory* (Springer Verlag, 2011).
6. R. Forman, Functional determinants and geometry. Invent. Math. **88**, 447 (1987).
7. R. Forman, Functional determinants and geometry (Erratum). Invent. Math. **108**, 453 (1992).
8. H. Braden, Periodic functional determinants. J. Phys. A **18,** 2127 (1985).
9. D. Burghelea, L. Friedlander, T. Kappeler, On the determinant of elliptic differential and finite difference operators in vector bundles over S^1. Commun. Math. Phys. **150**, 431 (1992).
10. D. Burghelea, L. Friedlander, T. Kappeler, On the determinant of elliptic boundary value problems on a line segment. Proc. Am. Math. Soc. **123**, 3027 (1995).
11. M. Lesch, J. Tolksdorf, On the determinant of one-dimensional elliptic boundary value problems. Commun. Math. Phys. **193**, 643 (1998), arXiv:dg-ga/9707022.
12. R. Rajaraman, in *Solitons and Instantons* (North Holland, Amsterdam, 1982).
13. M. Marino, in *Instantons and Large N* (Cambridge University Press, 2015).
14. A.J. McKane, M.B. Tarlie, Regularization of functional determinants using boundary perturbations. J. Phys. A **28**, 6931 (1995), arXiv:cond-mat/9509126.
15. K. Kirsten, A.J. McKane, Functional determinants by contour integration methods. Ann. Phys. **308**, 502 (2003), arXiv:math-ph/0305010.
16. K. Kirsten, A.J. McKane, Functional determinants for general Sturm-Liouville problems. J. Phys. A **37**, 4649 (2004), arXiv:math-ph/0403050.
17. K. Kirsten, P. Loya, Computation of determinants using contour integrals. Am. J. Phys. 76, 60 (2008), arXiv:0707.3755.
18. G.M. Falco, A.A. Fedorenko, On functional determinants of matrix differential operators with degenerate zero modes, arXiv:1703.07329.

Appendix C
Exact Spectrum for a Class of Fourth-Order Differential Operators

We examine the spectral properties of linear ordinary differential operators of the fourth order, whose coefficients are meromorphic and doubly periodic functions in one complex variable. They can be seen as higher-order generalization of the second-order Lamé equation. The exact solution of these operators has been found in the course of the study of the energy corrections for the rotating strings in Chap. 4.

In Appendix C.1 we write a generalization of the Floquet analysis—which is for homogeneous linear second-order operators with periodic coefficients [1] (also [2])—to the fourth-order operators (C.1) below. The Gel'fand-Yaglom method allows to express their determinants in terms of two *quasi-momenta* functions, which we are able to derive explicitly for a subclass of doubly periodic operators on the complex plane (C.17)–(C.18) in Appendices C.2–C.4. These results are the basis for the one-loop analysis in Chap. 4, which we supplement with some technical details in Appendices C.5–C.6.

C.1 Generalization of the Floquet-Bloch Theory to Periodic Fourth-Order Operators

Let us consider the fourth-order differential operator on the real line $x \in \mathbb{R}$[8]

$$\mathcal{O}^{(4)} = \frac{d^4}{dx^4} + v_1(x)\frac{d^2}{dx^2} + v_2(x)\frac{d}{dx} + v_3(x), \tag{C.1}$$

with coefficients having a *fundamental period* L,

$$v_i(x+L) = v_i(x)\ . \tag{C.2}$$

[8] We encourage the reader to consult in parallel Chap. 1 of [1] (also Sect. 4.1 [3]) for the original discussion of second-order Hill's operators.

E. Vescovi, *Perturbative and Non-perturbative Approaches to String Sigma-Models in AdS/CFT*, Springer Theses, DOI 10.1007/978-3-319-63420-3

The solution of the eigenvalue problem

$$\mathcal{O}^{(4)} f(x,\Lambda) = \Lambda f(x,\Lambda), \qquad f(x+L,\Lambda) = f(x,\Lambda) \tag{C.3}$$

consists of four independent functions $f_{(i)}(x,\Lambda)$ $(i = I, II, III, IV)$, which can be normalized

$$\begin{array}{llll} f_{(I)}(0,\Lambda) = 1, & f'_{(1)}(0,\Lambda) = 0, & f''_{(1)}(0,\Lambda) = 0, & f'''_{(1)}(0,\Lambda) = 0, \\ f_{(II)}(0,\Lambda) = 0, & f'_{(II)}(0,\Lambda) = 1, & f''_{(II)}(0,\Lambda) = 0, & f'''_{(II)}(0,\Lambda) = 0, \\ f_{(III)}(0,\Lambda) = 0, & f'_{(III)}(0,\Lambda) = 0, & f''_{(III)}(0,\Lambda) = 1, & f'''_{(III)}(0,\Lambda) = 0, \\ f_{(IV)}(0,\Lambda) = 0, & f'_{(IV)}(0,\Lambda) = 0, & f''_{(IV)}(0,\Lambda) = 0, & f'''_{(IV)}(0,\Lambda) = 1, \end{array} \tag{C.4}$$

such that the Wronskian determinant is one. The periodicity of the operator implies that also $f_{(i)}(x+L,\Lambda)$ is an eigenfunction with the same eigenvalue Λ, so it must be written as a linear combination of the $f_{(i)}(x,\Lambda)$:

$$f_{(i)}(x+L,\Lambda) = \sum_{j=I}^{IV} a_{ij} f_j(x,\Lambda)\,. \tag{C.5}$$

Setting $x = 0$ yields

$$a_{ij} = \frac{d^{(j-1)}}{dx^{(j-1)}} f_{(i)}(L,\Lambda)\,, \tag{C.6}$$

with $f^{(0)}_{(i)} \equiv f_{(i)}$. This shows that any four independent solutions at $x = L$ are related to those at $x = 0$ by the *monodromy matrix*

$$\mathcal{M}(\Lambda) \equiv \begin{pmatrix} f_{(I)}(L,\Lambda) & f'_{(I)}(L,\Lambda) & f''_{(I)}(L,\Lambda) & f'''_{(I)}(L,\Lambda) \\ f_{(II)}(L,\Lambda) & f'_{(II)}(L,\Lambda) & f''_{(II)}(L,\Lambda) & f'''_{(II)}(L,\Lambda) \\ f_{(III)}(L,\Lambda) & f'_{(III)}(L,\Lambda) & f''_{(III)}(L,\Lambda) & f'''_{(III)}(L,\Lambda) \\ f_{(IV)}(L,\Lambda) & f'_{(IV)}(L,\Lambda) & f''_{(IV)}(L,\Lambda) & f'''_{(IV)}(L,\Lambda) \end{pmatrix}. \tag{C.7}$$

After diagonalizing it one obtains a new set of four linear independent *Floquet* or *Bloch solutions* $\tilde{f}_i(x,\Lambda)$ with the property $\tilde{f}_i(x+L,\Lambda) = \rho \tilde{f}_i(x,\Lambda)$ $(\rho \in \mathbb{C})$, i.e. in matrix notation

$$[\mathcal{M}(\Lambda) - \rho\,\mathbb{I}_4] \begin{pmatrix} \tilde{f}_{(I)}(x,\Lambda) \\ \tilde{f}_{(II)}(x,\Lambda) \\ \tilde{f}_{(III)}(x,\Lambda) \\ \tilde{f}_{(IV)}(x,\Lambda) \end{pmatrix} = \begin{pmatrix} 0 \\ 0 \\ 0 \\ 0 \end{pmatrix}. \tag{C.8}$$

Setting $\rho = 1$ selects four independent periodic eigenfunctions with period L, while $\rho = -1$ four antiperiodic ones. We can restrict to the compact interval $x \in [0, L]$. The eigenvalues form a discrete set $\{\Lambda_i\}$: Λ is an allowed eigenvalue if we can find a non-trivial Bloch solution $\tilde{f}_{(i)}(x)$ from the algebraic system of equations (C.8), i.e. if there exists a ρ satisfying the *characteristic equation*

$$
\begin{aligned}
0 = \det(\mathcal{M}(\Lambda) - \rho\,\mathbb{I}) &= \rho^4 - (f_{(I)} + f'_{(II)} + f''_{(III)} + f'''_{(IV)})\rho^3 \\
&+ \Big[\det\begin{pmatrix} f_{(I)} & f'_{(I)} \\ f_{(II)} & f'_{(II)} \end{pmatrix} + \det\begin{pmatrix} f_{(I)} & f''_{(I)} \\ f_{(III)} & f''_{(III)} \end{pmatrix} + \det\begin{pmatrix} f_{(I)} & f'''_{(I)} \\ f_{(IV)} & f'''_{(IV)} \end{pmatrix} \\
&+ \det\begin{pmatrix} f'_{(II)} & f''_{(II)} \\ f'_{(III)} & f''_{(III)} \end{pmatrix} + \det\begin{pmatrix} f'_{(II)} & f'''_{(II)} \\ f'_{(IV)} & f'''_{(IV)} \end{pmatrix} + \det\begin{pmatrix} f''_{(III)} & f'''_{(III)} \\ f''_{(IV)} & f'''_{(IV)} \end{pmatrix} \Big] \rho^2 \\
&- \Big[\det\begin{pmatrix} f_{(I)} & f'_{(I)} & f''_{(I)} \\ f_{(II)} & f'_{(II)} & f''_{(II)} \\ f_{(III)} & f'_{(III)} & f''_{(III)} \end{pmatrix} + \det\begin{pmatrix} f_{(I)} & f'_{(I)} & f''_{(I)} \\ f_{(II)} & f'_{(II)} & f'''_{(II)} \\ f_{(IV)} & f'_{(IV)} & f'''_{(IV)} \end{pmatrix} \\
&+ \det\begin{pmatrix} f_{(I)} & f''_{(I)} & f'''_{(I)} \\ f_{(III)} & f''_{(III)} & f'''_{(III)} \\ f_{(IV)} & f''_{(IV)} & f'''_{(IV)} \end{pmatrix} + \det\begin{pmatrix} f'_{(II)} & f''_{(II)} & f'''_{(II)} \\ f'_{(III)} & f''_{(III)} & f'''_{(III)} \\ f'_{(IV)} & f''_{(IV)} & f'''_{(IV)} \end{pmatrix} \Big] \rho + 1 \,.
\end{aligned}
\tag{C.9}
$$

Let ρ_i $(i = 1, 2, 3, 4)$ be the roots of this polynomial equation. We pose

$$
\begin{aligned}
\det(\mathcal{M}(\Lambda) - \rho\,\mathbb{I}_4) \equiv \rho^4 &- (\rho_1 + \rho_2 + \rho_3 + \rho_4)\rho^3 \\
&+ (\rho_1\rho_2 + \rho_1\rho_3 + \rho_1\rho_4 + \rho_2\rho_3 + \rho_2\rho_4 + \rho_3\rho_4)\rho^2 \\
&- (\rho_1\rho_2\rho_3 + \rho_1\rho_2\rho_4 + \rho_1\rho_3\rho_4 + \rho_2\rho_3\rho_4)\rho + \rho_1\rho_2\rho_3\rho_4
\end{aligned}
\tag{C.10}
$$

and the comparison between (C.9) and (C.10) gives a condition on the *Floquet factors* ρ_i

$$\rho_1\rho_2\rho_3\rho_4 = 1. \tag{C.11}$$

The constraint implies that a general solution of (C.9) would require the introduction of three functions to parametrize the four roots. However, elaborating on the Floquet-Bloch theory [1], we will show that we can conveniently solve the equation introducing just *two quasi-momenta functions* $p_i(\Lambda)$ $(i = 1, 2)$, periodic with period L, from the definitions

$$
\begin{aligned}
\rho_1 &\equiv e^{ip_1(\Lambda)L}\,, \qquad & \rho_2 &\equiv e^{-ip_1(\Lambda)L}\,, \\
\rho_3 &\equiv e^{ip_2(\Lambda)L}\,, \qquad & \rho_4 &\equiv e^{-ip_2(\Lambda)L}\,.
\end{aligned}
\tag{C.12}
$$

We postpone the construction of the $p_i(\Lambda)$ in Appendix C.2 for a subclass of operators with coefficients (C.18).

The advantage of our method is that, given the quasi-momenta, we can immediately compute the functional determinant Det $\mathcal{O}^{(4)}$ without knowing the eigenvalues explicitly. We apply the corollary (B.11) of Gel'fand and Yaglom [4] for differential operators of even order [5–9]. The method relies on an integral identity to express the spectral zeta-function, which defines the finite part of the determinant in zeta-function regularization, in a contour integral representation. The proof parallels what was done for the Hill's operators [10], where the so-called *resolvent* function carries the spectral information to be plugged in such integral representation.

We shall use (B.27)–(B.28) and set the range of $x \in [a, b]$ to be the line segment $[0, L]$. The functional determinant can be eventually calculated in closed form[9]

$$\mathrm{Det}\,\mathcal{O}^{(4)} = \det\left(\mathcal{M}(\Lambda = 0) - \rho\,\mathbb{I}_4\right) . \tag{C.13}$$

The formula makes sense only considering the ratio of two determinants of operators in the form (C.1) with the same boundary conditions on a interval (i.e. same values for L and ρ). The division by a reference determinant ensures that the divergent part of the determinant (C.13) is properly subtracted.[10]

It is useful to spell out the important formulas for Sect. 4.2 for periodic (P, $\rho = 1$) or anti-periodic (AP, $\rho = -1$) boundary conditions at the endpoints.

- Periodic functions with period L
  Periodic eigenfunctions $f_i(x + L) = f_i(x)$ exist only for special values of Λ which are determined by setting $\rho = 1$ in (C.10) and using (C.12)–(C.13) we get

$$\left(\mathrm{Det}\,\mathcal{O}^{(4)}\right)_{P,L} = 16\sin^2\left(\frac{L}{2}p_1\right)\sin^2\left(\frac{L}{2}p_2\right) . \tag{C.14}$$

- Anti-periodic functions on an interval of length L
  We analogously get the determinant for antiperiodic eigenfunctions $f_i(x + L) = -f_i(x)$ by setting $\rho = -1$

$$\left(\mathrm{Det}\,\mathcal{O}^{(4)}\right)_{AP,L} = 16\cos^2\left(\frac{L}{2}p_1\right)\cos^2\left(\frac{L}{2}p_2\right) . \tag{C.15}$$

- Periodic functions with period $2L$
  Doubling the period $L \to 2L$ in (C.10), the case $f_i(x + 2L) = f_i(x)$ is the product of the previous matrix determinants:

$$\left(\mathrm{Det}\,\mathcal{O}^{(4)}\right)_{P,2L} = \left(\mathrm{Det}\,\mathcal{O}^{(4)}\right)_{P,L}\left(\mathrm{Det}\,\mathcal{O}^{(4)}\right)_{AP,L} = 16\sin^2(Lp_1)\sin^2(Lp_2) . \tag{C.16}$$

[9]Note that the monodromy matrix (C.7) at zero eigenvalue $\Lambda = 0$ is the transpose of the matrix of fundamental solutions (B.9) for operators of the fourth order ($n = 4$) acting on scalar functions ($r = 1$).

[10]In the example of the (J_1, J_2)-string of Sect. 4.2.1, the normalization for the Landau-Lifshitz determinant (4.48), parametrized by k (4.47), is provided in (4.50) by the same determinant with $k = 0$. A similar normalization for (4.63) of the (S, J)-string of Sect. 4.2.2 would be naturally provided by the determinants of fermionic degrees of freedom, which we are not able to solve at the moment.

C.2 Quasi-Momenta for Operators with Doubly Periodic Coefficients

In the last appendix we presented an algorithm to evaluate the determinant of (C.1) given an ansatz (C.12) that needs the input of two quasi-momenta functions. While the problem with arbitrary periodic coefficients in (C.1) is hard to address, here we proceed with giving explicit formulas for the quasi-momenta for a special choice of such coefficients. We will also find the Bloch solutions $\tilde{f}_{(i)}$ (C.8). A first attempt to study this kind of equations was carried out by Mittag-Leffler [11].[11] The differential operators of interest are of the type (C.1)

$$\mathcal{O}^{(4)} = \frac{d^4}{dx^4} + v_1(x)\frac{d^2}{dx^2} + v_2(x)\frac{d}{dx} + v_3(x) \tag{C.17}$$

with *potentials*

$$\begin{aligned} v_1(x) &= \alpha_0 + \alpha_1 k^2 \mathrm{sn}^2(x, k^2)\,, \\ v_2(x) &= \beta_0 + \beta_1 k^2 \mathrm{sn}^2(x, k^2) + 2\beta_2 k^2 \mathrm{sn}(x, k^2)\mathrm{cn}(x, k^2)\mathrm{dn}(x, k^2)\,, \\ v_3(x) &= \gamma_0 + 2\gamma_3 k^2 + (\gamma_1 - 4(1+k^2)\gamma_3)k^2 \mathrm{sn}^2(x, k^2) \\ &\quad + 2\gamma_2 k^2 \mathrm{sn}(x, k^2)\mathrm{cn}(x, k^2)\mathrm{dn}(x, k^2) + 6\gamma_3 k^4 \mathrm{sn}^4(x, k^2) \end{aligned} \tag{C.18}$$

given in terms of elliptic functions—summarized in Appendix A—*with only one regular singular pole*[12] at $x = i\mathbb{K}'$. We conventionally suppress the elliptic modulus k^2 in what follows. We will now find conditions on the parameters $\alpha_i, \beta_i, \gamma_i$ such that the eigenvalue equation

$$\mathcal{O}^{(4)} f(x, \Lambda) = \Lambda f(x, \Lambda), \qquad f(x + L, \Lambda) = f(x, \Lambda) \tag{C.19}$$

is solved by a *Hermite-Bethe-like ansatz* [12][13]

---

[11] Apparently with the help of the diligent student Stenberg mentioned in a footnote of this paper.

[12] Given an ordinary differential equation $\sum_{i=0}^{n} p_i(z)(d/dz)^i f(z) = 0$ in one complex variable z, at *ordinary points* the coefficients of the equation are analytic functions, whereas at *singular points* some of them diverge. The latter are classified into *regular singular points*, if $p_{n-i}(z)$ has a pole of order at most i, otherwise the point is an *irregular singularity*.

[13] The ansatz provides four linearly independent solutions—see for example (4.60)–(4.61) with (C.44)–(C.52), or Fig. 4.3 which gives a graphical representation of them. However, at the edges (a finite set of points) where the color lines meet, there can be a problem, since two or all four functions become linearly dependent. This is expected from the second-order case [1, 12], where the ansatz gives all two linear independent solutions, except for a finite number of problematic points (the *band edge solutions* for Lamé operators [2]). The missing solutions at those points may be found [13] and this is expected to be generalizable to our fourth-order case. For our purpose of evaluating a partition function (see below 4.49) we only need the solutions, the associated quasi-momenta and the relations (C.35) in the *physical region* $\Omega^2 < 0$, which is free from such problematic "edge" points.

$$f(x,\Lambda) = \prod_{r=1}^{n} \frac{H(x+\bar{\alpha}_r)}{\Theta(x)} e^{x\rho} e^{x\lambda} \tag{C.20}$$

where the Jacobi H, Θ and Z functions are defined in terms of the Jacobi θ functions in (A.13). The constants ρ and $\bar{\alpha}_r$ are determined as functions of the parameters of the eigenfunction α_i, β_i, γ_i and the eigenvalue Λ by analyticity constraints on the eigenfunction as follows. Let us introduce the function

$$F(x,\Lambda) \equiv \frac{1}{f(x,\Lambda)} \mathcal{O}^{(4)} f(x,\Lambda), \tag{C.21}$$

which is an elliptic function with periods $2\mathbb{K}$ and $2i\mathbb{K}'$ and a certain number of poles x_i of order p_i in the period-parallelogram. We rewrite the eigenvalue equation as

$$F(x,\Lambda) = \Lambda. \tag{C.22}$$

Now we use the Liouville's theorem in complex analysis [12] to state that that an elliptic function without any poles in a fundamental period parallelogram of the complex plane has to be constant. If $f(x)$ in (C.20) is a solution of the differential equation, then the elliptic function $F(x)$ should merely be a constant. Therefore, we have to impose that in the Laurent expansion of $F(x)$

$$F(x_i+\epsilon,\Lambda) = \frac{A_{i,p_i}}{\epsilon^{p_i}} + \frac{A_{i,p_i-1}}{\epsilon^{p_i-1}} + \cdots + \frac{A_{i,1}}{\epsilon} + a_{i,0} + a_{i,1}\epsilon + \ldots \tag{C.23}$$

all coefficients $A_{i,j}$ of the principal part vanish. This will constrain the free parameters in (C.18) and deliver the corresponding Bethe-ansatz equations for the spectral parameters $\bar{\alpha}_i$.

C.3 Pole Structure

In order to proceed we need to collect information about the pole structure of the functions appearing in (C.20)–(C.21). In the study of their analytic properties, it is useful to start with the *auxiliary function*

$$\Phi(x,\Lambda) \equiv \frac{1}{f(x,\Lambda)} \frac{df(x,\Lambda)}{dx} = \sum_{r=1}^{n} \left[Z(x+\bar{\alpha}_r+i\mathbb{K}') - Z(x) \right] + \rho + \lambda + \frac{n\pi i}{2\mathbb{K}} \tag{C.24}$$

that has $n+1$ poles at $x = i\mathbb{K}'$ and $x = -\bar{\alpha}_1, -\bar{\alpha}_2, \ldots, -\bar{\alpha}_n$, up to translations by the periods $2\mathbb{K}$ and $2i\mathbb{K}'$. We separately examine these two cases.

C.3.1 Expansion Around the Pole $x = i\mathbb{K}'$

The expansion of (C.24) around $i\mathbb{K}'$ produces

$$\begin{aligned}
\Phi(i\mathbb{K}' + \epsilon, \Lambda) &= \frac{A_1}{\epsilon} + a_0 + a_1\epsilon + a_2\epsilon^2 + a_3\epsilon^3 + \dots, \\
\Phi'(i\mathbb{K}' + \epsilon, \Lambda) &= -\frac{A_1}{\epsilon^2} + a_1 + 2a_2\epsilon + 3a_3\epsilon^2 + \dots, \\
\Phi''(i\mathbb{K}' + \epsilon, \Lambda) &= \frac{2A_1}{\epsilon^3} + 2a_2 + 6a_3\epsilon + \dots, \\
\Phi'''(i\mathbb{K}' + \epsilon, \Lambda) &= -\frac{6A_1}{\epsilon^4} + 6a_3 + \dots,
\end{aligned} \tag{C.25}$$

where we denoted

$$\begin{aligned}
A_1 &= -n, \qquad a_0 = \sum_{r=1}^{n} Z(\bar{\alpha}_r) + \rho + \lambda, \qquad a_1 = \frac{n}{3}(1+k^2) - k^2 \sum_{r=1}^{n} \mathrm{sn}^2(\bar{\alpha}_r), \\
a_2 &= -k^2 \sum_{r=1}^{n} \mathrm{sn}(\bar{\alpha}_r)\mathrm{cn}(\bar{\alpha}_r)\mathrm{dn}(\bar{\alpha}_r), \\
a_3 &= \frac{n}{45}(1 - 16k^2 + k^4) + \frac{2}{3}(1+k^2)k^2 \sum_{r=1}^{n} \mathrm{sn}^2(\bar{\alpha}_r) - k^4 \sum_{r=1}^{n} \mathrm{sn}^4(\bar{\alpha}_r).
\end{aligned} \tag{C.26}$$

The same procedure applied to the potentials (C.18) leads to the series

$$\begin{aligned}
v_1(i\mathbb{K}' + \epsilon) &= \frac{\alpha_1}{\epsilon^2} + \alpha_0 + \frac{\alpha_1}{3}(1+k^2) + \frac{\alpha_1}{15}(1 - k^2 + k^4)\epsilon^2 + 0 \cdot \epsilon^3 + \dots, \\
v_2(i\mathbb{K}' + \epsilon) &= -\frac{2\beta_2}{\epsilon^3} + \frac{\beta_1}{\epsilon^2} + \beta_0 + \frac{\beta_1}{3}(1+k^2) + \frac{2\beta_2}{15}(1 - k^2 + k^4)\epsilon \\
&\quad + \frac{\beta_1}{15}(1 - k^2 + k^4)\epsilon^2 + \dots, \\
v_3(i\mathbb{K}' + \epsilon) &= \frac{6\gamma_3}{\epsilon^4} - \frac{2\gamma_2}{\epsilon^3} + \frac{\gamma_1}{\epsilon^2} + \gamma_0 + \frac{\gamma_1}{3}(1+k^2) + \frac{2\gamma_3}{15}(1 - k^2 + k^4) \\
&\quad + \frac{2\gamma_2}{15}(1 - k^2 + k^4)\epsilon + \dots.
\end{aligned} \tag{C.27}$$

C.3.2 Expansion around the poles $x = -\bar{\alpha}_i$ $(i = 1, \dots n)$

The analysis carried out for the family of poles $-\bar{\alpha}_i$ yields

$$\Phi(-\bar{\alpha}_i + \epsilon, \Lambda) = \frac{1}{\epsilon} + b_{0,i} + b_{1,i}\epsilon + b_{2,i}\epsilon^2 + \dots \tag{C.28}$$

where we identify the ϵ-coefficients with

$$\begin{aligned}
b_{0,i} &= \sum_{r=1,\dots n\,,\,r\neq i} Z(\bar{\alpha}_r - \bar{\alpha}_i + i\mathbb{K}') + nZ(\bar{\alpha}_i) + \frac{i\pi(n-1)}{2\mathbb{K}} + \rho + \lambda\,, \\
b_{1,i} &= -\sum_{r=1,\dots n\,,\,r\neq i} \mathrm{cs}^2(\bar{\alpha}_i - \bar{\alpha}_r) + \frac{1}{3}(2-k^2) - n\mathrm{dn}^2(\bar{\alpha}_i)\,, \qquad (C.29) \\
b_{2,i} &= -\sum_{r=1,\dots n\,,\,r\neq i} \frac{\mathrm{cn}(\bar{\alpha}_i - \bar{\alpha}_r)\mathrm{dn}(\bar{\alpha}_i - \bar{\alpha}_r)}{\mathrm{sn}^3(\bar{\alpha}_i - \bar{\alpha}_r)} - nk^2\mathrm{sn}(\bar{\alpha}_i)\mathrm{cn}(\bar{\alpha}_i)\mathrm{dn}(\bar{\alpha}_i).
\end{aligned}$$

The potentials $v_j(-\bar{\alpha}_i)$ $(j = 1, 2, 3)$ are non-singular functions.

C.4 Consistency Equations

From the behaviour of the auxiliary function (C.24) and the potentials (C.18) around the singularities, it is now possible to reconstruct the pole structure of F, since the differential operators in (C.21) translate into combinations of the auxiliary function and its derivatives:

$$\begin{aligned}
\frac{1}{f(x,\Lambda)}\frac{d^2 f(x,\Lambda)}{dx^2} &= \Phi(x,\Lambda)^2 + \Phi'(x,\Lambda)\,, \\
\frac{1}{f(x,\Lambda)}\frac{d^3 f(x,\Lambda)}{dx^3} &= \Phi(x,\Lambda)^3 + 3\Phi(x,\Lambda)\Phi'(x,\Lambda) + \Phi''(x,\Lambda)\,, \qquad (C.30) \\
\frac{1}{f(x,\Lambda)}\frac{d^4 f(x,\Lambda)}{dx^4} &= \Phi(x,\Lambda)^4 + 6\Phi(x)^2\Phi'(x,\Lambda) + 4\Phi(x,\Lambda)\Phi''(x,\Lambda) \\
&\quad + 3\Phi'(x,\Lambda)^2 + \Phi'''(x,\Lambda).
\end{aligned}$$

The condition of vanishing Laurent coefficients of F at $x = i\mathbb{K}'$ gives constraining equations on the numerical parameters α_i, β_i and γ_i, provided we take into account (C.25)–(C.27):

$$
\begin{aligned}
&\rho = -\sum_{r=1}^{n} Z(\bar{\alpha}_r)\,,\\
&0 = n(n+1)(n+2)(n+3) + n(n+1)\alpha_1 + 2n\beta_2 + 6\gamma_3\,,\\
&0 = \lambda[4(n+2)(n+1)n + 2(n\alpha_1 + \beta_2)] + (n\beta_1 + 2\gamma_2)\,,\\
&0 = \lambda^2[6n(n+1) + \alpha_1] + \lambda\beta_1 + a_1[-2n(n+1)(2n+1) + \alpha_1(1-2n) - 2\beta_2]\\
&\quad +n(n+1)(\alpha_0 + \frac{1}{3}\alpha_1(1+k^2)) + \gamma_1\,,\\
&0 = 4\lambda^3 n - 2\lambda[a_1(6n^2+\alpha_1) - n(\alpha_0 + \frac{\alpha_1}{3}(1+k^2))]\\
&\quad -a_1\beta_1 + 2a_2[2n(1+n^2) + \alpha_1(n-1) + \beta_2] + n(\beta_0 + \frac{\beta_1}{3}(1+k^2))\,.
\end{aligned}
\tag{C.31}
$$

In particular, the term $O(\epsilon^0)$ in the Laurent expansion gives the relation between the eigenvalue parameter Λ and the spectral parameters $\bar{\alpha}_i$

$$
\begin{aligned}
\Lambda = {} & \lambda^4 + \lambda^2[6a_1(1-2n) + (\alpha_0 + \frac{\alpha_1}{3}(1+k^2))] + \lambda[a_2(4(2+3n(n-1)) + 2\alpha_1)\\
& +(\beta_0 + \frac{\beta_1}{3}(1+k^2))] + a_1^2[3(1-2n(1-n)) + \alpha_1]\\
& +a_1(1-2n)(\alpha_0 + \frac{\alpha_1}{3}(1+k^2)) + a_2\beta_1 +\\
& +a_3[2(2n-1)(n(1-n)-3) + \alpha_1(3-2n) - 2\beta_2] +\\
& +\frac{1}{15}(1-k^2+k^4)[n(n+1)\alpha_1 - 2n\beta_2 + 2\gamma_3] + \gamma_0 + \frac{\gamma_1}{3}(1+k^2)\,.
\end{aligned}
\tag{C.32}
$$

Finally, imposing that the $1/\epsilon$-coefficient around the poles $x = -\bar{\alpha}_i$ should vanish gives the Hermite-Bethe ansatz equations for the spectral parameters ($i = 1, \ldots n$)

$$
4b_{0,i}^3 + 2b_{0,i}(6b_{1,i} + \alpha_0 + \alpha_1 k^2 \mathrm{sn}^2(\bar{\alpha}_i)) + 8b_{2,i} + \beta_0 + \beta_1 k^2 \mathrm{sn}^2(\alpha_i) - 2\beta_2 k^2 \mathrm{sn}(\bar{\alpha}_i)\mathrm{cn}(\bar{\alpha}_i)\mathrm{dn}(\bar{\alpha}_i) = 0
\tag{C.33}
$$

where we used (C.28). In deriving these conditions, we have assumed that $\alpha_i \neq \alpha_j$ for any pair $i, j = 1, \ldots n$.

It is important to mention that in the examples of Chap. 4 we made use of the $n = 1$ consistency equations alone, which we report here separately for reader's convenience.

$$
\begin{aligned}
&0 = 12 + \alpha_1 + \beta_2 + 3\gamma_3\,,\\
&0 = \lambda(24 + 2(\alpha_1 + \beta_2)) + \beta_1 + 2\gamma_2\,,\\
&0 = \lambda^2(12+\alpha_1) - a_1(\alpha_1 + 12 + 2\beta_2) + \lambda\beta_1 + 2\left(\alpha_0 + \frac{1}{3}\alpha_1(1+k^2)\right) + \gamma_1\,,\\
&0 = -4\lambda^3 - 2\lambda\left(\alpha_0 - a_1(6+\alpha_1) + \frac{1}{3}\alpha_1(1+k^2)\right) + a_1\beta_1 - 2a_2(4+\beta_2)\\
&\quad -\beta_0 - \frac{1}{3}\beta_1(1+k^2)\,,
\end{aligned}
\tag{C.34}
$$

$$
\begin{aligned}
\Lambda = \lambda^4 &+ \lambda^2 [-6a_1 + \alpha_0 + \frac{1}{3}\alpha_1(1+k^2)] + \lambda[2a_2(\alpha_1+4) + \beta_0 + \frac{1}{3}\beta_1(1+k^2)] \\
&+ a_1^2(\alpha_1+3) - a_1(\alpha_0 + \frac{1}{3}\alpha_1(1+k^2)) + a_2\beta_1 + a_3(\alpha_1 - 6 - 2\beta_2) \\
&+ \frac{2}{15}(1-k^2+k^4)(\alpha_1 - \beta_2 + \gamma_3) + \gamma_0 + \frac{1}{3}\gamma_1(1+k^2)\,.
\end{aligned}
$$

The condition for the pole at $x = -\bar{\alpha}$ to vanish turns out to be equivalent to the fourth equation in (C.34) and therefore it does not give any further constraint. Our result for the consistency equations is in partial disagreement with [11]. However, the examples discussed in the main text and the numerical cross-checks of the provided solutions, as well as the study of the square of the Lamé operator in appendix A of [14], give support to the correctness of our procedure.

C.5 Spectral Domain for the (J_1, J_2)-String

In Sect. 4.2 the expressions of α_i in the different branches for real Ω^2 read as follows.

- Case $-\infty < \Omega^2 < 0$

$$
\begin{aligned}
\alpha_1(\Omega,k) &= u(\Omega,k) - iv(\Omega,k) \\
\alpha_2(\Omega,k) &= 2\mathbb{K} - u(\Omega,k) + iv(\Omega,k) \\
\alpha_3(\Omega,k) &= 2\mathbb{K} + u(\Omega,k) + iv(\Omega,k) \\
\alpha_4(\Omega,k) &= 2\mathbb{K} - u(\Omega,k) + 2i\mathbb{K}' - iv(\Omega,k)
\end{aligned}
\tag{C.35}
$$

with

$$
u(\Omega,k) = \mathrm{sn}^{-1}\left[\sqrt{\frac{2}{\Omega^2}(1-\sqrt{1-\Omega^2})},\ k\right] \tag{C.36}
$$

$$
v(\Omega,k) = \mathrm{sn}^{-1}\left[\sqrt{\frac{\Omega^2 - 2k^2 + 2k^2\sqrt{1-\Omega^2}}{\Omega^2 - 4k^2k'^2}},\ k'\right].
$$

- Case $0 < \Omega^2 < 4k^2k'^2$

$$
\begin{aligned}
\alpha_1(\Omega,k) &= 2\mathbb{K} - u_2(\Omega,k) - iv_2(\Omega,k) \\
\alpha_2(\Omega,k) &= u_2(\Omega,k) + iv_2(\Omega,k) \\
\alpha_3(\Omega,k) &= 2\mathbb{K} + u_2(\Omega,k) - iv_2(\Omega,k) \\
\alpha_4(\Omega,k) &= 2\mathbb{K} - u_2(\Omega,k) + 2i\mathbb{K}' + iv_2(\Omega,k)
\end{aligned}
\tag{C.37}
$$

with

$$u_2(\Omega, k) = \mathrm{sn}^{-1}\left[\frac{1}{k}\sqrt{1 - 2k'^2\frac{1-\sqrt{1-\Omega^2}}{\Omega^2}}\right]$$

$$v_2(\Omega, k) = \mathrm{sn}^{-1}\left[\frac{1}{\sqrt{2}k'}\sqrt{1-\sqrt{1-\Omega^2}}, k'\right]. \tag{C.38}$$

- For $4k^2k'^2 < \Omega^2 < \infty$ as

$$\begin{aligned} \alpha_3(\Omega, k) &= 2\mathbb{K} - i\mathbb{K}' + 2i\alpha_0(\Omega, k') \\ \alpha_4(\Omega, k) &= 2\mathbb{K} + 3i\mathbb{K}' - 2i\alpha_0(\Omega, k'). \end{aligned} \tag{C.39}$$

- Case $4k^2k'^2 < \Omega^2 < 1$

$$\alpha_1(\Omega, k) = 2\mathbb{K} - i\,\mathrm{sn}^{-1}\left[\sqrt{1 - \frac{4k^2}{\Omega^2}\left(1 - 2k^2 - \sqrt{\Omega^2 - 4k^2k'^2}\right)}, k'\right]$$

$$\alpha_2(\Omega, k) = i\,\mathrm{sn}^{-1}\left[\sqrt{1 - \frac{4k^2}{\Omega^2}\left(1 - 2k^2 - \sqrt{\Omega^2 - 4k^2k'^2}\right)}, k'\right]. \tag{C.40}$$

- Case $1 < \Omega^2 < \infty$

$$\begin{aligned} \alpha_1(\Omega, k) &= 2\mathbb{K} - i\mathbb{K}' - 2\alpha_0(\Omega, k) \\ \alpha_2(\Omega, k) &= i\mathbb{K}' + 2\alpha_0(\Omega, k) \end{aligned} \tag{C.41}$$

with

$$\alpha_0(\Omega, k) = \mathrm{sn}^{-1}\left[\sqrt{1 - \frac{\Omega}{2k^2} + \frac{1}{2k^2}\sqrt{\Omega^2 - 4k^2k'^2}},\ k\right]. \tag{C.42}$$

C.6 Spectral Domain for the (S, J)-String

For convenience we define

$$\chi_\pm(\bar{\Omega}) = \left(\kappa\,\bar{w} \pm \sqrt{(\bar{w}^2 - \nu^2)(\kappa^2 - \nu^2 + \nu^2\bar{\Omega}^2)}\right)^2 - \nu^4. \tag{C.43}$$

Then in Sect. 4.2.2 the expressions of α_i in the different branches for real Ω^2 read as follows.

- Case $-\infty < \bar{\Omega}^2 < -(\frac{\kappa^2}{\nu^2} - 1)$

$$\begin{aligned}\alpha_{1,2}(\Omega) = \operatorname{sn}^{-1} & \left[\sqrt{\frac{(\kappa^2-\nu^2)+(\bar{w}^2-\nu^2)(1-\bar{\Omega}^2)-\sqrt{((\kappa^2+\bar{w}^2)-(\bar{w}^2-\nu^2)\bar{\Omega}^2)^2-4\kappa^2\bar{w}^2}}{2(\kappa^2-\nu^2+\nu^2\bar{\Omega}^2)}}, k^2 \right] \\ & \pm i \operatorname{dn}^{-1} \left[\frac{k}{\kappa} \sqrt{\frac{(\bar{w}^2-\kappa^2)^2-(\bar{w}^2-\nu^2)(\kappa^2+\bar{w}^2)\bar{\Omega}^2+(\bar{w}^2-\kappa^2)\sqrt{((\kappa^2+\bar{w}^2)-(\bar{w}^2-\nu^2)\bar{\Omega}^2)^2-4\kappa^2\bar{w}^2}}{-2(\bar{w}^2-\nu^2)\bar{\Omega}^2}}, k' \right]\end{aligned} \quad \text{(C.44)}$$

- Case $-(\frac{\kappa^2}{\nu^2} - 1) < \bar{\Omega}^2 < -\frac{(\kappa^2-\nu^2)(\bar{w}^2-\kappa^2)}{(\bar{w}^2-\nu^2)\kappa^2}$

$$\alpha_1(\bar{\Omega}) = \mathbb{K} - \operatorname{sn}^{-1}\left[\sqrt{\frac{\chi_-(\bar{\Omega})+\nu^4\bar{\Omega}^2/k^2}{\chi_-(\bar{\Omega})+\nu^4\bar{\Omega}^2}}, k\right] \quad \text{(C.45)}$$

- Case $-\frac{(\kappa^2-\nu^2)(\bar{w}^2-\kappa^2)}{(\bar{w}^2-\nu^2)\kappa^2} < \bar{\Omega}^2 < 0$

$$\alpha_1(\bar{\Omega}) = 2\mathbb{K} - \operatorname{sn}^{-1}\left[\sqrt{\frac{-\nu^4\bar{\Omega}^2}{k^2\chi_-(\bar{\Omega})}}, k\right] \quad \text{(C.46)}$$

- Case $0 < \bar{\Omega}^2 < \frac{(\bar{w}-\kappa)^2}{\bar{w}^2-\nu^2}$

$$\alpha_1(\bar{\Omega}) = 2\mathbb{K} + i\,\operatorname{sn}^{-1}\left[\sqrt{\frac{\nu^4\bar{\Omega}^2}{k^2\chi_-(\bar{\Omega})+\nu^4\bar{\Omega}^2}}, k'\right] \quad \text{(C.47)}$$

- Case $\frac{(\bar{w}-\kappa)^2}{\bar{w}^2-\nu^2} < \bar{\Omega}^2 < 1 - \frac{\kappa^2}{\bar{w}^2}$

$$\alpha_1(\bar{\Omega}) = 2\mathbb{K} + i\mathbb{K}' - \operatorname{sn}^{-1}\left[\sqrt{\frac{\chi_-(\bar{\Omega})}{-\nu^4\bar{\Omega}^2}}, k\right] \quad \text{(C.48)}$$

- Case $1 - \frac{\kappa^2}{\bar{w}^2} < \bar{\Omega}^2 < \frac{(\bar{w}+\kappa)^2}{\bar{w}^2-\nu^2}$

$$\alpha_1(\bar{\Omega}) = \mathbb{K} + i\mathbb{K}' - \operatorname{sn}^{-1}\left[\sqrt{\frac{\chi_-(\bar{\Omega})+\nu^4\bar{\Omega}^2}{k^2\chi_-(\bar{\Omega})+\nu^4\bar{\Omega}^2}}, k\right] \quad \text{(C.49)}$$

- Case $\frac{(\bar{w}+\kappa)^2}{\bar{w}^2-\nu^2} < \bar{\Omega}^2 < \infty$

$$\alpha_1(\bar{\Omega}) = i\,\operatorname{sn}^{-1}\left[\sqrt{\frac{\nu^4\bar{\Omega}^2}{k^2\chi_-(\bar{\Omega})+\nu^4\bar{\Omega}^2}}, k'\right] \quad \text{(C.50)}$$

- Case $-(\frac{\kappa^2}{\nu^2} - 1) < \bar{\Omega}^2 < 0$

$$\alpha_2(\bar{\Omega}) = \operatorname{sn}^{-1}\left[\sqrt{-\frac{\nu^4\bar{\Omega}^2}{k^2\chi_+(\bar{\Omega})}}, k\right] \quad \text{(C.51)}$$

- Case $0 < \bar{\Omega}^2 < \infty$

$$\alpha_2(\bar{\Omega}) = i\,\mathrm{sn}^{-1}\left[\sqrt{\frac{\nu^4\bar{\Omega}^2}{k^2\chi_+(\bar{\Omega})+\nu^4\bar{\Omega}^2}}, k'\right] \tag{C.52}$$

In the main body we have used the following identity, which does not seem to be tabulated but can be easily checked to be true

$$Z(\alpha,k) = \frac{2}{1+\tilde{k}'}Z\left(\frac{\alpha}{1+\tilde{k}'}+\frac{i\mathbb{K}'}{1+\tilde{k}'},\tilde{k}\right) - \frac{1+\mathrm{cn}(\alpha,k)\mathrm{dn}(\alpha,k)}{\mathrm{sn}(\alpha,k)} + \frac{i\pi}{2\mathbb{K}}, \tag{C.53}$$

where $\tilde{k}$ is the *Landen transformed modulus*, i.e. $\tilde{k}^2 = 4k/(1+k)^2$.

Bibliography

1. W. Magnus, S. Winkler, in *Hill's Equation* (Dover Phoenix Editions, New York, 1966).
2. M. Beccaria, G.V. Dunne, V. Forini, M. Pawellek, A.A. Tseytlin, Exact computation of one-loop correction to energy of spinning folded string in $AdS_5 \times S^5$. J. Phys. A **43**, 165402 (2010), arXiv:1001.4018.
3. V. Forini, V.G.M. Puletti, O. Ohlsson Sax, The generalized cusp in $AdS_4 \times \mathbb{CP}^3$ and more one-loop results from semiclassical strings. J. Phys. A **46**, 115402 (2013), arXiv:1204.3302.
4. I.M. Gelfand, A.M. Yaglom, Integration in functional spaces and it applications in quantum physics. J. Math. Phys. **1**, 48 (1960).
5. A.J. McKane, M.B. Tarlie, Regularization of functional determinants using boundary perturbations. J. Phys. A **28**, 6931 (1995), arXiv:cond-mat/9509126.
6. K. Kirsten, A.J. McKane, Functional determinants by contour integration methods. Ann. Phys. **308**, 502 (2003), arXiv:math-ph/0305010.
7. K. Kirsten, A.J. McKane, Functional determinants for general Sturm-Liouville problems. J. Phys. A **37**, 4649 (2004), arXiv:math-ph/0403050.
8. K. Kirsten and P. Loya, Computation of determinants using contour integrals. Am. J. Phys. **76**, 60 (2008), arXiv:0707.3755.
9. G.V. Dunne, Functional determinants in quantum field theory. J. Phys. A **41**, 304006 (2008), arXiv:0711.1178 (in, *Proceedings, 5th International Symposium on Quantum Theory and Symmetries (QTS5)*, p. 304006.
10. H. Braden, Periodic functional determinants. J. Phys. A **18**, 2127 (1985).
11. M.G. Mittag-Leffler, Ueber die Integration der Hermiteschen Differentialgleichungen der dritten und vierten Ordnung, bei denen die Unendlichkeitsstellen der Integrale von der ersten Ordnung sind. Annali di Matematica **11**, 65 (1882) (in German).

12. G.N. Watson, E.T. Whittaker, *A Course of Modern Analysis* (Cambridge University Press, 1902).
13. E.L. Ince, in *Ordinary Differential Equations* (Dover, New York, 1956).
14. V. Forini, V.G.M. Puletti, M. Pawellek, E. Vescovi, One-loop spectroscopy of semiclassically quantized strings: bosonic sector. J. Phys. A **48**, 085401 (2015), arXiv:1409.8674.

Appendix D
Conventions for Worldsheet Geometry

Throughout Chaps. 3 and 5 we use the following index conventions, when not otherwise stated.

$$\begin{array}{ll} M, N, ... = 0, ... 9 & \text{curved target-space indices} \\ A, B, ... = 0, ... 9 & \text{flat target-space indices} \\ i, j, ... = 0, 1 = \tau, \sigma & \text{curved worldsheet indices} \\ a, b, ... = 0, 1 = \tau, \sigma & \text{flat worldsheet indices} \end{array} \tag{D.1}$$

We introduce the target-space metric G_{MN} and a set of vielbein E^A_M and its inverse E^M_A

$$G_{MN} = E^A_M E^B_N \eta_{AB}\,, \qquad E^A_M E^M_B = \delta^A_B\,, \qquad \eta_{AB} = \text{diag}(-1, 1, \dots 1)\,, \tag{D.2}$$

Given a classical surface Σ in Sect. 3.2, we have similar relations for the induced metric $h_{ij} \equiv G_{MN}\partial_i X^M \partial_j X^N$ and the zweibein e^a_i

$$h_{ij} = e^a_i e^b_j \eta_{ab}\,, \qquad e^a_i e^i_b = \delta^a_b\,, \qquad \eta_{ab} = \text{diag}(-1, 1)\,, \tag{D.3}$$

When not explicit, indices are raised/lowered via contractions with the metrics above. The 10d indices can be “pullbacked” onto the 2d embedding Σ using the derivatives of the background, namely $V_i \equiv \partial_i X^M V_M$.

We also need a pair of tangent vectors (labeled by the flat index $\tilde{a} = 0, 1$)

$$t^A_{\tilde{a}} = E^A_M e^i_{\tilde{a}} \partial_i X^M \tag{D.4}$$

and 8 vectors $N^A_{\underline{a}}$ orthogonal to Σ (labeled by the flat index $\underline{a} = 2, \dots 10$):

$$\begin{gathered} t^A_{\tilde{a}} t^B_{\tilde{b}} \eta_{AB} = \eta_{\tilde{a}\tilde{b}}\,, \qquad N^A_{\underline{a}} N^B_{\underline{b}} \eta_{AB} = \eta_{\underline{ab}}\,, \qquad t^A_{\tilde{a}} N^B_{\underline{b}} \eta_{AB} = 0\,, \\ t^A_{\tilde{a}} t^B_{\tilde{b}} \eta^{\tilde{a}\tilde{b}} + N^A_{\underline{a}} N^B_{\underline{b}} \eta^{\underline{a},\underline{b}} = \eta^{AB}\,. \end{gathered} \tag{D.5}$$

E. Vescovi, *Perturbative and Non-perturbative Approaches to String Sigma-Models in AdS/CFT*, Springer Theses, DOI 10.1007/978-3-319-63420-3

For convenience we often tolerate an abuse of notation that consists in adding a tilde over 2d curved indices, e.g. identifying $\tilde{i} = i$.

Flat and curved 32×32 Dirac matrices are respectively denoted by Γ_A and $\Gamma_M \equiv E^A_M \Gamma_A$ while the symbol $\mathbb{I}_n$ stands for the $n \times n$ identity matrix:

$$\{\Gamma_A, \Gamma_B\} = 2\eta_{AB}\mathbb{I}_{32}\,, \qquad \{\Gamma_M, \Gamma_N\} = 2G_{MN}\mathbb{I}_{32}\,. \tag{D.6}$$

The doublet of 10d spinors of type IIB string theory have the same chirality

$$\Gamma_{11}\Psi^I = \Psi^I \tag{D.7}$$

and labeled by capital letters $I, J, ... = 1, 2$.

D.1 Latitude Wilson Loops at Strong Coupling

In what follows we specialize to the analysis in Chap. 5. Here, G_{MN} is (5.28) upon Wick rotation $t \to it$ to Euclidean time. The chosen representation of the $SO(10)$ Dirac matrices reads ($\sigma_1, \sigma_2, \sigma_3$ are Pauli matrices)

$$\begin{array}{ll}
\Gamma_0 = i\,(\sigma_3 \otimes \sigma_2) \otimes \mathbb{I}_4 \otimes \sigma_1 & \Gamma_5 = \mathbb{I}_4 \otimes (\sigma_3 \otimes \sigma_2) \otimes \sigma_2 \\
\Gamma_1 = (\mathbb{I}_2 \otimes \sigma_1) \otimes \mathbb{I}_4 \otimes \sigma_1 & \Gamma_6 = \mathbb{I}_4 \otimes (\sigma_1 \otimes \sigma_2) \otimes \sigma_2 \\
\Gamma_2 = (\mathbb{I}_2 \otimes \sigma_3) \otimes \mathbb{I}_4 \otimes \sigma_1 & \Gamma_7 = \mathbb{I}_4 \otimes (-\sigma_2 \otimes \sigma_2) \otimes \sigma_2 \\
\Gamma_3 = (\sigma_1 \otimes \sigma_2) \otimes \mathbb{I}_4 \otimes \sigma_1 & \Gamma_8 = \mathbb{I}_4 \otimes (\mathbb{I}_2 \otimes \sigma_1) \otimes \sigma_2 \\
\Gamma_4 = (-\sigma_2 \otimes \sigma_2) \otimes \mathbb{I}_4 \otimes \sigma_1 & \Gamma_9 = \mathbb{I}_4 \otimes (\mathbb{I}_2 \otimes \sigma_3) \otimes \sigma_2
\end{array} \tag{D.8}$$

along with the chirality matrix

$$\Gamma_{11} = \Gamma_{0123456789} = -\mathbb{I}_4 \otimes \mathbb{I}_4 \otimes \sigma_3\,. \tag{D.9}$$

Appendix E
Details on the Null Cusp Fluctuation Lagrangian in $AdS_4 \times \mathbb{CP}^3$

E.1 Fluctuation Lagrangian in the Null Cusp Background

We present the expressions for the Feynman propagators and the interaction vertices of the fluctuation Lagrangian (6.17). We drop the tildes over the fluctuation fields for better readability. The bosonic propagators can be easily read off from the quadratic Lagrangian (6.24)

$$G_{\varphi\varphi}(p) = \frac{1}{p^2+1}\,, \qquad G_{z_a\bar{z}^b}(p) = \frac{2\,\delta^b_a}{p^2}\,, \qquad G_{x^1x^1}(p) = \frac{1}{p^2+\frac{1}{2}} \tag{E.1}$$

and the fermionic propagators from the inverse of the kinetic matrix K_F (6.27)

$$\begin{aligned} G_{\eta_4\bar{\eta}^4}(p) &= G_{\theta_4\bar{\theta}^4}(p) = \frac{p_0}{p^2}\,, & G_{\eta_4\bar{\theta}^4}(p) &= G_{\theta_4\bar{\eta}^4}(-p) = -\frac{p_1}{p^2}\,, \\ G_{\eta_a\bar{\eta}^b}(p) &= G_{\theta_a\bar{\theta}^b}(p) = \frac{p_0}{p^2+\frac{1}{4}}\delta^b_a\,, & G_{\eta_a\bar{\theta}^b}(p) &= G_{\theta_a\bar{\eta}^b}(-p) = -\frac{p_1+\frac{i}{2}}{p^2+\frac{1}{4}}\delta^b_a\,. \end{aligned} \tag{E.2}$$

The expansion of the exponentials of $\tilde{\phi}$ in (6.18)–(6.20) would bring an an infinite tower of interactions. Only terms with four fields at most are relevant for the two-loop computation in Sect. 6.3. Vertices bring along the factor $\frac{T}{2}$ from the overall coefficient in the action (6.17). We include the factor of $\frac{1}{2}$ but ignore the string tension T in the lists below.

The cubic interactions read

$$V_{\varphi x^1 x^1} = -4\varphi\left(\partial_s x^1 - x^1\right)^2, \quad V_{\varphi^3} = 2\varphi\left[(\partial_t\varphi)^2 - (\partial_s\varphi)^2\right], \quad V_{\varphi|z|^2} = 2\varphi\left[|\partial_t z|^2 - |\partial_s z|^2\right], \tag{E.3}$$

$$V_{z\eta\eta} = -\epsilon^{abc}\,\partial_t\bar{z}_a\eta_b\eta_c + \text{h.c.}, \qquad V_{z\eta\theta} = -2\,\epsilon^{abc}\bar{z}_a\eta_b(\partial_s\theta_c - \theta_c) - \text{h.c.}, \tag{E.4}$$

$$V_{\varphi\eta\theta} = -4\,i\,\varphi\,\eta_a(\partial_s\theta^a - \theta^a) - \text{h.c.}, \qquad V_{x^1\eta\eta} = -4\,i\,\eta^a\eta_a(\partial_s x^1 - x^1), \tag{E.5}$$

$$V_{z\eta_a\eta_4} = -2\,\partial_t z^a\eta_a\eta_4 + \text{h.c.}, \qquad V_{z\eta_a\theta_4} = 2\,\partial_s z^a\eta_a\theta_4 - \text{h.c.}, \tag{E.6}$$

E. Vescovi, *Perturbative and Non-perturbative Approaches to String Sigma-Models in AdS/CFT*, Springer Theses, DOI 10.1007/978-3-319-63420-3

$$V_{\varphi\eta_4\theta^4} = -2i\,\varphi\,(\theta^4\partial_s\eta_4 - \partial_s\theta^4\eta_4) - \text{h.c.},\quad V_{x^1\psi^4\psi_4} = -2i\,(\eta^4\eta_4 + \theta^4\theta_4)(\partial_s x^1 - x^1), \tag{E.7}$$

while the quartic vertices are

$$V_{z^4} = \frac{1}{6}\Big[\left(\bar{z}_a\partial_t z^a\right)^2 + \left(\bar{z}_a\partial_s z^a\right)^2 + \left(z^a\partial_t\bar{z}_a\right)^2 + \left(z^a\partial_s\bar{z}_a\right)^2 - |z|^2\left(|\partial_t z|^2 + |\partial_s z|^2\right) - \left|\bar{z}_a\partial_t z^a\right|^2 - \left|\bar{z}_a\partial_s z^a\right|^2\Big], \tag{E.8}$$

$$V_{\varphi^2 x^1 x^1} = 16\,\varphi^2\left[(\partial_s x^1 - x^1)\right]^2,\qquad V_{\varphi^4} = 4\,\varphi^2\left[(\partial_t\varphi)^2 + (\partial_s\varphi)^2 + \frac{2}{3}\varphi^2\right], \tag{E.9}$$

$$V_{\varphi^2|z|^2} = 4\,\varphi^2\left[|\partial_t z|^2 + |\partial_s z|^2\right],\qquad V_{\dot{z}\bar{z}\psi^4\psi_4} = -2i\,(\eta^4\eta_4 + \theta^4\theta_4)\bar{z}_b\partial_t z^b + \text{h.c.}, \tag{E.10}$$

$$V_{\eta^2\eta^4\eta_4} = 8\,\eta^4\eta_4\eta^a\eta_a,\qquad V_{z'\bar{z}\psi^4\psi_4} = -2i\,(\eta^4\theta_4 - \theta^4\eta_4)\bar{z}_b\partial_s z^b - \text{h.c.}, \tag{E.11}$$

$$V_{\eta^4} = 4(\eta^a\eta_a)^2,\qquad V_{\varphi^2\eta_4\theta^4} = 4i\,\varphi^2\,(\theta^4\partial_s\eta_4 - \partial_s\theta^4\eta_4) - \text{h.c.}, \tag{E.12}$$

$$V_{\eta_4\eta^4\theta_4\theta^4} = -8\,\eta^4\eta_4\theta^4\theta_4,\qquad V_{\varphi x^1\psi^4\psi_4} = 12i\,\varphi\,(\eta^4\eta_4 + \theta^4\theta_4)(\partial_s x^1 - x^1), \tag{E.13}$$

$$V_{\eta^3\eta_4} = 4\,\epsilon^{abc}\eta_a\eta_b\eta_c\eta_4 + \text{h.c.},\qquad V_{zz\eta^a\eta_4} = -2i\,\epsilon_{abc}\partial_t z^a z^b\eta^c\eta_4 + \text{h.c.}, \tag{E.14}$$

$$V_{\varphi z\eta_a\theta_4} = -8\,\varphi\,\partial_s z^a\eta_a\theta_4 - \text{h.c.},\qquad V_{\varphi z\eta\theta} = 8\,\varphi\epsilon^{abc}\bar{z}_a\eta_b(\partial_s\theta_c - \theta_c) - \text{h.c.}, \tag{E.15}$$

$$V_{zz\eta^a\theta_4} = 2i\,\epsilon_{abc}\partial_s z^a z^b\eta^c\theta_4 - \text{h.c.},\quad V_{zz\eta\eta} = -2i\,(\bar{z}_a\partial_t z^a\eta^b\eta_b - \bar{z}_b\partial_t z^a\eta^b\eta_a) + \text{h.c.}, \tag{E.16}$$

$$V_{\varphi x^1\eta\eta} = 24i\,\varphi\,\eta^a\eta_a(\partial_s x^1 - x^1),\qquad V_{zz\eta\theta} = -2i\,[|z|^2\eta_a(\partial_s\theta^a - \theta^a) + - \bar{z}_b z^a\eta_a(\partial_s\theta^b - \theta^b)] - \text{h.c.}, \tag{E.17}$$

$$V_{\varphi^2\eta\theta} = 8i\,\varphi^2\,\eta_a(\partial_s\theta^a - \theta^a) - \text{h.c.},\quad V_{x^1 z\eta\eta} = -4\,(\partial_s x^1 - x^1)\epsilon^{abc}\bar{z}_a\eta_b\eta_c - \text{h.c.}. \tag{E.18}$$

E.2 Two-Loop Integral Reductions

We list the relevant *Passarino-Veltmann reductions* of tensor integral appearing in the computation of the two-loop free energy. We first define the *master scalar integrals*

$$I\left(m^2\right) \equiv \int \frac{d^2p}{(2\pi)^2}\,\frac{1}{p^2 + m^2}, \tag{E.19}$$

$$I\left(m_1^2, m_2^2, m_3^2\right) \equiv \int \frac{d^2p\,d^2q\,d^2r}{(2\pi)^4}\,\frac{\delta^{(2)}(p+q+r)}{(p^2 + m_1^2)(q^2 + m_2^2)(r^2 + m_3^2)}, \tag{E.20}$$

in terms of which we can decompose the following integrals ($\mu, \nu, \ldots = 0, 1$ and $(2\pi)^{-4}$ is omitted in the integrands)

$$\int \frac{d^2p\, d^2q\, d^2r\; p^\mu q^\nu\, \delta^{(2)}(p+q+r)}{(p^2+m_1^2)(q^2+m_2^2)(r^2+m_3^2)} = \tag{E.21}$$

$$= \frac{\delta^{\mu\nu}}{4}\left[I(m_1^2)I(m_2^2) - I(m_1^2)I(m_3^2) - I(m_2^2)I(m_3^2) + (m_1^2+m_2^2-m_3^2)I(m_1^2, m_2^2; m_3^2)\right],$$

$$I(m_1^2, m_2^2; m_3^2) = \int \frac{d^2p\, d^2q\, d^2r\;(p\cdot q)\;\delta^{(2)}(p+q+r)}{(p^2+m_1^2)(q^2+m_2^2)(r^2+m_3^2)} = \tag{E.22}$$

$$= \frac{1}{2}\left[I(m_1^2)I(m_2^2) - I(m_1^2)I(m_3^2) - I(m_2^2)I(m_3^2) + (m_1^2+m_2^2-m_3^2)I(m_1^2, m_2^2; m_3^2)\right],$$

$$\int \frac{d^2p\, d^2q\, d^2r\; p^\mu\, p^\nu\, \delta^{(2)}(p+q+r)}{(p^2+m_1^2)(q^2+m_2^2)(r^2+m_3^2)} = \frac{\delta^{\mu\nu}}{2}\left[I(m_2^2)I(m_3^2) - m_1^2\, I(m_1^2, m_2^2; m_3^2)\right], \tag{E.23}$$

$$J \equiv \int \frac{d^2p\, d^2q\, d^2r\; p^2q^2\,\delta^{(2)}(p+q+r)}{(p^2+m_1^2)(q^2+m_2^2)(r^2+m_3^2)} = m_1^2m_2^2\, I(m_1^2, m_2^2; m_3^2) - m_1^2\, I(m_1^2)I(m_3^2) + \\ - m_2^2\, I(m_2^2)I(m_3^2), \tag{E.24}$$

$$K \equiv \int \frac{d^2p\, d^2q\, d^2r\,(p\cdot q)^2\,\delta^{(2)}(p+q+r)}{(p^2+m_1^2)(q^2+m_2^2)(r^2+m_3^2)} = \frac{1}{2}\Big[-m_2^2\, I(m_2^2)I(m_3^2) - m_1^2\, I(m_1^2)I(m_3^2) + \\ +(m_1^2+m_2^2-m_3^2)I^\mu_\mu(m_1^2, m_2^2; m_3^2)\Big], \tag{E.25}$$

$$\int \frac{d^2p\, d^2q\, d^2r\; p^\mu\, p^\nu\, q^\rho\, q^\sigma\,\delta^{(2)}(p+q+r)}{(p^2+m_1^2)(q^2+m_2^2)(r^2+m_3^2)} = \left(\frac{3}{8}J - \frac{1}{4}K\right)\delta^{\mu\nu}\delta^{\rho\sigma} + \\ + \left(\frac{1}{4}K - \frac{1}{8}J\right)\left(\delta^{\mu\rho}\delta^{\nu\sigma} + \delta^{\mu\sigma}\delta^{\nu\rho}\right), \tag{E.26}$$

$$\int \frac{d^2p\, d^2q\, d^2r\; p^\mu\, p^\nu\, p^\rho\, q^\sigma\,\delta^{(2)}(p+q+r)}{(p^2+m_1^2)(q^2+m_2^2)(r^2+m_3^2)} = \frac{1}{8}\left(\delta^{\mu\nu}\delta^{\rho\sigma} + \delta^{\mu\rho}\delta^{\nu\sigma} + \delta^{\mu\sigma}\delta^{\nu\rho}\right), \\ \left[m_2^2\, I(m_2^2)I(m_3^2) - m_1^2\, I^\mu_\mu(m_1^2, m_2^2; m_3^2)\right], \tag{E.27}$$

$$L \equiv \int \frac{d^2p\, d^2q\, d^2r\; p^2\;(q\cdot r)\;\delta^{(2)}(p+q+r)}{(p^2+m_1^2)(q^2+m_2^2)(r^2+m_3^2)} = -m_1^2\, I^\mu_\mu(m_3^2, m_2^2; m_1^2), \tag{E.28}$$

$$M \equiv \int \frac{d^2p\, d^2q\, d^2r\,(p\cdot q)(p\cdot r)\,\delta^{(2)}(p+q+r)}{(p^2+m_1^2)(q^2+m_2^2)(r^2+m_3^2)} = \frac{1}{2}\Big[(m_1^2+m_3^2-m_2^2)I^\mu_\mu(m_1^2, m_2^2; m_3^2) + \\ +m_1^2\, I(m_1^2)I(m_3^2) - m_2^2\, I(m_2^2)I(m_3^2)\Big], \tag{E.29}$$

$$\int \frac{d^2p\, d^2q\, d^2r\; p^\mu\, p^\nu\, q^\rho\, r^\sigma\,\delta^{(2)}(p+q+r)}{(p^2+m_1^2)(q^2+m_2^2)(r^2+m_3^2)} = \left(\frac{3}{8}L - \frac{1}{4}M\right)\delta^{\mu\nu}\delta^{\rho\sigma} + \\ + \left(\frac{1}{4}M - \frac{1}{8}L\right)\left(\delta^{\mu\rho}\delta^{\nu\sigma} + \delta^{\mu\sigma}\delta^{\nu\rho}\right). \tag{E.30}$$

Appendix F
Simulating Strings on the Lattice

This appendix serves as a complement to Chap. 7. Appendix F.1 presents the matrix algebra of the ρ-matrices. Appendices F.2 and F.3 are meant to provide the preliminaries of what we need to discretize and simulate our model and a short presentation of the issue represented by fermion doublers. While we do not to directly address the implementation of numerical routines, the Monte Carlo algorithm is illustrated in Appendix F.4. The reader can find a list of the relevant technical parameters that were used with the $SO(6)$-preserving fermion discretization (7.40)–(7.41) in the main text. We conclude in Appendix F.6 with a summary of the results descending from an alternative fermion discretization that breaks *both* the $U(1)$ and $SO(6)$ symmetry of the model. Further details can be found in [1].

The literature on numerical quantum field theory is extremely vast: we recommend the textbooks [2–4] for a comprehensive and coherent account of the subject.

F.1 $SO(6)$ Matrix Representation

We adopt the chiral representation for the 8×8 Euclidean Dirac matrices γ^M ($M = 1, \ldots 6$) in the fundamental representation of the $\mathfrak{so}(6)$ algebra:

$$\gamma^M \equiv \begin{pmatrix} 0 & \rho_M^\dagger \\ \rho_M & 0 \end{pmatrix} = \begin{pmatrix} 0 & \left(\rho^{\dagger M}\right)^{ij} \\ \left(\rho^M\right)_{ij} & 0 \end{pmatrix}, \qquad \left\{\gamma^M, \gamma^N\right\} = 2\delta^{MN}. \tag{F.1}$$

The two off-diagonal blocks, carrying upper and lower indices respectively, are related by $(\rho^M)^{ij} = -(\rho^M_{ij})^* \equiv (\rho^M_{ji})^*$, so that indeed the block with upper indices is the conjugate transpose of the one with lower indices. For these matrices we use the representation

E. Vescovi, *Perturbative and Non-perturbative Approaches to String Sigma-Models in AdS/CFT*, Springer Theses, DOI 10.1007/978-3-319-63420-3

$$\rho^1_{ij} = \begin{pmatrix} 0 & 1 & 0 & 0 \\ -1 & 0 & 0 & 0 \\ 0 & 0 & 0 & 1 \\ 0 & 0 & -1 & 0 \end{pmatrix}, \quad \rho^2_{ij} = \begin{pmatrix} 0 & i & 0 & 0 \\ -i & 0 & 0 & 0 \\ 0 & 0 & 0 & -i \\ 0 & 0 & i & 0 \end{pmatrix}, \quad \rho^1_{ij} = \begin{pmatrix} 0 & 0 & 0 & 1 \\ 0 & 0 & 1 & 0 \\ 0 & -1 & 0 & 0 \\ -1 & 0 & 0 & 0 \end{pmatrix}, \tag{F.2}$$

$$\rho^4_{ij} = \begin{pmatrix} 0 & 0 & 0 & -i \\ 0 & 0 & i & 0 \\ 0 & -i & 0 & 0 \\ i & 0 & 0 & 0 \end{pmatrix}, \quad \rho^5_{ij} = \begin{pmatrix} 0 & 0 & i & 0 \\ 0 & 0 & 0 & i \\ -i & 0 & 0 & 0 \\ 0 & -i & 0 & 0 \end{pmatrix}, \quad \rho^6_{ij} = \begin{pmatrix} 0 & 0 & 1 & 0 \\ 0 & 0 & 0 & -1 \\ -1 & 0 & 0 & 0 \\ 0 & 1 & 0 & 0 \end{pmatrix}.$$

The Clifford algebra (F.1) induces the anticommutation relations

$$\left(\rho^M\right)_{il}\left(\rho^{\dagger N}\right)^{lj} - \left(\rho^N\right)_{il}\left(\rho^{\dagger M}\right)^{lj} = \delta_i^j \delta^{MN}, \tag{F.3}$$

$$\left(\rho^{\dagger M}\right)^{il}\left(\rho^N\right)_{lj} - \left(\rho^{\dagger N}\right)^{il}\left(\rho^M\right)_{lj} = \delta^i_j \delta^{MN}. \tag{F.4}$$

We can also build the $SO(6)$ generators generators

$$\left(\rho^{MN}\right)_i^{\ j} \equiv \frac{1}{2}\left[\left(\rho^M\right)_{il}\left(\rho^{\dagger N}\right)^{lj} - \left(\rho^N\right)_{il}\left(\rho^{\dagger M}\right)^{lj}\right], \tag{F.5}$$

$$\left(\rho^{\dagger MN}\right)^i_{\ j} \equiv \frac{1}{2}\left[\left(\rho^{\dagger M}\right)^{il}\left(\rho^N\right)_{lj} - \left(\rho^{\dagger N}\right)^{il}\left(\rho^M\right)_{lj}\right], \tag{F.6}$$

for which it holds

$$\begin{aligned} &(\rho^{MN})^i_{\ j} = \frac{1}{2}(\rho^{M i\ell}\,\rho^N_{\ell j} - \rho^{N i\ell}\,\rho^M_{\ell j}) = \frac{1}{2}(\rho^M_{i\ell}\,\rho^{N\ell j} - \rho^N_{i\ell}\,\rho^{M\ell j})^* \equiv \left((\rho^{MN})_i^{\ j}\right)^* \\ &(\rho^{MN})^i_{\ j} = -(\rho^{MN})_j^{\ i} \qquad (\rho^{MN})_i^{\ j} = -(\rho^{MN})^j_{\ i}, \end{aligned} \tag{F.7}$$

where in the last equation we used that $\frac{1}{2}(\rho^{Mi\ell}\,\rho^N_{\ell j} - \rho^{Ni\ell}\,\rho^M_{\ell j}) = -\frac{1}{2}(\rho^M_{j\ell}\,\rho^{N\ell i} - \rho^N_{j\ell}\,\rho^{M\ell i})$.

Useful flipping rules are

$$\eta\,\rho^M\,\theta = \eta^i\,\rho^M_{ij}\,\theta^j = -\theta^j\,\rho^M_{ij}\,\eta^i = \theta^j\,\rho^M_{ji}\,\eta^i \equiv \theta^i\,\rho^M_{ij}\,\eta^j = \theta\,\rho^M\,\eta, \tag{F.8}$$

$$\eta^\dagger\,\rho^\dagger_M\,\theta^\dagger = \eta_i\,\rho^{Mij}\,\theta_j = -\theta_j\,\rho^{Mij}\,\eta_i = \theta_j\,\rho^{Mji}\,\eta_i \equiv \theta_i\,\rho^{Mij}\,\eta_j = \theta^\dagger\,\rho^\dagger_M\,\eta^\dagger, \tag{F.9}$$

$$\eta_i\,(\rho^{MN})^i_{\ j}\,\theta^j = -\theta^j\,(\rho^{MN})^i_{\ j}\,\eta_i = \theta^j\,(\rho^{MN})_j^{\ i}\,\eta_i \equiv \theta^i\,(\rho^{MN})_i^{\ j}\,\eta_j\,. \tag{F.10}$$

F.2 From the Worldsheet to the Lattice

The field theory corresponding to the model (7.24) lives on a two-dimensional Euclidean spacetime equipped with coordinates (t, s). To discretize it, the spacetime is replaced by a two-dimensional array of points Λ with lattice spacing a (Fig. F.1)

$$\Lambda = \{\xi^{\mu} = (a\, n_0, a\, n_1) : n_0 = 0, \ldots N_T - 1, \;\; n_1 = 0, \ldots N_L - 1\} \qquad \text{(F.11)}$$

so that a two-integer label (n_0, n_1) takes the place of the continuous coordinates (t, s). Above, ξ^{μ} is a discrete vector ($\mu = 0, 1$) while N_T and $N_L \equiv N$ are the number of lattice points in the time and space direction respectively. Let us call T and L the corresponding extensions of the grid in length units.[14] The total number of lattice points is $|\Lambda| = N_T N_L$ and we have the relations $T = N_T\, a$ and $L = N_L\, a$. In our simulations the lattice temporal extent T is always twice the spatial extent L.

Surface integration over the worldsheet is replaced by a discrete sum

$$\int dt ds \quad \to \quad a^2 \sum_{n_{\mu} \in \Lambda} . \qquad \text{(F.12)}$$

This formula shows that 2d "volume" V_2 of the classical worldsheet in (7.1) translates into the (dimensionful) lattice volume

$$V_2 \equiv \int dt ds\, 1 \quad \to \quad a^2 |\Lambda| = T\, L\, . \qquad \text{(F.13)}$$

We shall simplify the notation by referring to both sides in (F.13) as simply V_2 when the precise expression is clear from context. We run simulations on a lattice with $N_T = 2N_L$, i.e. $T = 2L \equiv 2aN$ in Tables F.1 and F.2, so we have $V_2 = TL = 2a^2 N^2$ in the main text.

The model (7.24) consists of 8 commuting (x, x^*, z^M with $M = 1, \ldots 6$) and 16 anti-commuting fields (seen as components of $\psi \equiv (\theta^i, \theta_i, \eta^i, \eta_i)^T$), supplemented by 7 commuting auxiliary fields (ϕ, ϕ^M with $M = 1, \ldots 6$). The latter ones are integrated out (7.27) and the resulting 16-component commuting variable ζ (7.29) will be treated in Appendix F.4.1 as a set of non-dynamical degrees of freedom. The discretization of the fields is straightforward because they are all scalars[15]: one assigns them, here generically called f, to each lattice site

$$f(t, s) \to f(n_0, n_1) \qquad \text{(F.14)}$$

---

[14] The time length T has not to be mistaken with the string tension in (7.3).

[15] This observation drastically simplifies the discretization of the field content. For instance, gauge fields would require the introduction of link variables between pairs of neighbouring sites with the net effect of driving up the computational cost.

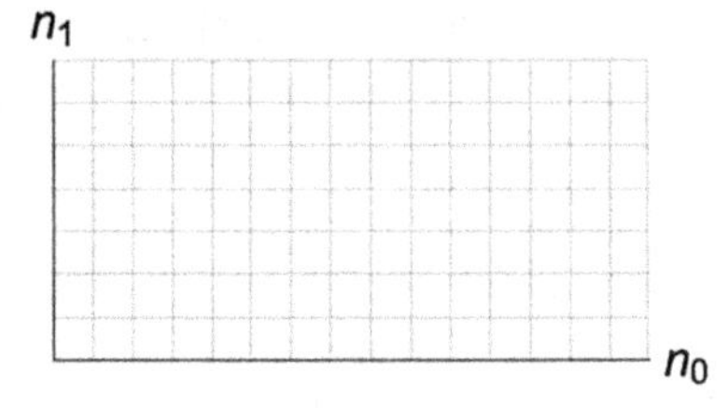

Fig. F.1 Rectangular lattice grid with $N_T = 2N_L = 16$. Scalar fields live at the intersection points

imposing periodic boundary conditions for all fields, save for antiperiodic temporal boundary conditions in the case of Grassmann-odd fields, i.e. in compact form borrowed form [4]

$$f(\xi + \vec{\mu}N_\mu) = e^{2\pi i \delta_\mu(f)} f(\xi), \quad \delta_\mu(f) = \begin{cases} 0 & \text{for } f = x,\, x^*,\, z^M,\, \phi,\, \phi^M \text{ and } \mu = 0, 1 \\ 0 & \text{for } f = \theta^i,\, \theta_i,\, \eta^i,\, \eta_i \text{ and } \mu = 1 \\ \frac{1}{2} & \text{for } f = \theta^i,\, \theta_i,\, \eta^i,\, \eta_i \text{ and } \mu = 0 \end{cases}, \tag{F.15}$$

where we identified $N_0 \equiv N_T$ and $N_1 \equiv N_L$.

The momentum space $\tilde{\Lambda}$ is the reciprocal lattice of Λ

$$\tilde{\Lambda} = \left\{ p_\mu = (p_0, p_1) :\, p_\mu = \frac{2\pi}{aN_\mu}(k_\mu + \delta_\mu(f)),\, k_\mu = -\frac{N_\mu}{2} + 1, \ldots \frac{N_\mu}{2} \right\} \tag{F.16}$$

where points are shifted by the phase $\delta_\mu(f)$ to ensure that plane waves $e^{ip_\mu n_\mu a}$ obey the boundary conditions (F.15) for any $p_\mu \in \tilde{\Lambda}$. Integration in this space becomes a summation

$$\int dp_0 dp_1 \quad \rightarrow \quad \frac{1}{|\Lambda|} \sum_{\vec{p} \in \tilde{\Lambda}} \tag{F.17}$$

and an integral in the infinite-volume limit $T, L \rightarrow \infty$

$$\int dp_0 dp_1 \quad \rightarrow \quad \frac{1}{(2\pi)^2} \int_{-\frac{\pi}{a}}^{\frac{\pi}{a}} \int_{-\frac{\pi}{a}}^{\frac{\pi}{a}} dp_0 dp_1 \,. \tag{F.18}$$

In the latter case, the momenta become continuous and the effect of the field-dependent phase $\delta_\mu(f)$ is washed out.

The lattice Fourier transform and its inverse [2] are defined by

$$\tilde{f}(p_0, p_1) = \sum_{\vec{n} \in \Lambda} f(n_0, n_1) e^{-ip^\mu n_\mu a} \,, \qquad f(n_0, n_1) = \frac{1}{|\Lambda|} \sum_{\vec{p} \in \tilde{\Lambda}} \tilde{f}(p_0, p_1) e^{ip^\mu n_\mu a} \,. \tag{F.19}$$

The discrete approximation of continuum derivatives are finite difference operators. There are different *two-point stencils* for the discretization of $\partial_\mu f(t,s)$: the forward derivative in the direction of the unit vector $\hat{\mu}$

$$\left(\Delta_\mu f\right)(\xi) \equiv \frac{f(\xi) - f(\xi - \hat{\mu})}{a}\,, \tag{F.20}$$

the backward derivative

$$\left(\Delta^*_\mu f\right)(\xi) \equiv \frac{f(\xi) - f(\xi - \hat{\mu})}{a}\,, \tag{F.21}$$

and the centered derivative

$$\left(\bar{\Delta}_\mu f\right)(\xi) \equiv \frac{\left(\Delta_\mu f + \Delta^*_\mu f\right)(\xi)}{2} = \frac{f(\xi + \hat{\mu}) - f(\xi - \hat{\mu})}{2a}\,. \tag{F.22}$$

The first two are accurate discretizations up to $O(a)$ corrections, and this improves to $O(a^2)$ in the third case. We always pick the centered derivative due to its smaller errors. It is also useful to write its expression in momentum space

$$\left(\tilde{\bar{\Delta}}_\mu\right)(p) = \frac{i}{a}\sin(p_\mu a) \equiv i\,\mathring{p}_\mu\,. \tag{F.23}$$

The wave operator $\Box f(\tau,\sigma) \equiv \partial_\mu\partial_\mu f(\tau,\sigma)$ acts on a lattice field as

$$(\Box f)(\xi) \equiv \left(\Delta^*_\mu \Delta_\mu f\right)(\xi) = \frac{1}{a^2}\sum_{\hat{\mu}}\left[f(\xi+\hat{\mu}) - 2f(\xi) + f(\xi - \hat{\mu})\right]. \tag{F.24}$$

and in Fourier space as

$$\left(\tilde{\Box}\right)(p) = \sum_{\mu=0,1}\left(\frac{2}{a}\sin\frac{p_\mu a}{2}\right)^2 \equiv \sum_{\mu=0,1}\hat{p}_\mu^2\,. \tag{F.25}$$

The path-integral (7.28) becomes a multi-dimensional integral[16]

$$\int\prod_{i=1}^{15}\mathcal{D}f_i\,(\mathrm{Det}(\hat{\mathcal{O}}_F\,\hat{\mathcal{O}}_F^\dagger)[\{f_i\}])^{\frac{1}{4}}\,e^{-S_B[\{f_i\}]}$$
$$\rightarrow \int\prod_{i=1}^{15}\prod_{n_0=0}^{N_T-1}\prod_{n_1=0}^{N_L-1}df_i(n_0,n_1)\,(\mathrm{Det}(\hat{\mathcal{O}}_F\,\hat{\mathcal{O}}_F^\dagger)[\{f_i\}])^{\frac{1}{4}}_{\mathrm{LAT}}\,e^{-S_{B,\mathrm{LAT}}[\{f_i\}]}\,. \tag{F.26}$$

[16]We will run phase-quenched simulations, see Sect. 7.6.3.

where the product over lattice sites is discrete and the infinite-volume limit will be taken afterwards. In the spirit of previous appendices, we have already specialized to the 2d model of interest with field content

$$f_i = (x, x^*, z^1, z^2, z^3, z^4, z^5, z^6, \phi, \phi^1, \phi^2, \phi^3, \phi^4, \phi^5, \phi^6)_i \qquad i = 1, \dots 15\,. \tag{F.27}$$

We also understood the various quantities labeled by "LAT" in (F.26) to be the discretizations of the their continuum counterparts according to the prescriptions above.

F.3 Fermion Doublers in the Standard Dirac Operator

This appendix briefly elucidates the origin of the *fermion doubling problem* raised at the beginning of Sect. 7.4. Doublers are identified when one computes the inverse lattice fermionic operator in momentum space for the case of free lattice fermions in d dimensions. In the *naive* discretization one replaces partial derivatives in the continuum action $S_{\rm ferm}$ (γ^μ are $SO(d)$ Dirac matrices)

$$S_{\rm ferm} = \int d^d x\, \bar{\psi}(x)(\gamma^\mu \partial_\mu + m)\psi(x) \tag{F.28}$$

with centered derivatives (F.22) to obtain

$$S_{\rm ferm,\,LAT} = a^d \sum_{\vec{n}=(n_1,\dots n_d)\in\Lambda} \bar{\psi}(\vec{n}) \sum_{\hat{\mu}} \left(\frac{\psi(\vec{n}+\vec{\mu}) - \psi(\vec{n}-\vec{\mu})}{2a} + m\,\psi(\vec{n}) \right)\,. \tag{F.29}$$

The inverse of the lattice Dirac operator in momentum space is (see equation (5.43) in [4])

$$\tilde{D}(p)^{-1} = \frac{m\,\mathbb{I} - \frac{i}{a}\sum_{\mu=1}^{d} \gamma_\mu \sin(p_\mu a)}{m^2 + \frac{1}{a^2}\sum_{\mu=1}^{d} \sin^2(p_\mu a)}\,, \qquad p^2 = \sum_{\mu=1}^{d} p_\mu^2\,. \tag{F.30}$$

In the continuum limit $a \to 0$ it has a pole in $p^2 = -m^2$ in correspondence to the value of the physical mass. The situation is different on the lattice because additional poles appear where one or more components p_μ are at the corners of the first Brillouin zone (i.e. equal to $\frac{\pi}{a}$). For a massless fermion this explicitly means that each component is either 0 or $\frac{\pi}{a}$:

$$p_\mu = \left(\frac{\pi}{a}, 0, \dots 0\right), \left(0, \frac{\pi}{a}, \dots 0\right), \dots \left(\frac{\pi}{a}, \frac{\pi}{a}, \dots \frac{\pi}{a}\right)\,. \tag{F.31}$$

When one of the components is close to $\frac{\pi}{a}$ the lattice fermion propagator (F.30) behaves in the same way as around the point $p^2 = -m^2$. One physical fermion is then accompanied by $2^d - 1$ unphysical excitations in the spectrum of the theory at finite a. In two-dimensions the operator (7.25) would produce a total number 16×3 doublers. This observation motivates the discussion of Sect. 7.4 to construct the doubler-free operator (7.40)–(7.41).

F.4 Monte Carlo Algorithm for the Worldsheet Model

Numerical simulations of lattice field theory can measure expectation values of functionals $F[\{f_i\}]$ of the field variables (F.27), providing a practical mean to measure the quantum mechanical average

$$\langle F \rangle \equiv \frac{1}{Z} \int \prod_{i=1}^{15} \mathcal{D} f_i \; F[\{f_i\}] \, (\mathrm{Det}(\hat{\mathcal{O}}_F \, \hat{\mathcal{O}}_F^\dagger)[\{f_i\}])^{\frac{1}{4}} \, e^{-S_B[\{f_i\}]} \tag{F.32}$$

$$\text{with } Z \equiv \int \prod_{i=1}^{15} \mathcal{D} f_i \; (\mathrm{Det}(\hat{\mathcal{O}}_F \, \hat{\mathcal{O}}_F^\dagger)[\{f_i\}])^{\frac{1}{4}} \, e^{-S_B[\{f_i\}]} \, .$$

We consider a bosonic path-integral where the fermionic factor $(\mathrm{Det}(\hat{\mathcal{O}}_F \, \hat{\mathcal{O}}_F^\dagger))^{\frac{1}{4}}$ is a real functional of the fields.[17] For us the functional of fields is $F = x x^*$ in Sect. 7.6.1 and $F = S_B$ in Sect. 7.6.2. The regularization imposed by the lattice in Appendix F.2 translates the expressions above into

$$\langle F_{\mathrm{LAT}} \rangle \equiv \frac{1}{Z_{\mathrm{LAT}}} \int \prod_{i=1}^{15} \prod_{n_0=0}^{N_T-1} \prod_{n_1=0}^{N_L-1} d f_i(n_0, n_1) \; F_{\mathrm{LAT}}[\{f_i\}] \, (\mathrm{Det}(\hat{\mathcal{O}}_F \, \hat{\mathcal{O}}_F^\dagger)[\{f_i\}])^{\frac{1}{4}}_{\mathrm{LAT}} \, e^{-S_{B,\mathrm{LAT}}[\{f_i\}]}$$

$$\text{with } Z_{\mathrm{LAT}} \equiv \int \prod_{i=1}^{15} \prod_{n_0=0}^{N_T-1} \prod_{n_1=0}^{N_L-1} d f_i(n_0, n_1) \; (\mathrm{Det}(\hat{\mathcal{O}}_F \, \hat{\mathcal{O}}_F^\dagger)[\{f_i\}])^{\frac{1}{4}}_{\mathrm{LAT}} \, e^{-S_{B,\mathrm{LAT}}[\{f_i\}]} \, . \tag{F.33}$$

We now concentrate on how to measure the lattice observables and leave the topic of their continuum extrapolation in Sect. 7.5.

The number of degrees of freedom (15 field variables for each of the $N_T \, N_L$ sites) increases rapidly with the lattice size and makes a deterministic numerical integration of (F.33) prohibitive in terms of computation time. The central idea of *Monte Carlo (MC) simulations* is to replace the quantum average with a time average over a sequence of states produced by a stochastic dynamics. One resorts to randomly generate a sequence of field configurations $\{f_i\}_j$ ($j = 1, \ldots N_{\mathrm{MC}}$) according to a statistical probability $P(\{f_i\}_j)$, evaluate the corresponding values of the observable and compute their arithmetic sum

[17] See comments in footnote 3.

$$\langle F_{\rm LAT}\rangle = \frac{1}{N_{\rm MC}} \sum_{j=1}^{N_{\rm MC}} F_{\rm LAT}[\{f_i\}_j] + O(N_{\rm MC}^{-1/2}), \qquad N_{\rm MC} \gg 1. \qquad (F.34)$$

The probability distribution is the *Gibbs measure* (modified by the inclusion of the fermionic determinant)

$$P(\{f_i\}_j) = \frac{1}{Z_{\rm LAT}} (\mathrm{Det}(\hat{\mathcal{O}}_F\, \hat{\mathcal{O}}_F^\dagger)[\{f_i\}])_{\rm LAT}^{\frac{1}{4}}\, e^{-S_{B,\rm LAT}[\{f_i\}]}, \qquad 0 < P(\{f_i\}_j) < 1, \qquad (F.35)$$

This method replaces the exact average by a sample over the phase space of the quantum fields. The *sample average* (F.34) for the Monte Carlo integration captures the expectation value of the observable in the lattice field theory (F.33) within (probabilistic) accuracy of order $N_{\rm MC}^{-1/2}$ by the central limit theorem.[18]

The MC method relies on an algorithm to find field configurations distributed according to the probability (F.35). To this end, one considers a *Markov chain* to construct a sequence of configurations from an initial arbitrary state $\{f_i\}_1$

$$\{f_i\}_1 \to \{f_i\}_2 \to \cdots \to \{f_i\}_j \to \cdots \to \{f_i\}_{N_{\rm MC}}, \qquad (F.36)$$

where the probability of being in the state $\{f_i\}$ is given by the same $P(\{f_i\})$ in Eq. (F.35). Any *Markov state* consists for us of $15N_T N_L$ field variables describing the values of the fields on the discretized spacetime (F.11). The update of a state to the next one is a *Monte Carlo step* and the length of the chain is $N_{\rm MC}$ *Monte Carlo units*.

The state of the system undergoes a stochastic process where each state $\{f_i\}$ evolves into a new one $\{f_i'\}$ with a conditional *transition probability* $P_M\left(\{f_i\} \to \{f_i'\}\right)$ that depends on the current state $\{f_i\}$ only. The latter property is what defines a *time-homogeneous* Markov chain. Together with the *detailed balance equation*

$$P(\{f_i\})\, P_M\left(\{f_i\} \to \{f_i'\}\right) = P_M(\{f_i'\})\, P\left(\{f_i'\} \to \{f_i\}\right), \qquad (F.37)$$

it guarantees that the *equilibrium distribution* of the chain exists and coincides with (F.35). This means that—after the information on the starting configuration is lost (*thermalization*)—we eventually reach the equilibrium distribution regardless of which state the process begins in. Hence we have a way to produce a canonical *ensemble* of configurations that appears in the chain (for sufficiently large j) with frequency (F.35) and can be included into the ensemble average (F.34). However, at equilibrium subsequent configurations are not completely uncorrelated. The typical

---

[18]The length of a typical *Monte Carlo history* in our simulations is limited to $N_{\rm MC} \approx 10^3$ by the available computer resources. The index j labels the configurations as they are subsequently generated. Since j is a measure of the "fictitious time" of the stochastic process, it is not to be mistaken with the Wick-rotated time t of the 2d worldsheet/lattice or the time variable $t_{\rm HMC}$ of the Hybrid Monte Carlo algorithm (see below F.40 in the next appendix).

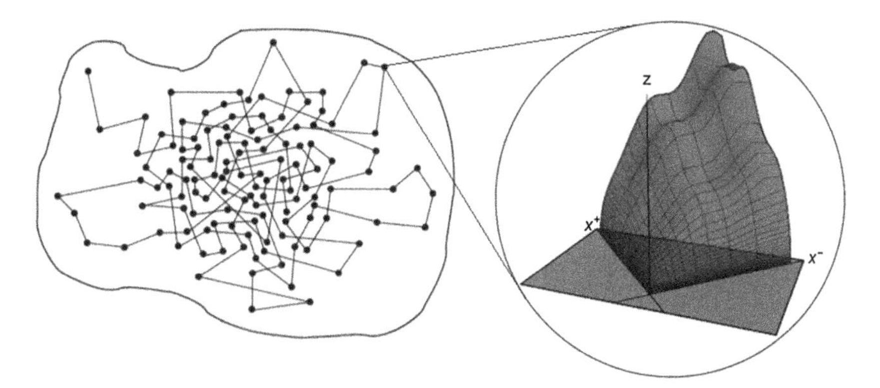

Fig. F.2 Sketch of a Markov chain in the space of field configurations. A *black dot* symbolizes a configuration of the physical x, x^*, z^M and auxiliary scalars ϕ, ϕ^M over the 2d lattice grid. In the inset it is graphically represented for one of them (z) as the "vertical" fluctuation of the string embedding in AdS_3. Each *black* segment is a state update occurred in the time of one Monte Carlo step. The *dots* tends to visit the region where the Boltzmann factor (F.35) is larger, here at the *center* of the blob. Figure adapted from [4]

length of a series of states that keeps "memory" of the first configuration is given by the *autocorrelation time* [2]. An optimal number of sample configurations must be much larger than this time scale (Fig. F.2).

Another important point to address is whether any configuration can be eventually reached in a finite number of steps. In our case, the condition $P(\{f_i\} \to \{f_i'\}) > 0$ for any pair of states guarantees that the stochastic process can explore the entire state space. An efficient algorithm to randomly generate new configurations takes also into account that a direct sampling of the entire state space proves to be computationally too difficult, while only a relatively small number of field configurations is statistically significant. The *importance sampling* involves visiting more frequently states with significant Boltzmann weight. This concept is incorporated in the type of MC algorithm that we present in the next paragraph.

F.4.1 Rational Hybrid Monte Carlo Algorithm

The lattice simulations in Chap. 7 are prepared with a type of *Rational Hybrid Monte Carlo* (RHMC) algorithm [5–7].

"Rational" refers to the fact that it differs from the standard HMC (described below) in the treatment of the fermion contribution to the action, as it uses a rational approximation of the fourth root of operators called *Remez algorithm* [7,8]

$$(M^{\dagger}M)^{-\frac{1}{4}} = \alpha_0 + \sum_{i=1}^{P} \frac{\alpha_i}{M^{\dagger}M + \beta_i}, \tag{F.38}$$

here with $M = \hat{O}_F^\dagger$ (7.40)–(7.41) and coefficients α_i and β_i tuned by the range of eigenvalues of M to optimize the rational approximation. More information is in Appendix F.5.

The term "hybrid" Monte Carlo [9] characterizes the Markov chain as an algorithm that uses deterministic molecular dynamics to generate new configurations, paired up with a Metropolis acceptance test[19] (see below).

We shall work with an equivalent form of (F.33) in terms of the 16 complex pseudo-fermions ζ_k ($k = 1, \dots 16$) in (7.29)

$$\begin{aligned}\langle F_{\mathrm{LAT}}\rangle &= \frac{1}{Z_{\mathrm{LAT}}}\int \prod_{n_0=0}^{N_T-1}\prod_{n_1=0}^{N_L-1}\left(\prod_{i=1}^{15} df_i(n_0,n_1)\right)\left(\prod_{k=1}^{16} d\zeta_k^\dagger(n_0,n_1)d\zeta_k(n_0,n_1)\right)\\ &\quad\times F_{\mathrm{LAT}}[\{f_i\}]\, e^{-S_{B,\mathrm{LAT}}[\{f_i\}]-S_{\zeta,\mathrm{LAT}}[\{f_i\}]} \qquad \text{(F.39)}\\ \text{with } Z_{\mathrm{LAT}} &= \int \prod_{n_0=0}^{N_T-1}\prod_{n_1=0}^{N_L-1}\left(\prod_{i=1}^{15} df_i(n_0,n_1)\right)\left(\prod_{k=1}^{16} d\zeta_k^\dagger(n_0,n_1)d\zeta_k(n_0,n_1)\right)\\ &\quad\times e^{-S_{B,\mathrm{LAT}}[\{f_i\}]-S_{\zeta,\mathrm{LAT}}[\{f_i\}]}\,.\end{aligned}$$

We also postulate the evolution Hamiltonian

$$H_{\mathrm{HMC}} = \frac{1}{2}\sum_{i=1}^{15}\pi_i^2 + S_{B,\mathrm{LAT}}[\{f_i\}] + S_{\zeta,\mathrm{LAT}}[\{\zeta_k\}] \qquad \text{(F.40)}$$

to describe the evolution of the fields in the "fictitious" time t_{HMC}. At variance with [12], the pseudo-fermions do not undergo a time evolution and we only need the canonically conjugate momenta π_i to the f_i fields.

A MC simulation begins with the initialization of the field variables $\{f_i\}_1$. The central part of the program is the updating subroutine to advance the Markov chain from $\{f_i\}_j$ to $\{f_i\}_{j+1}$ in succession for $j = 1, \dots N_{\mathrm{MC}}$.

1. Suppose the system is in a state $\{f_i\}$ at a certain j. At the beginning of each HMC trajectory the momenta π_i, the real and imaginary part of the pseudo-fermions ζ_k are randomly generated with Gaussian distribution centered at zero and with unit standard deviation.
2. To make a proposal for a new configuration, we numerically integrate the Hamilton's equations of motion

$$\frac{\partial \pi_i}{\partial t_{\mathrm{HMC}}} = -\frac{\partial H_{\mathrm{HMC}}}{\partial f_i} \equiv F_i\,, \qquad \frac{\partial f_i}{\partial t_{\mathrm{HMC}}} = \frac{\partial H_{\mathrm{HMC}}}{\partial \pi_i} = \pi_\phi\,, \qquad \text{(F.41)}$$

to evolve deterministically the state into $\{f_i'\}$ (Fig. F.3). The time is discretized in small steps of size δt and the F_i are the so-called *forces*. Formulas in (F.41)

[19] See [10] (also in Chap. 5 of [11]) for a detailed discussion of the algorithm and its variations.

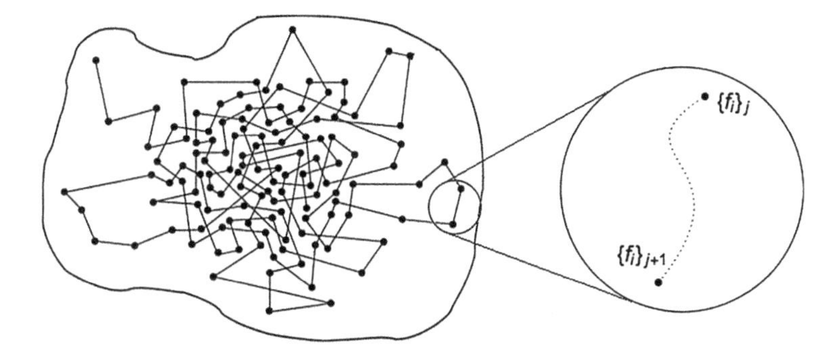

Fig. F.3 Sketch of a HMC trajectory. In the magnified area, an initial state is driven by the molecular dynamics (F.41) to a final configuration. The latter is depicted as passing the acceptance test and turning into the next chain state. The *dotted* trajectory lies on a hypersurface of constant energy within numerical errors. Figure adapted from [4]

are also known as *molecular dynamics equations* because of their evident resemblance with the differential equations governing the mechanical evolution of a classical system.

The numerical integration uses the *leapfrog scheme*: we update the values of fields and their conjugate momenta at staggered time steps as in Fig. F.4:

$$f_i(t_{\mathrm{HMC}} + \delta t/2) = f_i(t_{\mathrm{HMC}}) + \pi_i(t_{\mathrm{HMC}})\delta t\,, \tag{F.42}$$

$$\pi_i(t_{\mathrm{HMC}} + 3\delta t/2) = \pi_i(t_{\mathrm{HMC}} + \delta t/2) + F_i(t_{\mathrm{HMC}})\delta t\,. \tag{F.43}$$

It is a second-order method (the error per step is proportional to δt^2) in contrast to the Euler integration (linear in δt).

3. We obtain a *HMC trajectory* made of n_{HMC} time steps. The final configuration $\{f_i'\}$ is affected by numerical errors that make the energy only approximately conserved. One needs a quantitative way to test whether $\{f_i'\}$ can be taken as the result of a truly quantum mechanical evolution at $\delta t = 0$. Although one can extrapolate to vanishing δt, it is practically more convenient to introduce a corrective test called *Metropolis algorithm* [9]. The acceptance test renders the discrete Hamiltonian evolution exact for any small, yet finite, δt.
The candidate configuration undergoes the *acceptance test*: it is accepted with probability given by $\min\left(1, e^{-\delta H}\right)$ with $\delta H = H_{\mathrm{HMC}}[\{f_i'\}] - H_{\mathrm{HMC}}[\{f_i\}]$.
   (a) This means that the state is always updated if the change in the energy is negative, so driving the simulation towards states of higher probability;
   (b) otherwise, the test can still be passed if the condition $e^{-\delta H} > n$ is verified, where n is a number randomly generated in the interval $(0, 1)$ with flat distribution. The proposal is rejected if $e^{-\delta H} < n$.

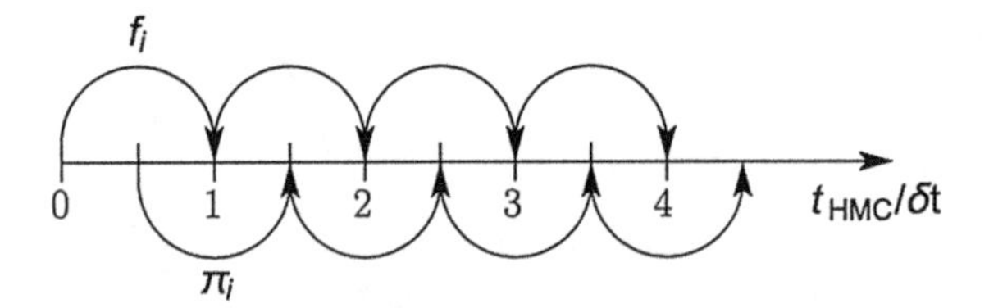

Fig. F.4 Leapfrog integration advances fields and momenta at evenly-spaced interleaved time points in a manner that they "leapfrog" over each other. Courtesy of [12]

Notice the if we could solve the equations of motion (F.41) for zero δt, the time evolution would be exactly energy-preserving ($\delta H = 0$) and any configuration would always be accepted, rendering the Metropolis test superfluous.

4. The chain is either updated with the candidate $\{f_i\}_{j+1} = \{f_i'\}$ or the starting configuration is considered as the new state $\{f_i\}_{j+1} = \{f_i\}_j$.
5. The process restarts from point 1 taking $\{f_i\}_{j+1}$ as the next configuration to evolve.

F.5 Subroutines and Simulation Parameters

The source code has been implemented[20] in Fortran 95 and uses the `ranlux` pseudo-random number generator [13].[21]

- The Remez algorithm (F.38) uses a rational approximation of degree $P = 15$ [14] and we checked for a subset of the generated configurations that its accuracy is always better than 10^{-3}.
- The right hand side of (F.38) enters the bilinear in the pseudo-fermionic Lagrangian in (7.29)

$$\zeta^\dagger(\hat{O}_F\hat{O}_F^\dagger)^{-\frac{1}{4}}\zeta = \alpha_0\,\zeta^\dagger\zeta + \sum_{i=1}^{P}\zeta^\dagger\frac{\alpha_i}{\hat{O}_F\hat{O}_F^\dagger + \beta_i}\zeta \tag{F.44}$$

and eventually appears in the bosonic forces (F.41) via (F.40). Defining $s_i \equiv \frac{1}{\hat{O}_F\hat{O}_F^\dagger+\beta_i}\zeta$ suggests to evaluate the summand in (F.44) by solving a system of P matrix equations

$$\left(\hat{O}_F\hat{O}_F^\dagger + \beta_i\right)s_i = \zeta \qquad i = 1, \ldots P\,. \tag{F.45}$$

The equations are solved simultaneously at the cost of solving one single equation using a *multi-mass conjugate gradient* solver [15]. This is an iterative algorithm with a total computational cost given by the number of iterations required to solve the slowest-convergent equation in (F.45).

[20]We are very thankful to Mattia Bruno for having played a key role in setting up the code in an early stage of the project.

[21]The implementation in C language is available at http://luscher.web.cern.ch/luscher/ranlux/index.html.

- Each HMC trajectory stops after a time span of $t_{\mathrm{HMC}} = 0.5$. The typical number of integration steps n_{HMC} ranges from 50 to 100, which means a time step δt between 0.005 and 0.01. In the simulation runs of Table F.1 the acceptance probability of the Metropolis test is above 90% in order to quickly sample a large region of the configuration space.

Table F.1 summarizes the other relevant parameters: the coupling g, the temporal and spatial extent of the lattice $N_T \times N_L = \frac{T}{a} \times \frac{L}{a}$ in units of the lattice spacing a, the line of constant physics fixed by $L\,m$ and the mass parameter $M = a\,m$.

- The size of the statistics after thermalization is given in the last column in terms of Molecular Dynamic Units (MDU). One unit of the time variable t_{HMC} defines one MDU and since a HMC trajectory lasts up to $t_{\mathrm{HMC}} = 0.5$, a statistics of n MDU equals a Monte Carlo history of $2n$ HMC trajectories.

  In the case of multiple replica the statistics for each replica is given separately.
- The auto-correlation times τ of our main observables are given in MDU.

F.6 $SO(6)$-Breaking Wilson-Like Term

We collect the results of the simulations performed employing an alternative discretization with fermionic operator

$$\widetilde{O}_F = \begin{pmatrix} \widetilde{W}_+ & -\mathring{p}_0 \mathbb{I}_4 & (\mathring{p}_1 - i\,\frac{m}{2})\rho^M \frac{z^M}{z^3} & 0 \\ -\mathring{p}_0 \mathbb{I}_4 & -\widetilde{W}_+^\dagger & 0 & \rho_M^\dagger (\mathring{p}_1 - i\,\frac{m}{2}) \frac{z^M}{z^3} \\ -(\mathring{p}_1 + i\,\frac{m}{2})\rho^M \frac{z^M}{z^3} & 0 & 2\frac{z^M}{z^4}\rho^M (\partial_s x - m\,\frac{x}{2}) + \widetilde{W}_- & -\mathring{p}_0 \mathbb{I}_4 - A^T \\ 0 & -\rho_M^\dagger (\mathring{p}_1 + i\,\frac{m}{2}) \frac{z^M}{z^3} & -\mathring{p}_0 \mathbb{I}_4 + A & -2\frac{z^M}{z^4}\rho_M^\dagger (\partial_s x^* - m\,\frac{x}{2}^*) - \widetilde{W}_-^\dagger \end{pmatrix} \tag{F.46}$$

where the only change with respect to (7.40) is in the Wilson term adopted here (Fig. F.5)

$$\widetilde{W}_\pm = \frac{r}{2\,z^3}\left(\hat{p}_0^2 \pm i\,\hat{p}_1^2\right)\left(\rho_6 \sum_{M=1}^{5} z^M z^M + \rho_1\,(z^6)^2\right). \tag{F.47}$$

We employed this discretization to present our preliminary results in [16] with the simulation parameters in Table F.2. It is consistent with lattice perturbation theory performed around a vacuum with all six entries of u^M vanishing but one (see (7.16)). It satisfies all requirements in the bulleted list of Sect. 7.4.1, except for the $SO(6)$- and $U(1)$-invariance, the latter being broken by (7.40) as well. One can compare the continuum extrapolations of the two observables under investigation in the

Table F.1 Parameters of the simulation runs with the discretization (7.40)–(7.41)

g	$T/a \times L/a$	$L\,m$	$a\,m$	τ_{int}^{S}	$\tau_{\text{int}}^{m_x}$	Statistics (MDU)
5	16×8	4	0.50000	0.8	2.2	900
	20×10	4	0.40000	0.9	2.6	900
	24×12	4	0.33333	0.7	4.6	900, 1000
	32×16	4	0.25000	0.7	4.4	850, 1000
	48×24	4	0.16667	1.1	3.0	92, 265
10	16×8	4	0.50000	0.9	2.1	1000
	20×10	4	0.40000	0.9	2.1	1000
	24×12	4	0.33333	1.0	2.5	1000, 1000
	32×16	4	0.25000	1.0	2.7	900, 1000
	48×24	4	0.16667	1.1	3.9	594, 564
20	16×8	4	0.50000	5.4	1.9	1000
	20×10	4	0.40000	9.9	1.8	1000
	24×12	4	0.33333	4.4	2.0	850
	32×16	4	0.25000	7.4	2.3	850, 1000
	48×24	4	0.16667	8.4	3.6	264, 580
30	20×10	6	0.60000	1.3	2.9	950
	24×12	6	0.50000	1.3	2.4	950
	32×16	6	0.37500	1.7	2.3	975
	48×24	6	0.25000	1.5	2.3	533, 652
	16×8	4	0.50000	1.4	1.9	1000
	20×10	4	0.40000	1.2	2.7	950
	24×12	4	0.33333	1.2	2.1	900
	32×16	4	0.25000	1.3	1.8	900, 1000
	48×24	4	0.16667	1.3	4.3	150
50	16×8	4	0.50000	1.1	1.8	1000
	20×10	4	0.40000	1.2	1.8	1000
	24×12	4	0.33333	0.8	2.0	1000
	32×16	4	0.25000	1.3	2.0	900, 1000
	48×24	4	0.16667	1.2	2.3	412
100	16×8	4	0.50000	1.4	2.7	1000
	20×10	4	0.40000	1.4	4.2	1000
	24×12	4	0.33333	1.3	1.8	1000
	32×16	4	0.25000	1.3	2.0	950, 1000
	48×24	4	0.16667	1.4	2.4	541

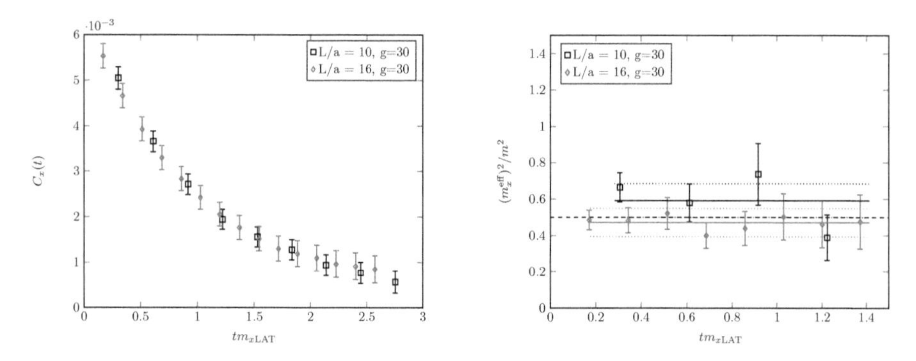

Fig. F.5 Correlator and mass for the x field, realized here using the $SO(6)$-breaking discretization (F.46)–(F.47). Detailed explanation and comments as in Fig. 7.5

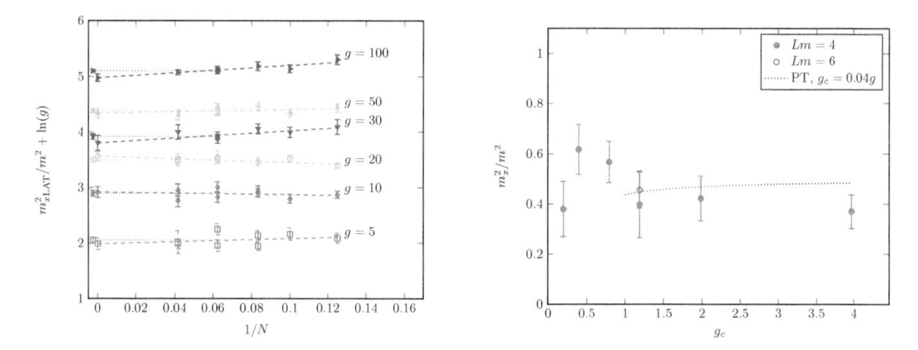

Fig. F.6 Plot of $m^2_{\mathrm{xLAT}}(N, g)/m^2 = m_x(g) + \mathcal{O}(1/N)$ and its continuum extrapolation, realized here using the $SO(6)$-breaking discretization (F.46)–(F.47). Detailed explanation and comments as in Fig. 7.6

two discretizations, namely Fig. 7.6 with Fig. F.6 for the x-mass and Fig. 7.10 with Fig. F.10 for the action. They agree within errors, which strongly suggests that the two discretizations lead to the same continuum limit (Figs. F.7, F.8, F.9 and F.10).

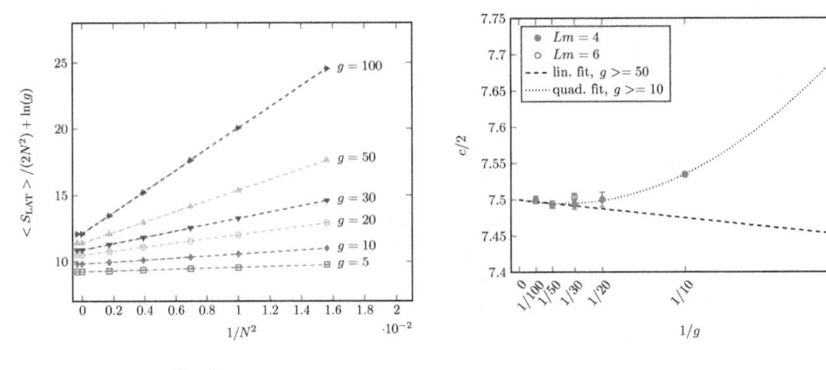

Fig. F.7 Plots of $\frac{\langle S_{\rm LAT}\rangle}{2N^2}$ and its continuum extrapolation to determine $c/2$, realized here using the $SO(6)$-breaking discretization (F.46)–(F.47). Detailed explanation and comments as in Fig. 7.7

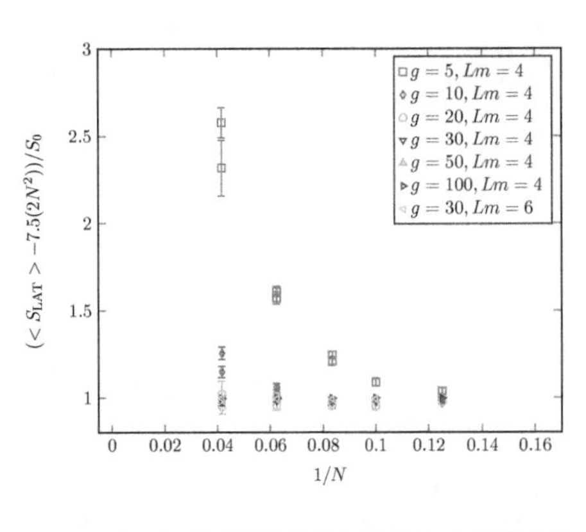

Fig. F.8 Plot of the ratio $\frac{\langle S_{\rm LAT}\rangle - \frac{c}{2}(2N^2)}{S_0} \equiv \frac{f'(g)}{4}$, realized here using the $SO(6)$-breaking discretization (F.46)–(F.47). Detailed explanation and comments as in Fig. 7.8

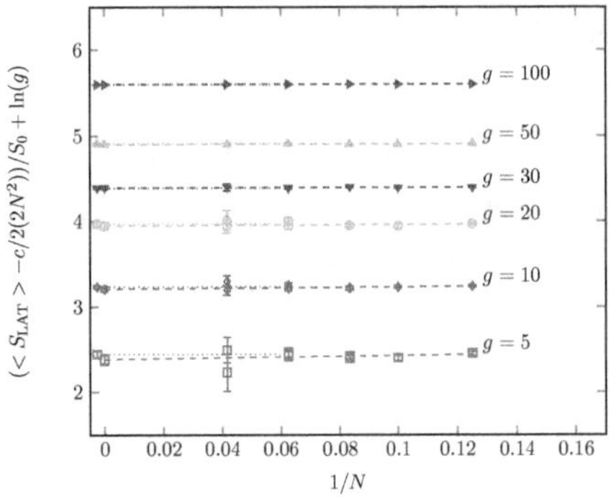

Fig. F.9 Plots for the ratio $\frac{\langle S_{\rm LAT}\rangle - \frac{c}{2}(2N^2)}{S_0} + \ln g$ as a function of $1/N$, realized here using the $SO(6)$-breaking discretization (F.46)–(F.47). Detailed explanation and comments as in Fig. 7.9

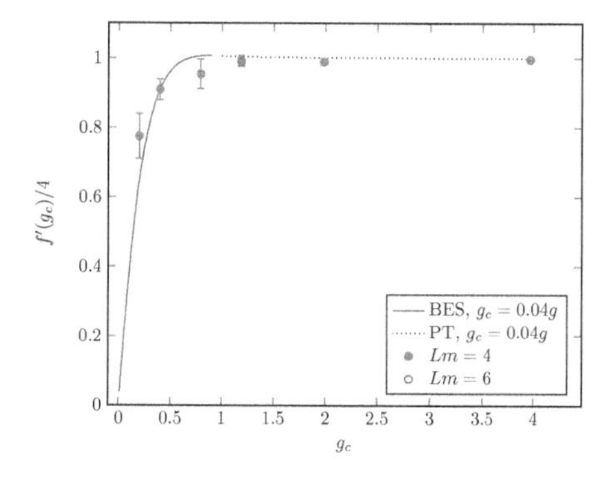

Fig. F.10 Plot for $f'(g)/4$ as determined from the $N \to \infty$ extrapolation of (7.52), realized here using the $SO(6)$-breaking discretization (F.46)–(F.47). Detailed explanation and comments as in Fig. 7.10

Table F.2 Parameters of the simulation runs with the discretization (F.46)–(F.47). See Appendix F.5 for explanations

g	$T/a \times L/a$	Lm	am	τ_{int}^{S}	$\tau_{\text{int}}^{m_x}$	Statistics (MDU)
5	16×8	4	0.50000	0.8	2.7	900
	20×10	4	0.40000	0.8	2.8	900
	32×16	4	0.25000	2.0	8.1	950, 950
10	20×10	8	0.80000	1.1	2.2	900
	24×12	8	0.66667	1.4	2.5	900
	32×1	8	0.50000	2.4	5.8	750, 750
	40×20	8	0.40000	5.8	10.6	900, 900
	16×8	4	0.50000	0.8	1.9	900
	20×10	4	0.40000	1.0	2.2	900
	24×12	4	0.33333	1.1	2.6	900, 900
	32×16	4	0.25000	1.9	5.0	925, 925
	40×20	4	0.20000	7.8	11.7	925, 925
20	16×8	4	0.50000	8.7	2.7	1000
	20×10	4	0.40000	10.9	2.3	1000
	24×12	4	0.33333	4.7	2.0	1000
	32×16	4	0.25000	6.5	3.3	850
	48×24	4	0.16667	6.2	3.2	918
30	16×8	4	0.50000	1.3	2.0	800
	20×10	4	0.40000	1.2	2.1	800
	24×12	4	0.33333	1.7	2.9	900
	32×16	4	0.25000	2.7	4.1	950, 950
	40×20	4	0.20000	3.7	11.0	950, 900
	64×32	4	0.12500	6.9	31.1	579, 900
100	16×8	4	0.50000	1.6	3.3	900
	20×10	4	0.40000	2.0	3.8	750
	32×16	4	0.25000	2.8	3.8	900, 900
	40×20	4	0.20000	6.2	10.4	900, 900

Bibliography

1. P. Topfer, Lattice discretisation of the Green-Schwarz superstring and AdS/CFT. Master Thesis, https://github.com/3uler/master-thesis/blob/master/main/main.pdf.
2. I. Montvay, G. Muenster, in *Quantum Fields on a Lattice* (Cambridge University Press, 1994).
3. H. Rothe, in *Lattice Gauge Theories: An Introduction* (World Scientific, 2005).
4. C. Gattringer, C. Lang, in *Quantum Chromodynamics on the Lattice: An Introductory Presentation* (Springer Berlin Heidelberg, 2009).
5. A.D. Kennedy, I. Horvath, S. Sint, A New exact method for dynamical fermion computations with nonlocal actions. Nucl. Phys. Proc. Suppl. **73**, 834 (1999), arXiv:hep-lat/9809092 (in, *Lattice Field Theory. Proceedings: 16th International Symposium, Lattice'98*, Boulder, USA, Jul 13–18, 1998, pp. 834–836).
6. M.A. Clark, A.D. Kennedy, The RHMC algorithm for two flavors of dynamical staggered fermions. Nucl. Phys. Proc. Suppl. **129**, 850 (2004), arXiv:hep-lat/0309084 (in, *Lattice field theory. Proceedings, 21st International Symposium, Lattice 2003, Tsukuba, Japan, July 15–19, 2003*, pp. 850–852, [850(2003)]).
7. M.A. Clark, The Rational Hybrid Monte Carlo Algorithm. PoS LAT2006, 004 (2006), arXiv:hep-lat/0610048, (in, *Proceedings, 24th International Symposium on Lattice Field Theory (Lattice 2006)*, p. 004
8. E.Y. Remez, Sur le calcul effectif des polynomes d'approximation de Tschebyscheff. C.R. Acad. Sci. Paris **199**, 337 (1934).
9. S. Duane, A.D. Kennedy, B.J. Pendleton, D. Roweth, Hybrid Monte Carlo. Phys. Lett. B **195**, 216 (1987).
10. R.M. Neal, MCMC using Hamiltonian dynamics, arXiv:1206.1901.
11. S. Brooks, A. Gelman, G.L. Jones and X.-L. Meng, in *Handbook of Markov Chain Monte Carlo* (Chapman & Hall/CRC Press, 2010).
12. R.W. McKeown, R. Roiban, The quantum $AdS_5 \times S^5$ superstring at finite coupling, arXiv:1308.4875.
13. M. Luscher, A Portable high quality random number generator for lattice field theory simulations. Comput. Phys. Commun. **79**, 100 (1994), arXiv:hep-lat/9309020.
14. S. Catterall, A. Joseph, An Object oriented code for simulating supersymmetric Yang-Mills theories. Comput. Phys. Commun. **183**, 1336 (2012), arXiv:1108.1503.
15. E. Hestenes, M.R.; Stiefel, Methods of conjugate gradients for solving linear systems. J. Res. Natl. Bur. Stand. **49**, 409 (1982).
16. V. Forini, L. Bianchi, M.S. Bianchi, B. Leder, E. Vescovi, Lattice and string worldsheet in AdS/CFT: a numerical study. PoS LATTICE2015, 244 (2015), arXiv:1601.04670 (in, *Proceedings, 33rd International Symposium on Lattice Field Theory (Lattice 2015)*, p. 244.

Curriculum Vitae

up-to-date: April 2017

Personal Data

Name: Edoardo Vescovi
Permanent email: edoardo (dot) vescovi (dot) 1989 (at) gmail (dot) com

Academic Education

2016–present Postdoctoral researcher
Institute of Physics, University of São Paulo, Brazil

2013–2016 PhD in Theoretical Physics
Institute of Physics, Humboldt University Berlin, Germany
Independent Emmy Noether Research Group "Gauge Fields from Strings" and Graduiertenkolleg "Masse Spektrum Symmetrie" (GRK 1504)
Thesis: Perturbative and non-perturbative approaches to string sigma-models in AdS/CFT
Advisor: Dr. Valentina Forini
Final grade: *summa cum laude* (highest honors)

2011–2013 Master of Science in Theoretical Physics
Department of Physics, Parma University, Italy
Thesis: 5D SUSY gauge theories on compact spaces and localization
Advisor: Prof. Luca Griguolo
Final grade: 110/110 *cum laude* (highest honors)

2008–2011 Bachelor of Science in Physics
Department of Physics, Parma University, Italy
Thesis: Analytic study of a quantum model for a Bose-Einstein condensate in a double-well potential
Advisors: Prof. Raffaella Burioni, Dr. Pierfrancesco Buonsante,

E. Vescovi, *Perturbative and Non-perturbative Approaches to String Sigma-Models in AdS/CFT*, Springer Theses, DOI 10.1007/978-3-319-63420-3

Dr. Alessandro Vezzani
Final grade: 110/110 *cum laude* (highest honors)

Research Interests

- Integrability in AdS_4/CFT_3 and AdS_5/CFT_4 correspondences, string sigma-model perturbation theory in AdS backgrounds for strong-coupling Wilson loops and correlators with local operators, methods for functional determinants (Gel'fand-Yaglom, heat kernel).
- Localization methods for partition functions and Wilson loops in supersymmetric 3D Chern-Simons matter theories, $\mathcal{N} = 4$ super Yang-Mills on S^4, $\mathcal{N} = 1$ super Yang-Mills with matter on S^5, with related one-/multi-cut solutions of matrix models.
- Non-perturbative, numerical study of lattice discretizations of the $AdS_5 \times S^5$ Green-Schwarz string sigma-model action to extract finite-coupling results for observables in the guage/gravity duality.
- Weak-coupling computation of supersymmetric Wilson loops.

Publications

arXiv profile: http://arxiv.org/a/vescovi_e_1
Inspire profile: http://inspirehep.net/author/profile/E.Vescovi.1

Peer-Reviewed Publications

1. V. Forini, A.A. Tseytlin, E. Vescovi, Perturbative computation of string one-loop corrections to Wilson loop minimal surfaces in $AdS_5 \times S^5$, JHEP **1703**, 003 (2017), arXiv:1702.02164
2. L. Bianchi, M.S. Bianchi, V. Forini, B. Leder and E. Vescovi, Green-Schwarz superstring on the lattice, JHEP **1607**, 014 (2016), arXiv:1605.01726
3. V. Forini, V.G.M. Puletti, L. Griguolo, D. Seminara, E. Vescovi, Precision calculation of 1/4-BPS Wilson loops in $AdS_5 \times S^5$, JHEP **1602**, 105 (2016), arXiv:1512.00841.
4. V. Forini, V.G.M. Puletti, L. Griguolo, D. Seminara, E. Vescovi, Remarks on the geometrical properties of semiclassically quantized strings, J. Phys. A **48**, 475401 (2015), arXiv:1507.01883.
5. V. Forini, V.G.M. Puletti, M. Pawellek, E. Vescovi, One-loop spectroscopy of semiclassically quantized strings: bosonic sector. J. Phys. A **48**, 085401 (2015), arXiv:1409.8674.
   Selected by the Editors of *Journal of Physics A: Mathematical and Theoretical* for the *IOP select collection* and the *Journal of Physics A Highlights of 2015 collection*.
6. L. Bianchi, M.S. Bianchi, A. Brès, V. Forini, E. Vescovi, Two-loop cusp anomaly in ABJM at strong coupling. JHEP **1410**, 13 (2014), arXiv:1407.4788.
7. P. Buonsante, R. Burioni, E. Vescovi and A. Vezzani, Quantum criticality in a bosonic Josephson junction. Phys. Rev. A **85**, 043625 (2012), arXiv:1112.3816.

Conference Proceedings

1. V. Forini, L. Bianchi, B. Leder, P. Toepfer and E. Vescovi, Strings on the lattice and AdS/CFT. PoS LATTICE2015 (2016) 206, arXiv:1702.02005.
2. V. Forini, L. Bianchi, M.S. Bianchi, B. Leder, E. Vescovi, Lattice and string worldsheet in AdS/CFT: a numerical study, PoS LATTICE2015 (2015) 244, arXiv:1601.04670.

Grants and Awards

2017	Springer Thesis Award for the doctoral thesis and 500 EUR prize
2016	Full financial support of IRIS fellowship of Humboldt University Berlin
2015	Full financial support of Graduiertenkolleg “Masse Spektrum Symmetrie” (GRK 1504)
2008–2010	Scholarship (12 000 EUR) of the Italian Physical Society (SIF)

CPSIA information can be obtained
at www.ICGtesting.com
Printed in the USA
LVHW061629170219
607791LV00043BA/712/P

9 783319 87551